METHODS IN MOLECULAR BIOLOGY™

Series Editor
John M. Walker
School of Life Sciences
University of Hertfordshire
Hatfield, Hertfordshire, AL10 9AB, UK

For further volumes:
http://www.springer.com/series/7651

Microsatellites

Methods and Protocols

Edited by

Stella K. Kantartzi

Department of Plant, Soil, and Agricultural Systems,
Southern Illinois University at Carbondale, Carbondale, IL, USA

Editor
Stella K. Kantartzi
Department of Plant, Soil, and Agricultural Systems
Southern Illinois University at Carbondale
Carbondale, IL, USA

ISSN 1064-3745 ISSN 1940-6029 (electronic)
ISBN 978-1-62703-388-6 ISBN 978-1-62703-389-3 (eBook)
DOI 10.1007/978-1-62703-389-3
Springer New York Heidelberg Dordrecht London

Library of Congress Control Number: 2013934703

Printed on acid-free paper

Humana Press is a brand of Springer
Springer is part of Springer Science+Business Media (www.springer.com)

Preface

Current developments in genetic studies and decreasing cost of genotyping have resulted in the rapid growth of the use of molecular markers. Microsatellites or simple sequence repeats (SSR) have become the markers of choice for a variety of molecular studies because of their versatility, operational flexibility, and lower cost than other marker systems.

This volume contains 21 chapters divided into 4 parts. Part I (seven chapters) presents and describes classical and modern methods for the discovery and development of microsatellite markers. Part II (four chapters) gives a description of amplification and visualization of SSRs. In Part III (four chapters), the use of four different automated capillary sequencers that are widely used for fragment analysis is presented. The last part (Part IV, five chapters) presents a variety of methods for the analysis of data obtained by the use of microsatellites. This book is aimed at new scientists who need detailed protocols for incorporating microsatellite markers into their projects and expert scientists who want to expand their knowledge of SSR discovery, use, and analysis.

I take this opportunity to thank my family (Kostas, Ioanna, and Manos) for their unconditional love and support.

Carbondale, IL, USA *Stella K. Kantartzi*

Contents

Contributors

J. Orville Adams • *Azco Biotech Inc., San Diego, CA, USA*
James Anderson • *Department of Plant, Soil, and General Agriculture, Southern Illinois University at Carbondale, Carbondale, IL, USA*
Marinês Bastianel • *Instituto Agronômico, Centro APTA Citros Sylvio Moreira, Cordeirópolis, SP, Brazil*
Ramesh Bhonde • *Manipal Institute of Regenerative Medicine, Manipal University, Bangalore, India*
Padmakar Bommisetty • *Division of Biotechnology, Indian Institute of Horticultural Research, Bangalore, India*
Michelle C. Brandi • *Department of Parasitology, Institute of Biomedical Sciences, University of São Paulo, São Paulo, SP, Brazil*
Ángeles Bueno • *Departamento de Biología Vegetal I, Facultad de CC Biológicas, Universidad Complutense, Madrid, Spain*
Pamela A. Burger • *Department of Biomedical Sciences, Institute of Population Genetics, University of Veterinary Medicine Vienna, Vienna, Austria*
Song-Bin Chang • *Department of Life Science, Institute of Biodiversity, National Cheng Kung University, Tainan, Taiwan*
Mariângela Cristofani-Yaly • *Instituto Agronômico, Centro APTA Citros Sylvio Moreira, Cordeirópolis, SP, Brazil*
Anjan Kumar Das • *Stem Cells and Regenerative Medicine, Stempeutics Research Malaysia Sdn Bhd, Kuala Lumpur, Malaysia*
Deborah A. Dean • *Entomology and Plant Pathology, University of Tennessee, Knoxville, TN, USA*
Marcelo U. Ferreira • *Department of Parasitology, Institute of Biomedical Sciences, University of São Paulo, São Paulo, SP, Brazil*
Iria Fernandez-Silva • *Hawai'i Institute of Marine Biology, University of Hawai'I, Kāne'ohe, HI, USA*
Lluvia Flores-Rentería • *Department of Biological Sciences, Merriam-Powell Center for Environmental Research, Northern Arizona University, Flagstaff, AZ , USA*
Ioannis Ganopoulos • *Department of Genetics and Plant Breeding, School of Agriculture, Aristotle University of Thessaloniki, Thessaloniki, Greece*
Sana Ghaffari • *Dry Land Farming and Oasis Cropping Laboratory, Arid Land Institute, Medenine, Tunisia*
Marc Ghislain • *International Potato Center, Nairobi, Kenya*
Melissa Giresi • *Center for Biosystematics and Biodiversity, Texas A&M University, College Station, TX, USA*
Arancha Gómez-Garay • *Departamento de Biología Vegetal I, Facultad de CC Biológicas, Universidad Complutense, Madrid, Spain*

ELENA G. GONZALEZ • *Departamento de Biodiversidad y Biología Evolutiva, Museo Nacional de Ciencias Naturales, MNCN-CSIC, Madrid, Spain; CCMAR, Universidade do Algarve, Faro, Portugal*

DENITA HADZIABDIC • *Entomology and Plant Pathology, University of Tennessee, Knoxville, TN, USA*

NEJIB HASNAOUI • *Industrial Biologic Chemistry Unit, Agro-Bio Tech - University of Liège, Gembloux, Belgium*

MARIA DEL ROSARIO HERRERA • *International Potato Center, Lima, Peru*

MUHAMMAD J. IQBAL • *Department of Plant, Soil and General Agriculture, Center of Excellence in Soybean Research, Teaching and Outreach, Southern Illinois University at Carbondale, Carbondale, IL, USA; Department of Crop Science, North Dakota State University, Fargo, ND, USA*

KYUNG SEOK KIM • *College of Veterinary Medicine, Seoul National University, Seoul, South Korea*

ANDREW KROHN • *Department of Biological Sciences, Merriam-Powell Center for Environmental Research, Northern Arizona University, Flagstaff, AZ , USA*

DAVID A. LIGHTFOOT • *Department of Plant, Soil and General Agriculture, Center of Excellence in Soybean Research, Teaching and Outreach, Southern Illinois University at Carbondale, Carbondale, IL, USA*

HENG-SHENG LIN • *Department of Life Science, Institute of Biodiversity, National Cheng Kung University, Tainan, Taiwan*

MARCOS A. MACHADO • *Instituto Agronômico, Centro APTA Citros Sylvio Moreira, Cordeirópolis, SP, Brazil*

PANAGIOTIS MADESIS • *Institute of Agrobiotechnology, CERTH, Thessaloniki, Greece*

KHALID MEKSEM • *Department of Plant, Soil, and General Agriculture, Southern Illinois University at Carbondale, Carbondale, IL, USA*

MURALI KRISHNA MAMIDI • *Manipal Institute of Regenerative Medicine, Manipal University, Bangalore, India*

VALDENICE M. NOVELLI • *Instituto Agronômico, Centro APTA Citros Sylvio Moreira, Cordeirópolis, SP, Brazil*

PAMELA ORJUELA-SÁNCHEZ • *Department of Parasitology, Institute of Biomedical Sciences, University of São Paulo, São Paulo, SP, Brazil; La Jolla Bioengineering Institute, San Diego, CA, USA*

RAJARSHI PAL • *Manipal Institute of Regenerative Medicine, Manipal University, Bangalore, India*

DARIO A. PALMIERI • *FCL-UNESP-Assis, SP, Brazil*

BEATRIZ PINTOS • *Departamento de Biología Vegetal I, Facultad de CC Biológicas, Universidad Complutense, Madrid, Spain*

JONATHAN B. PURITZ • *Hawaiʻi Institute of Marine Biology, University of Hawaiʻ I, Kāneʻohe, HI, USA; Department of Wildlife and Fisheries Sciences, Texas A&M University, College Station, TX, USA*

MAHENDRA RAO • *NIH Center for Regenerative Medicine, National Institutes of Health, Bethesda, MD, USA*

KUNDAPURA V. RAVISHANKAR • *Division of Biotechnology, Indian Institute of Horticultural Research, Bangalore, India*

MARK A. RENSHAW • *Department of Biological Sciences, University of Notre Dame, Notre Dame, IN, USA*

M. HUMBERTO REYES-VALDÉS • *Department of Plant Breeding, Universidad Autónoma Agraria Antonio Narro, Saltillo, Coahuila, Mexico*

KAISA RIKALAINEN • *Department of Biological and Environmental Science, University of Jyväskylä, Jyväskylä, Finland*

THOMAS W. SAPPINGTON • *Genetics Laboratory, Corn Insects and Crop Genetics Research Unit, USDA-ARS, Iowa State University, Ames, IA, USA*

ROBERT J. TOONEN • *Hawaiʻi Institute of Marine Biology, University of Hawaiʻi, Kāneʻohe, HI, USA*

ROBERT N. TRIGIANO • *Entomology and Plant Pathology, University of Tennessee, Knoxville, TN, USA*

ATHANASIOS TSAFTARIS • *Institute of Agrobiotechnology, CERTH, Thessaloniki, Greece; Department of Genetics and Plant Breeding, School of Agriculture, Aristotle University of Thessaloniki, Thessaloniki, Greece*

PHILLIP A. WADL • *Entomology and Plant Pathology, University of Tennessee, Knoxville, TN, USA*

XINWANG WANG • *Department of Horticultural Sciences, Texas A&M AgriLife Research and Extension Center at Dallas, Texas A&M University, Dallas, TX, USA*

DREW WRIGHT • *Department of Plant, Soil, and General Agriculture, Southern Illinois University at Carbondale, Carbondale, IL, USA*

RAFAEL ZARDOYA • *Departamento de Biodiversidad y Biología Evolutiva, Museo Nacional de Ciencias Naturales, MNCN-CSIC, Madrid, Spain*

Chapter 1

Microsatellites: Evolution and Contribution

Panagiotis Madesis, Ioannis Ganopoulos, and Athanasios Tsaftaris

Abstract

Microsatellites are codominant molecular genetic markers, which are universally dispersed within genomes. These markers are highly popular because of their high level of polymorphism, relatively small size, and rapid detection protocols. They are widely used in a variety of fundamental and applied fields of biological sciences for plants and animal studies. Microsatellites are also extensively used in the field of agriculture, where they are used in characterizing genetic materials, plant selection, constructing dense linkage maps, mapping economically important quantitative traits, identifying genes responsible for these traits. In addition microsatellites are used for marker-assisted selection in breeding programs, thus speeding up the process. In this chapter, genomic distribution, evolution, and practical applications of microsatellites are considered, with special emphasis on plant breeding and agriculture. Moreover, novel advances in microsatellite technologies are also discussed.

Key words Microsatellites, Inter simple sequence repeats, Simple sequence repeats, High-resolution melting analysis

1 Introduction

Assessing genetic variation is an important parameter in genetic studies, in studying biodiversity, germplasm characterization and generation of genetic variability in plant breeding. Moreover, estimation of genetic variation is important for the selection of desirable genotypes. The understanding that high amounts of genetic variation is not expressed on the phenotype rendered obvious that new methods should be developed to estimate and use this variation in favor of a breeding program. Recently, due to several technical advances made in molecular genetics, genetic variation could be measured at the DNA level by developing different molecular markers. Microsatellites which are the subject of this chapter are based on different repetitive sequences present in the genome. Marker assisted selection could help breeders avoid the traditional phenotype based selections in the field, thus speeding up the breeding programs and maximizing its progress. By definition, molecular marker is any site (locus) in the genome of an organism where the

Stella K. Kantartzi (ed.), *Microsatellites: Methods and Protocols*, Methods in Molecular Biology, vol. 1006,
DOI 10.1007/978-1-62703-389-3_1, © Springer Science+Business Media, LLC 2013

DNA base sequence differs among the individuals of a population. The arrival of genetic tools like restriction enzymes and the polymerase chain reaction plus the growing abundance of DNA sequence data, coupled with automated high-throughput assays, have revealed several classes of molecular markers, including restriction fragment length polymorphisms (RFLPs), variable number tandem repeats (VNTRs), microsatellite DNA, and single nucleotide polymorphisms (SNPs) ("molecular marker," http://www.encyclopedia.com/doc/1O6-molecularmarker.html). The genome of higher organisms contains three types of simple repetitive DNA sequences (satellite DNAs, minisatellites, and microsatellites), organized in clusters of differing sizes (1, 2). Microsatellites first described by Litt and Luty (3) can be found under the term simple sequence repeats (SSRs), simple repetitive sequences (SRS), or simple tandem repeats (STRs) (4) (SSLPs) (5).

2 Microsatellites

Microsatellites are tandem repeats of very short 1–6 bp patterns which are not repeated many times at a particular locus but are distributed relatively evenly at many different genomic loci (6). Other scientists define microsatellites as 2–8 or even 1–5 bp repeats (7, 8). The most abundant patterns found in the plant genome are (AT)n, (GA)n, and (GAA)n where n refers to the total number of repeats, usually ranging from 10 to 100. In addition, mononucleotide repeats consisting of A/T repeats are also present in the chloroplast genome. Increased intra- and inter-genetic variation is observed when the number of repeats is increasing (9). Between the two types of mononucleotide repeats, A/T was the most abundant in all plant species, while G/C was comparatively limited. In the mononucleotide repeats category, the maximum (99 %) A/T repeats have been found in the *Arabidopsis* genome and the minimum (78 %) in the *Brachypodium* genome. In the dinucleotide repeat category, the distribution of SSRs in different motif types was not uniform and the most frequent motif type was different for each plant species. For example, AG/CT repeats were more frequent in *Brachypodium* and rice, with 50.7 % and 41.9 % frequency respectively, whereas AT/AT repeats were more frequent in *Populus* (60.5 %) and *Medicago* (59.9 %). In rice, both AG/CT and AT/AT repeats were the most abundant among the other dinucleotide repeats. Interestingly, the CG/CG motif contributed less than 0.5 % in dicots, whereas it was 3.1–7.0 % in all dinucleotide repeats identified in the monocots. The analysis of mononucleotide and dinucleotide repeats concluded that CG-rich motifs were least preferred in both monocot and dicot genomes. However, for trinucleotide repeats the AGC/CGT, AGG/CCT, and CCG/CGG were observed more frequently in all monocot species, whereas A/T-rich repeats, such as AAC/GTT, AAG/CTT, and AAT/ATT, were more frequent in dicots. The frequency of tetranucleotide, pentanucleotide, and hexanucleotide repeats was very low in all the plant genomes (10). Overall, there are 501 possibilities of nonredundant monomeric to hexameric repeats. In plants, the frequency and number of microsatellites have been estimated for a number of species and results indicate that the most frequent microsatellite is (GT)*n*, while in mammals is (AT)*n* (10).

Microsatellites are sometimes associated with other genomic repeats, especially transposable elements. In humans, microsatellites are associated with repetitive DNA, especially non-LTR retrotransposons (11, 12). In plants, however, including *Arabidopsis*, rice, soybean, maize, and wheat, microsatellites are preferentially located in nonrepetitive DNA regions, which indicates that they reside in regions predating genome expansion (13, 14).

Nevertheless, microsatellites are evenly spaced in the genome, although they are highly variable in number of repeat units among individuals (7). Because unique DNA sequences flank individual microsatellites they could be genotyped via STS (Sequence Tagged Site) PCR (15). In species with low levels of genetic diversity identification of a fast mutating locus would be the optimal resource for the development of markers, markers which would thus be ideal for breeding programs. At the moment there are a vast number of SSR markers publicly available for research concerning the most important agricultural crops (16–19).

3 Generation of Microsatellite Diversity

The mutational rate for unique eukaryotic sequences is of approximately 10^{-9} per nucleotide per generation (20). Moreover, the mutation rate differs between species but also differs to a great extent within species (between loci) with long loci mutating more (21). The rate at which it's SSR loci mutate varies and depends on repeated motif, GC content in flanking DNA, allele size, chromosome position, cell division (mitotic vs. meiotic), sex, age, repeat type, and genotype (2, 8, 22). The differences are mainly observed as changes in the number of SSR repeats. These observations have significant implications for the development of molecular markers, as these differences can be visualized and can facilitate plant breeding. The main mutation event is gain and loss of entire repeat units, which suggest a specific mutational mechanism called replication slippage. As microsatellites mutate at such a high rate one would expect the microsatellite size to increase over time, yet this does not happen, probably because a point mutation breaks the perfect repeats of a microsatellite and, as has been shown, imperfect repeats have a reduced slippage mutation rate (23). Two mutational mechanisms can be used to explain such high rates of mutation. The first involves

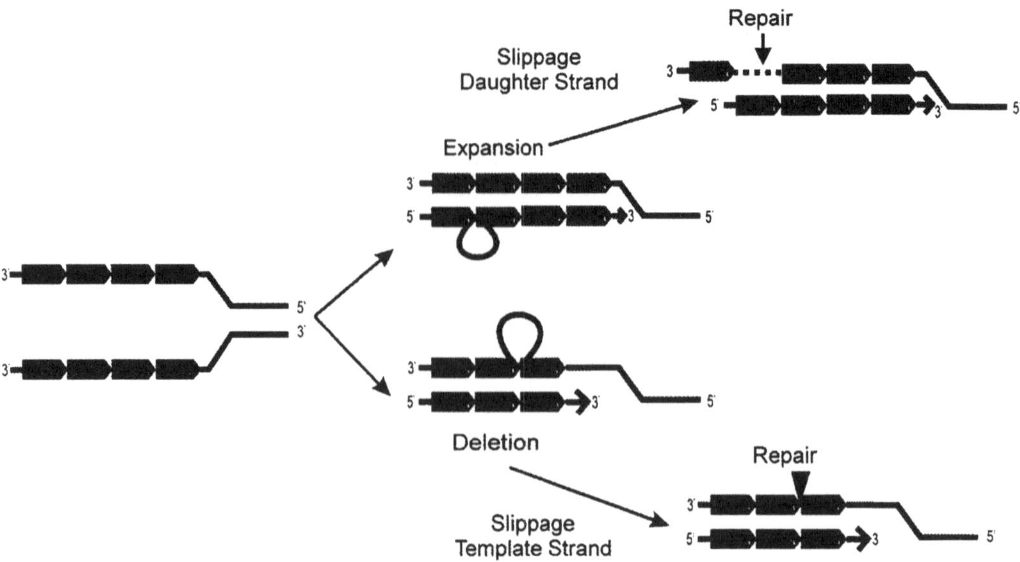

Fig. 1 Replication leads to new alleles with less (deletion) or more (expansion) repeats depending on the strand containing the error

DNA slippage during DNA replication (24), caused by mismatches between DNA strands when they are being replicated during meiosis (25). It has been estimated that replication slippage at each microsatellite occurs about once per 1,000 generations (26). The second involves recombination between DNA strands (27).

4 Replication Slippage

Replication slippage accounts for many mutations at SSR loci (28). This type of mutation occurs when one DNA strand is mispairing (slip strand) during DNA replication. The mispairing refers to a repeat unit hybridizing to a repeat in such a way where a loop is formed in the nascent strand resulting in the addition of a repeat (22). If the loop occurs in the template strand, then there will be a decrease in the number of units (29). These can lead to gain or loss of certain repeats. The mismatch repair mechanism and exonuclease activity of polymerase corrects a number of errors but many escape and become mutations (Fig. 1).

5 Recombination

Another mechanism of mutation is the recombination process which could change the SSR length by asymmetrical crossing over or by gene conversion (2, 30–33). Asymmetric exchanges, random

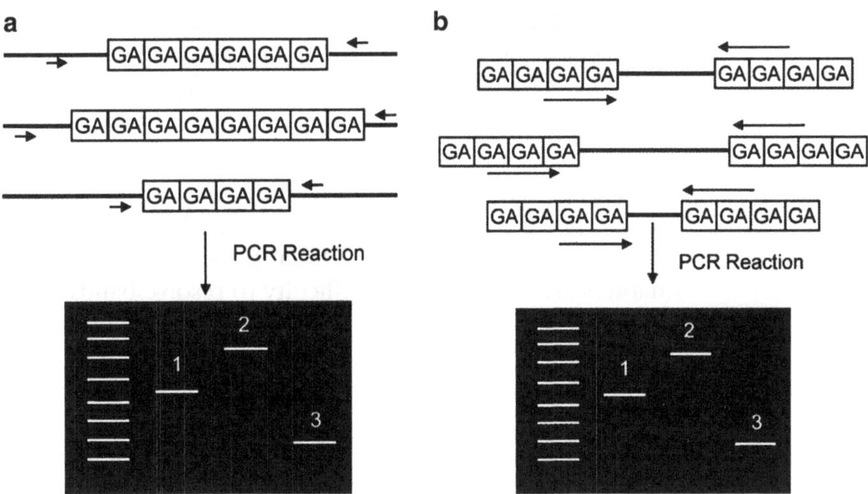

Fig. 2 (**a**) Amplification of microsatellites using a pair of SSR markers. PCR products are analyzed on polyacrylamide gels. (**b**) Amplification of microsatellites using one ISSR markers. PCR products are analyzed on polyacrylamide gels or simple 1.5 % agarose gels

genetic drift and selection can have a significant effect on the accumulation of tandem-repetitive sequences in the genome (34). Non reciprocal recombination also mutates tandem repeat number (for both microsatellites and minisatellites) (32, 33).

6 Infrastructure and Methods for the Study of Microsatellites

Molecular markers using microsatellites as targeting sequence polymorphisms can either multiply a DNA region containing the microsatellites, as in the case of SSR markers. This type of markers recognizes the microsatellite flanking sequences (using one pair of specific primers) (Fig. 2a). Another type of microsatellite markers (ISSR) bind on the microsatellite using only one primer and multiply the region between two microsatellites (Fig. 2b). PCR fragments are usually separated on polyacrylamide gels in combination with $AgNO_3$ staining (SSR primers) or on simple agarose gels (1.5 %), ISSR respectively. However, the development of microsatellite SSR primers for a new species is difficult, laborious, and expensive, although the genomic era could facilitates this process. Several protocols have been developed so far for the development of SSR markers (5, 35).

7 Technical Problems and Difficulties in Studying Microsatellites

Although microsatellites are extremely useful for genetic analysis, mapping, etc., there are certain difficulties concerning their use. They are expensive to develop, as a large number of sequences must be cloned and only a small number of these will be useful for the development of the SSR markers. Moreover, only a number of these markers will give informative results, especially for species with large genomes (36–38). In addition, problems that might occur are, for instance, as follows: (a) the primer may not amplify any PCR product; (b) the primer may produce very complex, weak, or nonspecific amplification patterns; (c) the amplification product may not be polymorphic. Other possible problems using SSR markers are as follows: the difficulty to resolve bands differing only in one or two base pairs, the cost of polyacrylamide gels and labeled primers, and the differences in identifying band size and their calling between laboratories, making comparisons between results very hard.

Yet, despite any problems, SSRs are now the marker of choice in many areas of molecular genetics due to their codominant and polymorphic nature, even between closely related lines, their requirement for low amounts of DNA, and the possibility of being automated for high-throughput screening make them attractive. In addition they can be easily exchanged between laboratories, and are highly transferable between populations (39). For example, a total of 18,828 SSR sequences have been detected in the rice genome (40), of which only 10–15 % have yet been used, suggesting the high potential available for such marker systems. SSRs are mostly codominant markers and are indeed excellent for studies of population genetics and mapping (31, 41). Another technical development like the use of fluorescent primers in combination with automatic capillary or gel-based DNA sequencers has facilitated the detection of bands and their analysis.

8 Advances in Microsatellites

Although microsatellites mainly occur in noncoding sequences, the development of EST databases revealed that microsatellite repetitive sequences also occur inside coding sequences (42–44). The information obtained by EST libraries has been recently used for the development of SSR markers (45–49). Microsatellites designed EST are expected to be slightly less polymorphic than genomic library derived SSRs, as there is selection pressure for sequence conservation in coding regions (50–53). This also explains why the most abundant microsatellites in genes are trinucleotides and hexanucleotides and the less frequent are mononucleotides and dinucleotides, as these types cause frame shift and most probably premature stop codons. While technology progresses and new genomes and EST libraries become available with the help of bioinformatics approaches, the development of SSR markers based on EST's through data mining has become a fast, efficient and relatively inexpensive, compared to development of genomic SSRs (54). However, these approaches require the existence of sequence information.

9 The Advances of High-Resolution Melting Analysis in Microsatellite Studies

Generally, laborious polyacrylamide gels followed by silver staining or, for better resolution, fluorescently labeled PCR products and automated sequencers are needed for microsatellite analysis. Moreover, post-PCR handling and dilution steps as well as fluorescently labeled primers for each microsatellite, are required by this method, resulting in increases in time and cost of the analysis. Over the last few years, Real-time PCR is often used to analyze amplified DNA and identify viruses and pathogens. In addition, it can also be used as an extremely quick analysis for reactions that do not require subsequent use of the amplified DNA.

Lately, high-resolution melting (HRM), a sensitive mutation detecting method has been introduced, extending the possibilities of analyzing the DNA melting curves which was a standard diagnostic feature in qPCR (55). HRM analysis is rapidly gaining in popularity as a cost-effective and faster alternative to traditional post-PCR genotyping methods such as single-stranded conformation polymorphism, denaturing high-performance liquid chromatography, and restriction fragment length polymorphism. The determination of the T_m values distance can be used for identifying the targeted amplicon among the nonspecific products (55) (Fig. 3a). The HRM curves obtained are highly specific for each amplicon and depend on the GC content, amplicon length and sequence (56).

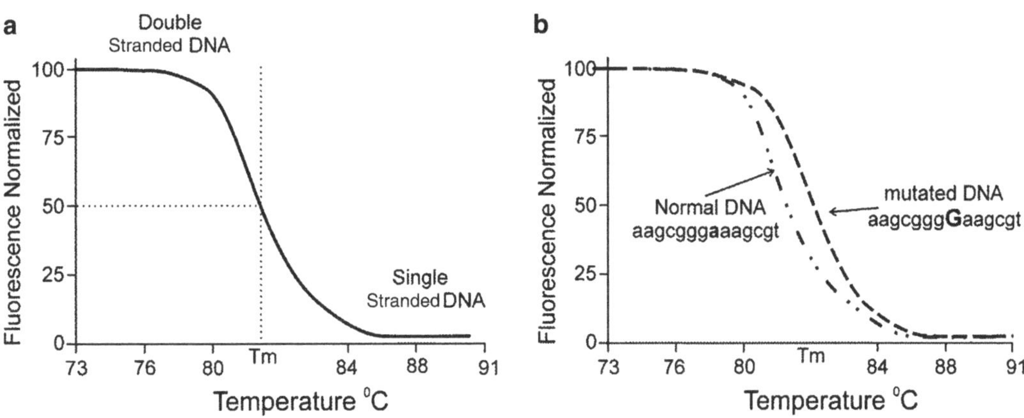

Fig. 3 (**a**) A double stranded DNA fragment melts at a specific temperature (T_m) which is specific for each DNA fragment. The highest rate of fluorescence decrease is generally at the melting temperature of the DNA sample (T_m). The T_m is defined as the temperature at which 50 % of the DNA sample is double stranded and 50 % is single stranded. (**b**) Different PCR products in size have different melting curves and can be distinguished having even one single point mutation

HRM analysis is based on the fact that although DNA melting curves are used primarily for the determination of the melting temperature (T_m) of amplified double-stranded DNA, the precise shape of a melting curve is typical of each DNA sequence (57). By making very small temperature size steps, accurate melting curves can be produced, while normalization and comparison of the melting curves can clarify whether different amplicons have the same or different sequences (58). Different amplicons can have the same T_m values but the advantage of the HRM method is that these different amplicons can be distinguished by the shape of HRM melting curves (59) (Fig. 3b). Even not being strictly a banding pattern-based method, HRM analysis is categorized as such because it relies on PCR amplification and detection of sequence variants without sequencing or hybridization procedures (60).

HRM could be used as an alternative method to detect microsatellites, especially for those laboratories that do not have immediate access to capillary sequencers (61). Its main advantage is the fast, accurate, and closed-tube determination of SNPs and sequence variations (62). The sensitivity of HRM analysis has already been broadly verified (63). Besides numerous applications in clinical mutation screening, HRM was suggested as another population genetics genotyping system (64) and has been used to discriminate closely related plant cultivars (61, 65–69). Mader et al. (69) proved the ability of HRM application to SSR analysis in principle, but also its limitations in comparison to CE (Capillary Electrophoresis). Specifically, only low-complexity SSRs with a few alleles in a population can be fully detected with HRM. The need for production of unknown PCR products artificial mixtures with already genotyped standards makes the procedure more complex and labor intensive. It may therefore be unlikely that HRM will replace CE for genotyping the highly complex SSRs typically used in population genetics. Ganopoulos et al. (67) suggested that HRM is able to detect and screen single locus markers without the need of labeled primers, product fractionation, DNA restriction or individual sequence analyses. This makes the technique ideal for cultivar identification studies where large populations are to be scored with numerous SSR loci.

There are numerous advantages of the HRM method of scoring SNPs/microsatellites comparing to existing systems that are based on high-resolution gel or CE (70). First, there is no a priori requirement to identify the position or identity of the SNP/microsatellite; any SNP or length polymorphism giving rise to a melt polymorphism can be scored without characterization. Second, there are no additional reagent costs for labeled primers. Third, the capacity to perform HRM directly after PCR makes the need for further handling of samples unnecessary. The capacity of HRM analysis instruments to perform more assays in the same time means that more data points can be generated within the 15–20 min required to perform an HRM following the end of PCR, thus increasing the overall throughput. Finally, the fact that melting curves shapes depends not only to amplified size fragments but also to base composition and SNP position, is more sensitive to distinguish closely related genotypes such as cultivars of the same species.

10 Applications of Microsatellites

Microsatellites have become a marker of choice for a huge range of applications in plants with a vast literature; refer all this literature is beyond the scope of this article. SSR markers are useful for a variety of applications in plant genetics and breeding because of their reproducibility, multiallelic nature, codominant inheritance, relative abundance, and good genome coverage (71). Furthermore, SSR markers have been useful for integrating the genetic physical and sequenced based physical maps in plant species and at the same time they have provided breeders and geneticists with an efficient tool to link phenotypic and genotypic variation (72). Microsatellites are also used in order to estimate genetic variation at molecular level in a germplasm collection which will help towards the correct choice of parents for crosses in a breeding program (i.e., hybrid breeding), mapping and tagging of genes or QTLs (quantitative trait loci) for agronomic and disease resistance traits, genome mapping, MAS of promising lines and Marker Assisted Backcrossing (MAB) during breeding programs, gender identification, studying the population structure and taxonomic and phylogenetic relationships.

In addition, the knowledge of genetic variation is mostly useful for characterization of accessions in plant germplasm collections and taxonomic studies and phylogenetic studies (73). In phylogenetic studies organelle specific markers (i.e., cpSSR and mtSSR) have also been used making great impact on the determination of structure and variation within a natural population too. Organelle microsatellites are attractive targets for phylogenetic studies or evolution studies and even migration histories due to uniparental mode of inheritance, conserved gene order and lack of heteroplasmy and recombination of organelle (74).

Microsatellites have also been used for hybrid determination and characterization of allelic contribution of each parent (71). Moreover, microsatellites have been used for mapping of specific genomic regions responsible for agronomic traits or mapping of specific genes (75).

11 Conclusions

Ever since their development, microsatellite markers are constantly being isolated and characterized in a wide range of plants including cereals, legumes, vegetables, forest trees, fruit plants, conifers, and other economically important plant species. Arrival of new technologies did not eliminate the use of microsatellites instead they have rendered microsatellites a useful multi-tool in plant breeding. Microsatellites are still the method of choice for marker assisted selection, population genetics, estimation of genetic diversity, fingerprinting, mapping, and gene association studies. SSR based association mapping holds a great promise for exploiting genetic diversity, characterizing accumulated phenotypic variation, and associating markers with traits in plant germplasm especially with the progress made in the genome programs. They owe their broad use to their cost-effectiveness easy to use and their excellent results.

Microsatellite markers not only are involved in genetic diversity studies, and evolutionary studies, but are also being used in fundamental research like genome analysis, gene mapping, marker-assisted selection, etc., yet there are several limitations limiting their use like the need to isolate them de novo although genome projects are expected to solve this problem, the presence of stutter bands, null alleles, and heterologous amplicons (76, 77). In conclusion, genomic progress and advancement in microsatellites markers will make their use even more attractive for molecular breeding and plant genetics and eventually they will have great contribution in major crop improvement.

References

1. Armour J et al (1999) Minisatellites and mutation processes in tandemly repetitive DNA. Oxford University Press, Oxford
2. Hancock JM (1999) Microsatellites and other simple sequences: genomic context and mutational mechanisms. Oxford University Press, Oxford
3. Litt M, Luty JA (1989) A hypervariable microsatellite revealed by in vitro amplification of a dinucleotide repeat within the cardiac muscle actin gene. Am J Hum Genet 44: 397–401
4. Tautz D (1989) Hypervariabflity of simple sequences as a general source for polymorphic DNA markers. Nucleic Acids Res 17: 6463–6471
5. McDonald DB, Potts WK (1997) DNA microsatellites as genetic markers for several scales. Academic, New York
6. Tautz D, Renz M (1984) Simple sequences are ubiquitous repetitive components of eukaryotic genomes. Nucleic Acids Res 12:4127–4138
7. Goldstein DB, Pollock DD (1997) Launching microsatellites: a review of mutation processes and methods of phylogenetic inference. J Hered 88:335–342
8. Schlötterer C (1998) Microsatellites. IRL, Oxford
9. Queller DC et al (1993) Microsatellites and kinship. Trends Ecol Evol 8:285–288
10. Sonah H et al (2011) Genome-wide distribution and organization of microsatellites in plants: an insight into marker development in *Brachypodium*. PLoS One 6:e21298
11. Kelkar YD et al (2011) A matter of life or death: how microsatellites emerge in and vanish from the human genome. Genome Res 21:2038–2048

12. Nadir E et al (1996) Microsatellite spreading in the human genome: evolutionary mechanisms and structural implications. Proc Natl Acad Sci 93:6470–6475
13. Morgante M et al (2002) Microsatellites are preferentially associated with nonrepetitive DNA in plant genomes. Nat Genet 30:194–200
14. Temnykh S et al (2001) Computational and experimental analysis of microsatellites in rice (*Oryza sativa* L.): frequency, length variation, transposon associations, and genetic marker potential. Genome Res 11:1441–1452
15. Weber J, May P (1989) Abundant class of human DNA polymorphisms which can be typed using the polymerase chain reaction. Am J Hum Genet 44:388–396
16. Milbourne D et al (1998) Isolation, characterisation and mapping of simple sequence repeat loci in potato. Mol Gen Genet 259:233–245
17. Sharopova N et al (2002) Development and mapping of SSR markers for maize. Plant Mol Biol 48:463–481
18. Song QJ et al (2002) Characterization of trinucleotide SSR motifs in wheat. Theor Appl Genet 104:286–293
19. Temnykh S et al (2000) Mapping and genome organization of microsatellite sequences in rice (*Oryza sativa* L.). Theor Appl Genet 100: 697–712
20. Crow J (1993) How much do we know about spontaneous human mutation rates? Environ Mol Mutagen 21:122–129
21. Zhu Y et al (2000) A phylogenetic perspective on sequence evolution in microsatellite loci. J Mol Evol 50:324–338
22. Ellegren H (2000) Microsatellite mutations in the germline: implications for evolutionary inference. Trends Genet 16:551–558
23. Jin L et al (1996) Mutation rate varies among alleles at a microsatellite locus:Phylogenetic evidence. Proc Natl Acad Sci 93: 15285–15288
24. Tachida H, Iizuka M (1992) Persistence of repeated sequences that evolve by replication slippage. Genetics 131:471–478
25. Tautz D, Schlötterer C (1994) Simple sequences. Curr Opin Genet Dev 4:832–837
26. Weber JL, Wong C (1993) Mutation of human short tandem repeats. Hum Mol Genet 2: 1123–1128
27. Harding RM et al (1992) The evolution of tandemly repetitive DNA: recombination rules. Genetics 132:847–859
28. Levinson G, Gutman GA (1987) Slipped-strand mispairing: a major mechanism for DNA sequence evolution. Mol Biol Evol 4: 203–221
29. Eisen J (1999) Mechanistic basis for microsatellite instability. Oxford University Press, Oxford
30. Brohede J, Ellegren H (1999) Microsatellite evolution: polarity of substitutions within repeats and neutrality of flanking sequences. Proc Biol Sci 266:825–833
31. Goldstein D, Schlotterer C (1999) Microsatellites, evolution and applications. Oxford University Press, Oxford
32. Jakupciak JP, Wells RD (1999) Genetic instabilities in (CTGB·CAG) repeats occur by recombination. J Biol Chem 274:23468–23479
33. Richard GF, Paques F (2000) Mini- and microsatellite expansions: the recombination connection. EMBO Rep 1:122–126
34. Charlesworth B et al (1994) The evolutionary dynamics of repetitive DNA in eukaryotes. Nature 371:215–220
35. Bruford M et al (1996) Microsatellites and their application to conservation genetics. Oxford University Press, Oxford
36. Kostia S et al (1995) Microsatellite sequences in a conifer, Pinus sylvestris. Genome 38: 1244–1248
37. Röder MS et al (1995) Abundance, variability and chromosomal location of microsatellites in wheat. Mol Gen Genet 246:327–333
38. Smith DN, Devey ME (1994) Occurrence and inheritance of microsatellites in Pinus radiata. Genome 37:977–983
39. Gupta PK et al (1999) Molecular markers and their applications in wheat breeding. Plant Breed 118:369–390
40. International Rice Genome Sequencing Project (2005) The map-based sequence of the rice genome Nature 436:793–800
41. Jarne P, Lagoda PJL (1996) Microsatellites, from molecules to populations and back. Trends Ecol Evol 11:424–429
42. Eujayl I et al (2004) *Medicago truncatula* EST-SSRs reveal cross-species genetic markers for *Medicago* spp. Theor Appl Genet 108:414–422
43. Hackauf B, Wehling P (2002) Identification of microsatellite polymorphisms in an expressed portion of the rye genome. Plant Breed 121:17–25
44. Thiel TT et al (2003) Exploiting EST databases for the development and characterization of gene-derived SSR-markers in barley (Hordeum vulgare). Theor Appl Genet 106: 411–422
45. Chapman M et al (2009) Development, polymorphism, and cross-taxon utility of EST–SSR markers from safflower (*Carthamus tinctorius* L.). Theor Appl Genet 120:85–91

46. Choudhary S et al (2009) Development of chickpea EST-SSR markers and analysis of allelic variation across related species. Theor Appl Genet 118:591–608
47. Gadaleta A et al (2010) Development and characterization of EST-derived SSRs from a 'totipotent' cDNA library of durum wheat. Plant Breed 129:715–717
48. Nunome T et al (2009) Development of SSR markers derived from SSR-enriched genomic library of eggplant (*Solanum melongena* L.). Theor Appl Genet 119:1143–1153
49. Wei W et al (2011) Characterization of the sesame (Sesamum indicum L.) global transcriptome using Illumina paired-end sequencing and development of EST-SSR markers. BMC Genomics 12:451
50. Chabane K et al (2005) EST versus genomic derived microsatellite markers for genotyping wild and cultivated barley. Genet Resour Crop Evol 52:903–909
51. Cho YG et al (2000) Diversity of microsatellites derived from genomic libraries and GenBank sequences in rice (*Oryza sativa* L.). Theor Appl Genet 100:713–722
52. Eujayl I et al (2001) Assessment of genotypic variation among cultivated durum wheat based on EST-SSRS and genomic SSRS. Euphytica 119:39–43
53. Scott KD et al (2000) Analysis of SSRs derived from grape ESTs. Theor Appl Genet 100: 723–726
54. Gupta PK et al (2003) Transferable EST-SSR markers for the study of polymorphism and genetic diversity in bread wheat. Mol Genet Genomics 270:315–323
55. Wilhelm J et al (2003) Validation of an algorithm for automatic quantification of nucleic acid copy numbers by real-time polymerase chain reaction. Anal Biochem 317:218–225
56. Wittwer CT (2009) High-resolution DNA melting analysis: advancements and limitations. Hum Mutat 30:857–859
57. Vossen RHAM et al (2009) High-resolution melting analysis (HRMA)—more than just sequence variant screening. Hum Mutat 30: 860–866
58. Wittwer CT et al (2003) High-resolution genotyping by amplicon melting analysis using LCGreen. Clin Chem 49:853–860
59. Stephens AJ et al (2008) High-resolution melting analysis of the spa repeat region of Staphylococcus aureus. Clin Chem 54: 432–436
60. Tindall EA et al (2009) Assessing high-resolution melt curve analysis for accurate detection of gene variants in complex DNA fragments. Hum Mutat 30:876–883
61. Mackay JF et al (2008) A new approach to varietal identification in plants by microsatellite high resolution melting analysis: application to the verification of grapevine and olive cultivars. Plant Meth 4:8
62. Wu SB et al (2008) High resolution melting analysis of almond SNPs derived from ESTs. Theor Appl Genet 118:1–14
63. Reed GH, Wittwer CT (2004) Sensitivity and specificity of single-nucleotide polymorphism scanning by high-resolution melting analysis. Clin Chem 50:1748–1754
64. Smith BL et al (2010) High-resolution melting analysis (HRMA): a highly sensitive inexpensive genotyping alternative for population studies. Mol Ecol Resour 10:193–196
65. Bosmali I et al (2012) Microsatellite and DNA-barcode regions typing combined with high resolution melting (HRM) analysis for food forensic uses: a case study on lentils (Lens culinaris). Food Res Int 46:141–147
66. Ganopoulos I et al (2011) Adulterations in Basmati rice detected quantitatively by combined use of microsatellite and fragrance typing with high resolution melting (HRM) analysis. Food Chem 129:652–659
67. Ganopoulos I et al (2011) Microsatellite high resolution melting (SSR-HRM) analysis for authenticity testing of protected designation of origin (PDO) sweet cherry products. Food Contr 22:532–541
68. Ganopoulos I et al (2012) Microsatellite genotyping with HRM (high resolution melting) analysis for identification of the PGI common bean variety Plake Megalosperma Prespon. Eur Food Res Tech 234:501–508
69. Mader E et al (2008) A strategy to setup codominant microsatellite analysis for high-resolution-melting-curve-analysis (HRM). BMC Genet 9:69
70. Reed GH et al (2007) High-resolution DNA melting analysis for simple and efficient molecular diagnostics. Pharmacogenomics 8:597–608
71. Powell W et al (1996) The comparison of RFLP, RAPD, AFLP and SSR (microsatellite) markers for germplasm analysis. Mol Breed 2:225–238
72. Gupta PK, Varshney RK (2000) The development and use of microsatellite markers for genetic analysis and plant breeding with emphasis on bread wheat. Euphytica 113: 163–185
73. Joshi SP et al (1999) Molecular markers in plant genome analysis. Curr Sci 77:230–240

74. Provan J et al (2001) Chloroplast microsatellites: new tools for studies in plant ecology and evolution. Trends Ecol Evol 16:142–147
75. Neeraja C et al (2007) A marker-assisted backcross approach for developing submergence-tolerant rice cultivars. Theor Appl Genet 115:767–776
76. Kalia R et al (2011) Microsatellite markers: an overview of the recent progress in plants. Euphytica 177:309–334
77. Wang M et al (2009) Microsatellite markers in plants and insects. Part I: applications of biotechnology. Genes Genomes Genomics 3: 54–67

Part I

Discovery and Development of Microsatellites

Chapter 2

Screening of Genomic Libraries

Valdenice M. Novelli, Mariângela Cristofani-Yaly, Marinês Bastianel, Dario A. Palmieri, and Marcos A. Machado

Abstract

Microsatellites, or simple sequence repeats (SSRs), have proven to be an important molecular marker in plant genetics and breeding research. The main strategies to obtain these markers can be through genomic DNA and from expressed sequence tags (ESTs) from mRNA/cDNA libraries. Genetic studies using microsatellite markers have increased rapidly because they can be highly polymorphic, codominant markers and they show heterozygous conserved sequences. Here, we describe a methodology to obtain microsatellite using the enrichment library of DNA genomic sequences. This method is highly efficient to development microsatellite markers especially in plants that do not have available ESTs or genome databases. This methodology has been used to enrich SSR marker libraries in *Citrus* spp., an important tool to genotype germplasm, to select zygotic hybrids, and to saturate genetic maps in breeding programs.

Key words Microsatellites, Molecular markers, Enrichment methods

1 Introduction

Microsatellites, or simple sequence repeats (SSR), are arrays of hypervariable short (1–5 bp) repeat motifs that can be found in both coding and noncoding DNA sequence of organisms. These single-locus markers are mainly characterized by high frequency, Mendelian inheritance, and co-dominance. Microsatellites have proven to be important molecular markers in plant genetics and breeding, because of their variability, detection based on DNA amplification, accessibility of detection, and reproducibility (1). Microsatellites are polymerase chain reaction (PCR) based, requiring previous sequence identification, primer designing for the conserved flanking regions, and amplification of the target repeat (2). The availability of microsatellite markers has been limited in a great number of species. Construction of genomic libraries for microsatellite markers' development has been an effective way to obtain polymorphic markers very useful to characterize germplasm

Stella K. Kantartzi (ed.), *Microsatellites: Methods and Protocols*, Methods in Molecular Biology, vol. 1006, DOI 10.1007/978-1-62703-389-3_2, © Springer Science+Business Media, LLC 2013

collections or as molecular tool for genetic mapping in breeding programs. Genomic libraries allow the screening of an entire genome (or a collection of genomes) by digesting genomic DNA (gDNA), cloning into vectors, and transforming bacterial cells that can be screened for a desired phenotype, i.e., clones containing DNA fragments with repeat motifs (3).

Despite their great utility for a broad range of plant species, especially in economically important crops, the number of polymorphic markers obtained using this strategy has been limited, requiring an intensive labor to generate an appropriate set of useful markers.

In the last 20 years, the number of methods and strategies to development genomic libraries for microsatellite isolation has evolved considerably, from traditional library screening and development of enriched libraries (reviewed by 1) to mining genomic and EST databases (4), and high-throughput identification from next-generation sequencing data (5). These strategies were able to identify and obtain a great number of markers rapidly and cost-effective, including high-quality genetic markers in non-model and understudied plant species (6).

2 Materials and Methods

Here we describe the main steps for the development and selection of microsatellite from genomic sequences from citrus DNA using the procedure of library enrichment. The first step is to obtain a DNA with high quality and purity, followed by digestion with restriction enzymes. Sau3AI has been chosen for citrus SSR library and the restriction fragments were ligated by corresponding adapters and amplified. The biotinylated SSR probes were used to hybridize the denatured pre-amplified fragments. The hybridized mixture was added to streptavidin-coated paramagnetic beads. The DNA-probe hybrids were incubated at room temperature, and a magnetic field was applied to precipitate the beads, which were attached by fragments containing SSR that hybridized to biotinylated probes. The SSR-enriched fragments were amplified by polymerase chain reaction (PCR); products were cloned into the pGEM®-T Easy Vector Systems, transformed into competent *Escherichia coli*, and plated onto Luria-Bertani medium (LB medium) with antibiotic selection. Single colonies were selected and they were grown overnight in LB. Plasmids were purified and the insert sequenced. All the steps for construction of genomic libraries of citrus are detailed below (Fig. 1).

2.1 DNA Extraction

Adapted from Murray and Thompson (7): Grind the sample (1 g) with liquid nitrogen to a powder. Transfer the sample to tube, add 20 mL CTAB–Sarkosyl Buffer (1 M Tris–HCl pH 7.5, 0.5 M EDTA, 5 M NaCl, 5 % CTAB, 10 % Sarkosyl, 140 mM

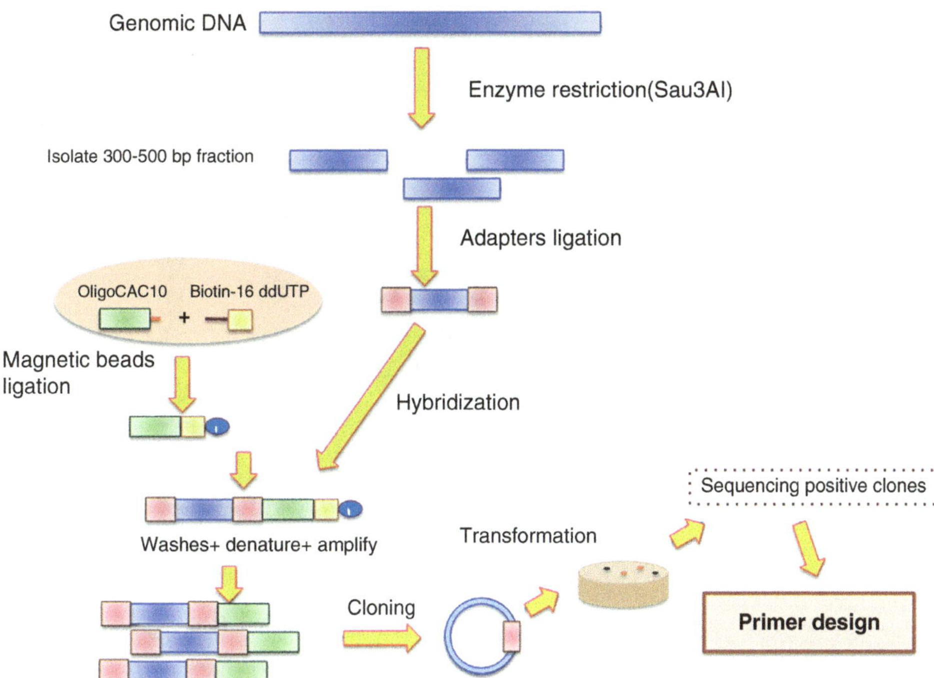

Fig. 1 Schematic representation for microsatellites' isolation from enriched genomic libraries

β-mercaptoethanol), homogenize and incubate at 60 °C for 10 min. Add 1 volume chloroform:isoamyl alcohol (24:1), mix by gentle inversion, and centrifuge at 1,900 × *g* for 8 min. Carefully remove the aqueous (top) layer to a fresh tube. Add 0.1 volume of a solution of 10 % CTAB, 5 M NaCl and mix carefully. Extract with an equal volume of chloroform:isoamyl alcohol (24:1), mix by gentle inversion, and centrifuge at 1,900 × *g* for 8 min. Transfer 15 mL of the aqueous (top) layer to a fresh tube, add 1 volume of precipitation buffer CTAB (1 % CTAB, 1 M Tris–HCl pH 7.5, 0.5 M EDTA), mix gently, and incubate at room temperature for 30 min. Centrifuge at 9,600 × *g* for 5 min. Discard the supernatant and dissolve the pellet in 4 mL of TE high salt (1 M Tris–HCl pH 7.5, 0.5 M EDTA pH 8.0, 5 M NaCl), and incubate at 65 °C for 10 min to total dissolution. Precipitate DNA by adding 2 volumes of cold (−20 °C) absolute ethanol and mix by gentle inversion and centrifuge at 3,500 × *g* for 6 min. Discard the supernatant and wash the pellet with 7 mL of cold (−20 °C) 70 % ethanol and centrifuge at 3,500 × *g* for 6 min. Remove the supernatant, add 4 mL of cold (−20 °C) absolute ethanol, and centrifuge at 3,500 × *g* for 6 min. Carefully remove the supernatant and incubate at room temperature for 20 min or until DNA is completely dry. Dissolve the pellet in 100 μL of TE 1/10 plus RNAse and incubate at 37 °C for 2 h, and after estimate the DNA concentration.

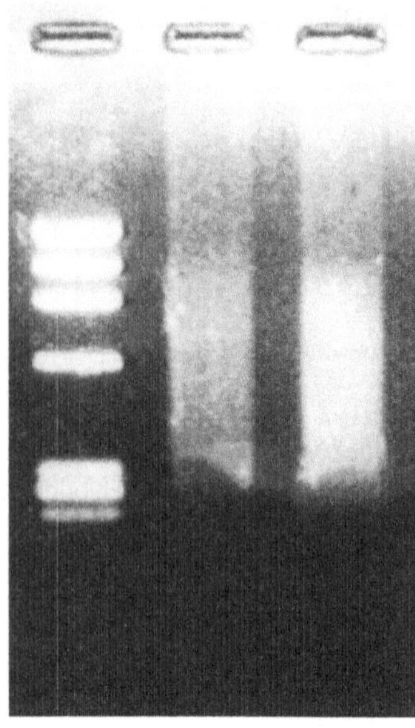

Fig. 2 Electrophoresis of fragment DNA obtained from digestion by Sau3AI

2.2 Genomic DNA Digestion

Digest the total genomic DNA (50 μg) using the blunt end-generating restriction endonuclease Sau3AI (250 U), Sau3AI buffer 10× (20 μL) and add water to a volume of 200 μL. Incubate at 37 °C overnight. Check digestion quality using the digested DNA (10 μL) and ΦX174 (50 ng/μL) as molecular weight standard, by electrophoresis through a 1.5 % agarose gel in 1× TAE buffer (40 mM Tris, 20 mM acetic acid, and 1 mM EDTA).

2.3 Gel-Fractionate to Isolate DNA Fragments

Perform electrophoresis through a 0.8 % low melting point agarose in 1× TAE buffer, using 190 μL of DNA digestion. Excise the 234–872 bp fraction from the gel and transfer to 1.5 mL microcentrifuge tube (Fig. 2). Add 3 volumes of TE buffer, and incubate at 65 °C for 5 min until complete agarose dissolution. Add an equal volume of TE-saturated phenol to the DNA sample, mix by vortex, and centrifuge. Remove about 90 % of the upper aqueous layer to a clean tube, carefully avoiding proteins at the aqueous:phenol interface. Extract a second time with an equal volume of 1:1 TE-saturated phenol:chloroform:isoamyl alcohol, centrifuge at 3,500 ×*g* for 5 min, and repeat the extraction using just chloroform. Transfer the supernatant and precipitate in 1:10 acetate sodium (3 M) and 3 volumes of absolute ethanol; incubate at −20 °C overnight. Carefully mix and centrifuge at 3,500 ×*g* for 30 min. Remove the supernatant, add 500 μL of 70 % ethanol, and centrifuge at 3,500 ×*g* for 20 min. Discard the supernatant, dry the precipitate for 5 min, and suspend in 50 μL water. Estimate the DNA concentration.

2.4 Ligation of Adapters

After confirming digestion on agarose gel electrophoresis and excise the 234–872 bp fraction from the gel, the Sau3AI adapters are ligated to the genomic fragments. Fragments of genomic DNA (10 μg) were ligated to adapters (200 μM) using T4 DNA ligase (400 U/μL) at 16 °C overnight. The adapters oligo sequences used were shorter adapter (5′CAG CCT AGA GCC GAA TTC ACC3′) and longer adapter (5′GAT CG GTG AAT TCG GCT CTA GGC TG3′).

2.5 Biotin-Labeled Oligonucleotide

Mix 100 pmol/μL of oligoprobe (for example, CAC_{10}), 5× terminal transferase buffer, Biotin-16 ddUTP (2 μL), terminal transferase (30 U), and water to 40 μL final volume. Incubate at 37 °C for 30 min and then add 0.5 M EDTA (4 μL) to enzyme inactivation. Precipitate with 2.5× volume of 100 % ETOH (±110 μL) incubating at −20 °C overnight. Centrifuge at 4 °C for 30 min at 13,800×*g*, washing twice with 100 μL 70 % ETOH, centrifuge at 4 °C for 10 min at 13,800×*g*, drying under vacuum and suspend in 30 μL of water. Incubate at refrigerator.

2.6 Preparation of Magnetic Beads

Use 1 mg Beads (Beads Streptavidin—Dynal S/A) (100 μg/μL) for each hybridization. Take out 100 μL and wash twice in PBS buffer (137 mM NaCl, 2.7 mM KCl, 10 mM NaH_2PO_4, 2 mM KH_2PO_4, HCl to pH 7.4) plus 1 % BSA (400 μL). Place tube in magnet stand for 1–2 min to allow beads to migrate to the side of the tube. Remove supernatant by aspiration with a pipette. Remove tube from magnet stand. Wash once in 400 μL 1× BEW buffer (1 M Tris–HCl pH 7.5, 100 mM EDTA pH 7.6, 5 M NaCl). Repeat the magnetic separation. Suspend gently in 2× BEW (200 μL), add 170 μL of H_2O and 30 μL of the biotin-labeled oligonucleotide. Shake at room temperature for 60 min. Wash twice in 1× BEW (400 μL) and wash once in 5× SSPE (20× SSPE = 0.2 M NaH_2PO_4, 3.0 M NaCl, 0.02 M EDTA, NaOH to pH 7.4) plus 400 μL 0.1 % SDS. Suspend in 10× SSPE + 0.2 % SDS (150 μL) pre-warmed at 65 °C. Save at 65 °C until hybridization.

2.7 Hybridization

To allow the biotinylated probe to hybridize to the target DNA, denature DNA plus adaptors heating at 95 °C for 10 min in 150 μL (before this, make the 1:1,000 dilution in water and save to carry out the PCR control). Transfer the tube to ice. Then, add the DNA + adaptors to the beads incubating at 65 °C for 1 h and 30 min and shake in each 10 min. After that, recover the hybridization solution in a new tube. To capture the fragments hybridized to the probe, it was used the affinity of the biotin in the probe for the streptavidin-coated magnetic beads. Then proceed washes: (a) twice in 2× SSPE + 0.1 % SDS (400 μL) for 5 min at room temperature, (b) once in 2× SSPE + 0.1 % SDS (400 μL) for 15 min at 65 °C, and (c) rinse the beads in 2× SSPE solution (400 μL). Suspend the beads in 200 μL water. Recover solutions after each wash.

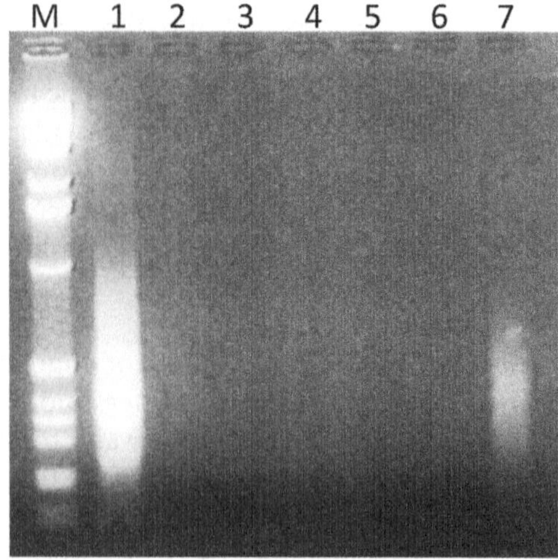

Fig. 3 Electrophoresis of products from enrichment procedure. M = ladder 1 kb, 1 = DNA + adaptor (1:1,000), 2 = hybridization solutions, 3–6 = wash solutions, 7 = DNA adsorbed to beads

2.8 PCR Control for Enrichment Procedure and Chemiluminescent Probe Detection

Mix 10× PCR buffer (2.5 μL), 2 mM dNTPs, 0.01 M adaptor primer, 1.5 U Taq DNA polymerase, and 3 μL of sample* and complete at 25 μL final volume [*samples of DNA-adaptor (1:1,000), hybridization solutions, wash solutions (first to fourth), and DNA adsorbed to beads]. PCR conditions of 95 °C for 3 min (hot start), 94 °C for 2 min, before 25 cycles of 94 °C for 45 s, 56 °C for 45 s, and 72 °C for 2 min followed by 7 min at 72 °C for final extension.

2.9 Hybridization

PCR products (25 μL) are electrophoresed in 2 % agarose using TAE buffer, stained with ethidium bromide (Fig. 3). After electrophoresis, treat gel with denature solution (NaOH 0.5 M + NaCl 1.5 M) with constant agitation for 30 min. Rinse in deionized water, add neutralization solution (NaCl 0.5 M + 0.5 M Tris–HCl pH 8.0), and shake for 30 min. Transfer the DNA by capillary using Hybond-N$^+$ membrane and allow the transfer for 8–12 h. Expose the membrane to a source of UV irradiation (254 nm) and neutralize by washing, twice for 5 min each, in 2× SSPE and pre-hybridize at 65 °C for 3 h. Then, discard the pre-hybridization solution, add the probe (3 μg/μL) previously denatured in 2× SSPE (95 °C for 20 min), and incubate overnight. Remove unbound biotinylated probe by washing 2× for 5 min each in 2× SSC (1× SSC = 150 mM NaCl, 15 mM sodium citrate) + 0.1 % SDS followed by 15 min in 0.1 % SSC + 0.1 % SDS. After rinse, detect the site of biotinylated probe by chemiluminescence using a digoxigenin-labeled nucleic acids and CSPD substrate.

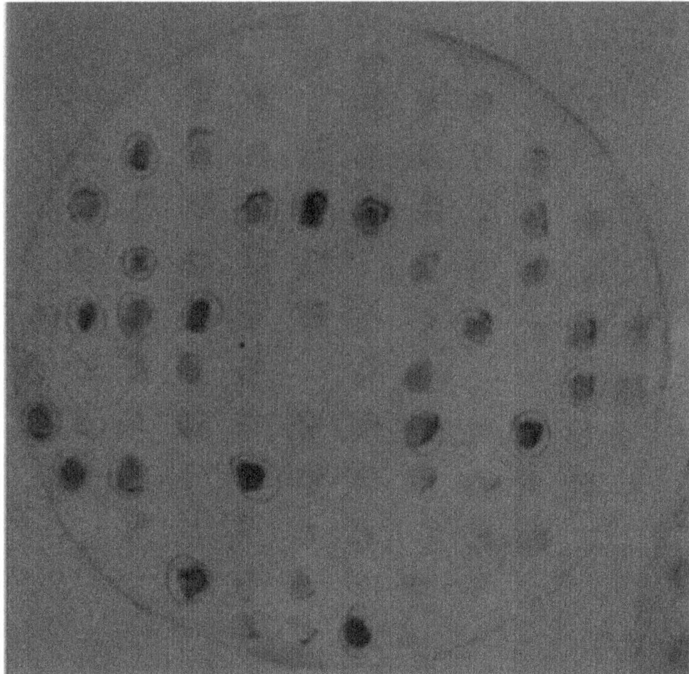

Fig. 4 Positive clones to microsatellite sequences (GT) after enrichment procedure

2.10 Cloning of PCR Fragments

The DNA recovered from the enriched library is PCR-amplified in a 25 μL reaction with ten replicates. Purify the PCR products are purified using a Gel and PCR Clean-Up System. Then, cloning the fragments into a vector using 10× T4 Buffer (1 μL), vector (1 μL), T4 DNA ligase (1 U); complete the volume with water to 10 μL, and incubate overnight at 4 °C. Transform recombinants into competent Dh 5α or JM109 *E. coli* cells. Transfer the resulting colonies to Hybond-N$^+$ membrane. Expose the membrane to a source of UV irradiation (254 nm) and neutralize by washing 2 × 5 min each in 2× SSPE and pre-hybridize at 65 °C for 3 h. Then, discard the pre-hybridization solution, add the biotinylated probe (3 μg/μL) previously denatured in 2× SSPE (95 °C for 20 min), and incubate overnight. Remove unbound biotinylated probe by washing 2× for 5 min each in 2× SSC (1× SSC = 150 mM NaCl, 15 mM sodium citrate) + 0.1 % SDS followed by 15 min in 0.1 % SSC + 0.1 % SDS. After rinse, detect the site of the biotinylated probe by chemiluminescence using a digoxigenin-labeled nucleic acids and CSPD substrate (Fig. 4).

In the next and last step, plasmid DNA is isolated from the selected positive clones and they are sequenced using standard protocol and primers flanking the repeated sequences are designed.

Acknowledgments

The authors gratefully acknowledge the financial support by INCT Project, CNPq, and FAPESP. V.M.N., M.C.Y., M.B., and M.A.M. are recipients of research fellowships from CNPq.

References

1. Zane L, Bargelloni L, Patarnello T (2002) Strategies for microsatellite isolation: a review. Mol Ecol 11:1–16
2. Palmieri DA, Novelli VM, Bastianel M, Cristofani-Yaly M, Astúa-Monge G, Carlos EF, Carlos de Oliveira A, Machado MA (2007) Frequency and distribution of microsatellites from ESTs of citrus. Genet Mol Biol 30:1009–1018
3. Nicolaou SA, Gaida SM, Papoutsakis ET (2011) Coexisting/Coexpressing Genomic Libraries (CoGeL) identify interactions among distantly located genetic loci for developing complex microbial phenotypes. Nucleic Acids Res 39:e152
4. Victoria FC, da Maia LC, de Oliveira AC (2011) In silico comparative analysis of SSR markers in plants. BMC Plant Biol 11:15
5. Egan AN, Schlueter J, Spooner DM (2012) Applications of next-generation sequencing in plant biology. Am J Bot 99:175–185
6. Zalapa JE, Cuevas H, Zhu H, Steffan S, Senalik D, Zeldin E, Mccown B, Harbut R, Simon P (2012) Using next-generation sequencing approaches to isolate simple sequence repeat (SSR) loci in the plant sciences. Am J Bot 99:193–208
7. Murray MG, Thompson WF (1980) Rapid isolation of high molecular weight plant DNA. Nucleic Acids Res 8:4321–4325

Chapter 3

PCR-Based Isolation of Microsatellite Arrays (PIMA)

Heng-Sheng Lin and Song-Bin Chang

Abstract

Microsatellite is one of the most high-speed developing genetic markers for its wide application in molecular biology researches. It is proved to be a powerful marker-assisted tool in genetic relationship identification, the inheritance breeding, the population genetics, the physical map construction, the management and security of germplasm. These short tandem repeats loci are distributed throughout the eukaryotic genome. They represent not only highly conservative trait but also significant differentiation properties between individuals, making it advantageous over other molecular markers. Traditionally, hard labor is required for isolating these loci and the flanking sequences, including small fragment DNA library construction, DNA cloning, radioactive hybridization, sequencing, and microsatellite test. PIMA is a relatively simple microsatellite isolation technique which avoids not only library construction but also radioactivity manipulation. This approach builds on random amplified polymorphic DNA (RAPD) process but investigates microsatellite arrays by repeat-specific PCR rather than radioactive hybridization. PIMA screening microsatellites use one repeat-specific and two vector primers to run PCR. A number of useful vectors are widely circulated and the repeat-specific primer is easy to obtain. The advantages of obtaining both flank sequences simultaneously, no need of specific sequencing primers, the ease of operation, and well amplification of bacterial colonies persuade us of its high value. It prevails other tools because of its traits of cheaper, high-efficient, and relatively lower requirement of specialized equipment tool. Since no protocol is universal and perfect for every species, it is recommended that modification should be made according to the objective of the experiments. Existing examples serve as good sources of future works.

Key words Microsatellite, PIMA, PCR-based isolation microsatellite array, RAPD, Repeat-specific primer

1 Introduction

Polymorphic microsatellite repeat arrays have become one of the most powerful molecular tools used with applications in a wide range of different fields. Microsatellites are loci with short repeating nucleotide throughout the eukaryotic genome. Microsatellite was manipulated as a powerful tool applied in many fields, such as genetic relationship identification, the population genetics, physical map construction, the management and security of germplasm, the marker-assisted breeding, and many other biological

Stella K. Kantartzi (ed.), *Microsatellites: Methods and Protocols*, Methods in Molecular Biology, vol. 1006, DOI 10.1007/978-1-62703-389-3_3, © Springer Science+Business Media, LLC 2013

applications (1–3). It prevails other tools because of its highly conservative traits and significant differentiation properties between individuals. Although development of the microsatellite markers was time and cost consuming, the application, convenience, and high information content are the best among molecular tools (4).

Before using microsatellites as markers, DNA sequences flanking the repeat motifs must be found. Traditional strategy for microsatellite isolation contains the construction of a small fragment DNA library, in which fragmented genomic DNA is inserted into a vector. These small DNA fragments are then used as templates for primer extension reactions, primed with repeat-specific oligonucleotides. Theoretically, this library construction operation only generates a double-stranded product from vectors containing the desired repeat. Only a limited investigated genome is cloned; therefore, the population of inserts undergoes the problem that rare repeat motifs will lose a lot (5). Additionally, microsatellite methods including the step of genomic library construction will prolong the processes by several months and the relatively long procedures might explain their limited application (3). To avoid the lengthy procedures of library construction, several straightforward approaches relating to RAPD modification have been proposed (6).

This PCR isolation of microsatellite arrays (PIMA) approach was proposed by Lunt et al. (7). They found out that RAPD profile contained many distinctive microsatellite loci and because of the unique cloned products can exceed the number of visible bands, which seems to strengthen the sensitivity during the stages of cloning and screening. In addition, microsatellites were reported to be present in RAPD bands (8). These use the fact that the RAPD fragments contain microsatellite repeats more frequently than traditional genomic clones (9, 10). Therefore, RAPD-based techniques are suitable for microsatellite isolation. Other than PIMA, random amplified hybridization microsatellites (RAHM) is also a process that could be operated simply through Southern hybridization of RAPD profiles with repeat-containing probes, followed by the selective cloning of positive bands (9, 10). These approaches with operation of radioactivity are generally more sensitive, but the need for dedicated equipment and laboratory space for the manipulation of radionucleotides might be the limitations for researchers that could not access to these equipment. Moreover, the short life of radioisotopes also makes radiolabeled probes of limited use. Recently, nonradioactive labeling techniques have greatly improved the efficiency, which allow comfortable and safer working conditions. Additional bonus of these techniques is the long-term storage of the probes. To identify the repeat-containing clones, specific primers are designed and the PCR conditions are optimized to allow the amplification of each locus from different individuals of a population (3). In view of this, PIMA seems to be much safer. It could be accessed through the cloning of all the RAPD products and then screening of arrayed clones using PCR (7).

Polymerase chain reaction (PCR) is the most convenient technique that is used in molecular research by scientists. PCR-based identification of microsatellite arrays (PIMA) was a relatively simple technique to isolate microsatellite, which avoids not only library construction but also radioactivity manipulation. This approach starts on random amplified polymorphic DNA (RAPD) process but investigates microsatellite arrays by repeat-specific PCR rather than radioactive hybridization. In the beginning, RAPD primers are manipulated to obtain randomly amplified fragments from the target species genome. These fragments are cloned by using a T-vector and arrayed clones are screened using one repeat-specific and two vector primers by PCR. Positive bands could be recognized by clones with one or more bands comparing to standard reaction (only two vector primers were used in PCR). Clones with positive results were cultivated and the colony DNA were then sequenced (7). This PCR-based microsatellite isolation strategy is widely accepted due to its easy operation by PCR, yet the procedure of clone production is still essential.

Actually, a similar strategy using vector- and repeat-specific primers has been established (11, 12), which skips the procedure of cloning but screens the fragment-containing vectors by PCR directly. These strategies seem to have higher efficiency of screening; however, several benefits persuade scientists to choose PIMA technique because of its ability to obtain both flanking regions at the same time. Theoretically, higher screening speed would increase the percentage of false positives; the advantages of obtaining both flanks simultaneously, the lack of a need for specific sequencing primers, the ease of operation, and well amplification of bacterial colonies of PIMA demonstrated its high value (7). PIMA is proved to be cheaper, with high efficiency, and with the advantage that it requires a minimum of specialized equipment (7). In recent years, PIMA had been reported to successfully isolate microsatellites in many species, including gymnosperm, monocotyledon, dicotyledon, pteridophyte, mammals, fish, birds, reptiles, and arthropods (Table 1). We introduce this popular and convenient technique in this chapter.

Theoretically, most isolated microsatellite repeats could successfully be used. However, some criteria for microsatellite isolation are still followed by our laboratories. For example, the size of expected microsatellite lower than 250 bp accessed by forward and reverse primers through PCR would be favored for its ease of observation. According to the experiences in our laboratory, larger sizes of microsatellite fragments will lower the resolution of recognition in gel electrophoresis. The expected sizes of most microsatellites developed are smaller than 250 bp. However, no matter how

Table 1
Different species, vectors, and competent cells using PIMA to isolate microsatellites

	Species	Vector system	Competent cell (*E. coli*)	DNA extraction strategy	References
Gymnosperm	*Taxus sumatrana*	pGEM-T Easy Vector	–	CTAB method (16)	(47)
	Cycas hainanensis	pGEM-T vector	DH5α	CTAB method (16)	(49)
Monocotyledon	*Setaria italica*	pGEM-T Easy Vector	DH5α	DNeasy Plant Mini Kit (Qiagen, Hilden, Germany)	(33)
	Miscanthus sinensis	pGEM-T Easy Vector	–	CTAB method (16)	(64)
Dicotyledon	*Linum usitatissimum*	pMD18-T vector	Top10	TIAN Gel Midi Purification Kit (Tiangen, China)	(32)
	Acer opalus	Dephosphorylated *BamHI*-digested pBluescript II SK(+/−) plasmid	XL1–blue	CTAB method (16)	(50)
	Camellia sinensis	pGEM-T Easy Vector	DH5α	CTAB method (16)	(65)
	Pedicularis verticillata	pGEM-T Easy Vector	DH5α	CTAB method (16)	(66)
	Suzukia shikikunensis	pGEM-T Easy Vector	DH5α	CTAB method (16)	(67)
	Euphrasia nankotaizanensis	pGEM-T Easy Vector	DH5α	CTAB method (16)	(42)
	Ludwigia polycarpa	pGEM-T Easy Vector	DH5α	CTAB method (16)	(68)
	Ajuga taiwanensis	pGEM-T Easy Vector	DH5α	CTAB method (16)	(63)
	Fatsia polycarpa	pGEM-T Easy Vector	DH5α	CTAB method (16)	(69)
Pteridophyte	*Lycopodium fordii* Bak.	pGEM-T Easy Vector	DH5α	CTAB method (16)	(70)
Mammalia	*Bubalus bubalis*	PTZ57R TA cloning vector	DH5α	Salting out procedure (16)	(19)
	Ailuropoda melanoleuca	pMD18-T vector	Top10	Standard phenol–chloroform procedures (19)	(71)
	Ursus thibetanus	pMD18-T vector	Top10	Standard phenol–chloroform procedures (19)	(72)
	Ailurus fulgens	pMD18-T vector	Top10	Standard phenol–chloroform procedures (19)	(73)
	Apodemus agrarius	pMD18-T vector	Top10	Standard phenol–chloroform procedures (19)	(37)
	Apodemus draco	pMD18-T vector	Top10	Standard phenol–chloroform procedures (19)	(38)
	Megaderma lyra	pTZ57R/T cloning vector	Top10	Standard phenol–chloroform procedures (19)	(74)

Fish	*Cynoglossus semilaevis*	T-vector	–	Standard protocol described by Strauss (20)	(75)
	Scophthalmus maximus	T-vector	–	Standard protocol described by Strauss (20)	(40)
	Lates calcarifer	pGEM-T vector	JM109	Phenol–chloroform extraction methods (21)	(34)
	Pararasbora moltrechti	pGEM-T vector	–	Standard phenol–chloroform procedures (19)	(76)
	Varicorhinus alticorpus	pGEM-T vector	–	Standard phenol–chloroform procedures (19)	(77)
	Coilia mystus	pGEM-T vector	–	Standard phenol–chloroform procedures (19)	(78)
	Candidia barbata	pGEM-T vector	–	Standard proteinase K-SDS digestion followed phenol–chloroform extraction (22)	(79)
	Acrossocheilus paradoxus	pT7 Blue T-vector	–	Standard proteinase K-SDS digestion, phenol–chloroform extraction (22)	(80)
	Hemibarbus labeo	pT7 Blue T-vector	–	Standard proteinase K-SDS digestion, phenol–chloroform extraction (22)	(36)
	Centropomus undecimalis	Bluescript PBC KS-plasmid vectors	–	–	(81)
	Squalidus argentatus	pMD19-T vector	DH5α	Standard phenol–chloroform procedures (19)	(82)
Bird	*Garrulax morrisonianus*	pGEM-T Easy Vector	DH5α	Genomic DNA Mini Kit (Geneaid, Taipei, Taiwan)	(60)
Reptiles	*Coronella austriaca*	pGEM-T Easy Vector	JM109	Vertebrate genome extraction (23)	(83)
Insects	*Lysandra bellargus*	pGEM-T Easy Vector	JM109	–	(35)
Arthropods	*Penaeus vannamei*	pGEM-T vector	DH5α	Standard phenol–chloroform procedures (19)	(51)
	Austinogebia edulis	pGEM-T vector	DH5α	Standard proteinase K-SDS digestion followed by phenol–chloroform extraction (22)	(84)
	Caridina gracilipes	pGEM-T vector	DH5α	Standard phenol–chloroform procedures (19)	(85)
	Scylla paramamosain	pMD19-T vector	Top10	Standard proteinase K-SDS digestion followed by phenol–chloroform extraction (22)	(46)
	Tetranychus urticae	pGEM-T Easy Vector	DH5α	Salting out protocol (23)	(30)

larger the microsatellite size should be selected, polymorphism level is still the primary consideration since the applications of microsatellites span over different areas ranging from ancient and forensic DNA studies to population genetics and conservation/management of biological resources (1). Therefore, in regard to the versatility of microsatellite, it is nearly impossible to find criteria or protocol that could be suitable for all purposes. It is good to modify the existing isolation approaches or to establish new ones for specific purposes as demands arise.

2 Materials (Table 2)

2.1 PCR-Based Identification of Microsatellite Arrays

Polymerase chain reaction (PCR) is the most convenient technique that is used in molecular research by scientists. During PCR, oligonucleotide primer molecules are bound at low temperature to templates of heat-denatured DNA and extended on their 3′ end using a thermostable DNA polymerase. Three steps including DNA denaturation, primer annealing, and extension are repeated several times under program control to amplify a large number of identical DNA sequence copies between the primers (13) (see Note 1). The following are several basic equipment used in PCR process:

1. Thermocycler, e.g., MyCycler™ Thermal Cycler, Bio-Rad, USA.
2. DNA template.
3. Two primers.
4. *Taq* polymerase.
5. Deoxynucleoside triphosphates/dNTPs (10 mM).
6. Buffer solution including divalent cations, and monovalent cations (10× PCR buffer: 500 mM KCl, 100 mM Tris–HCl (pH 8.3), and 15 mM $MgCl_2$) (14).
7. The PCR is commonly carried out in a reaction volume of 10–200 μl in small reaction tubes in a thermal cycler (15). The volume depends on different species.

2.2 Preparation for Genomic DNA of Target Species

1. DNA isolation is a widely used procedure to obtain DNA for further molecular studies or analysis. Several steps are operated in a DNA extraction. Although some steps are optional such as lipid and protein removal, higher quality of DNA is favored. Theoretically, lower chances of any unexpected interference may increase the efficiency of experiments (see Notes 2 and 3).
2. Cell disruption or cell lysis, which means to break the cells in open condition. To expose the inside DNA could be accessed by chemical and physical methods such as blending, grinding, or sonicating the sample tissues. For cell disruption, mortar with pestle, liquid nitrogen, or −80 °C freezer is needed.

Table 2
List of materials used in PIMA

Steps of PIMA		Materials needed
Basic equipment for labs using PIMA		1. Thermocycler 2. DNA template 3. Two primers 4. Taq polymerase 5. Deoxynucleoside triphosphates/dNTPs (10 mM) 6. Buffer solution: including divalent cations, and monovalent cations (10× PCR buffer: 500 mM KCl, 100 mM Tris–HCl (pH 8.3), and 15 mM MgCl2) 7. 10–200 μl in small reaction tubes 8. Personal protection equipment (lab coat, gloves, goggles)
Genomic DNA isolation	CTAB method	1. Water bath 2. CTAB buffer: 100 ml 1 M Tris–HCl pH 8.0, 280 ml 5 M NaCl, 40 ml of 0.5 M EDTA, 20 g of CTAB (cetyltrimethylammonium bromide), bring total volume to 1 l with ddH2O 3. Polyvinylpyrrolidone (PVP) and β-mercaptoethanol: prior to starting extraction, add 0.8 g polyvinylpyrrolidone (PVP) and 100 μl β-mercaptoethanol into 20 ml CTAB buffer. Put the solution in the water bath for 10–20 min to dissolve the PVP 4. –80 °C Freezer/liquid nitrogen 5. Mortar and pestles 6. RNase A (100 mg/ml) 7. Ammonium acetate 8. Isopropanol 9. 70 % Ethanol (EtOH) 10. 95 % Ethanol (EtOH) 11. TE buffer: 1 mM EDTA, ethylenediaminetetraacetic acid (pH 8.0) and 10 mM Tris–HCl (pH 7.5)
	DNeasy Plant Mini Kit	1. Mortar and pestle 2. Centrifuge for microcentrifuge tubes 3. Buffer AP1(disruption buffer) 4. Buffer AP2 (acetic acid) 5. Buffer AP3/E (guanidine hydrochloride) 6. Buffer AW (wash buffer) 7. Buffer AE (10 mM Tris–Cl, 0.5 mM EDTA, pH 9.0)

(continued)

Table 2
(continued)

Steps of PIMA			Materials needed
Fragmentation	RAPD primers	Purchase/ generate and synthesis	1. Purchase the arbitrary nucleotide sequences (e.g., 5′-GTTTC GCTCC-3¢, 100 mM) 2. RAPD-primer generator free online (http://www2.uni-jena.de/biologie/mikrobio/tipps/rapd.html; J. Wöstemeyer, Institute of General Microbiology and Microbial Genetics, Germany)
	Gel extraction		Gel extraction kit of Geneaid, Taipei, Taiwan. (Cat. No. DF-100)
Ligation	T4 DNA ligase		T4 DNA ligase: 1 U/μl, 50 μl (Invitrogen, USA)
	Vector		1. pGEM-T Vector, pGEM-T Easy Vector (Promega, Madison, Wisconsin, USA) 2. pMD18-T vector (TaKaRa, Japan) 3. PTZ57R TA cloning vector (Fermentas, USA) 4. pT7 Blue T-vector (Novagen, USA) 5. Plasmid vectors (Bluescript PBC KS-, Stratagene, UK)
Transformation	Competent cell	*Escherichia coli*	1. *E. coli* Top10 competent cells (Invitrogen, USA), E. coli DH5α competent cells (Takara), XL1–blue E. coli competent cells (Stratagene, USA), E. coli JM109 competent cells (Promega, USA) 2. LB (Luria–Bertani) medium: 1,000 ml deionized water, 10 g Bactotryptone, 5 g Bacto yeast, 5 g NaCl, 1 ml 5 M NaOH, 1 ml 1 M HCl 3. LB plate: 1,000 ml deionized water, 10 g Bactotryptone, 5 g Bacto yeast, 5 g NaCl, 1 ml 5 M NaOH, 1 ml 1 M HCl, 15 g agar, 1 mg 1,000× ampicillin (for 40 plate) 4. Ampicillin 5. IPTG 6. X-gal 7. LB plates with ampicillin/IPTG/X-gal and SOC medium 8. SOC medium (20 g Tryptone, 5 g Yeast Extract, 0.5 g NaCl, 10 ml 250 mM KCl, adjust volume to 1 l with ddH_2O)
Screening/ selection	Blue–white screen		LB plates with ampicillin/IPTG/X-gal and SOC medium
	Colony preparation	Liquid/plate	A single colony of *E. coli* DH5-α, maintained on a fresh LB agar plate or was inoculated into 5 ml of LB medium and incubated at 37 °C with shaking at 200 rpm for 16 h
	PCR screening	Plasmid isolation/ direct PCR	1. Using kits, which are available from varying manufacturers to purify plasmid DNA (different types of plasmid isolation kits are named by size of bacterial culture and corresponding plasmid yield) 2. Alternatively, the bacterial colonies could be regarded as the NA template for PCR screening, this could be used for screening roughly and rapidly

	Microsatellite detection	Repeat-specific primer	1. Dinucleotide repeat-specific primer reported by Lunt et al. (7) 2. Deng et al. (32) choose trinucleotide repeat-specific primer (TTC and ATC) to find trinucleotide repeat microsatellites (see Note 11) 3. The following are examples of the repeat-specific primers: (a) TG-repeat primer (5¢-TGTGGCGG CCGC(TG)8V-3¢) as the repeat-specific primer (7) (b) Microsatellite-specific primer E (VRV $(TTC)_{10}$) or primer F (VRV $(ATC)_{10}$) for microsatellite isolation (32) (c) Repeat-specific primers including $(AC)_5$, $(AG)_5$, $(AT)_5$, $(CG)_5$, $(CT)_5$, and $(GT)_5$ and 2 vector primers including forward M13 and reverse M13 primers for operating the clone screening (33) (d) RAPD-based library for screening with three repeat-specific primers, namely, 5′-(GA)7H-3′, 5′-(CA)7D-3′, and 5′-(TG)7V-3′, where H=A/C, D=A/G, and V=A/C/G (34) (e) (CA)*n* repeats were designed to screen the colonies for using a colony-PCR-based approach (PIMA). The DNA from each colony was amplified using three primers: M13 forward and M13 reverse primers, plus a (CA)*n*-specific oligonucleotide (5′-TGTGGCGGCCGC(TG)8V-3′) (35)
Colony sequencing			1. Plasmid isolation: using kits, which are available from varying manufacturers to purify plasmid DNA (different types of plasmid isolation kits are named by size of bacterial culture and corresponding plasmid yield) 2. These years, many companies provide the services of sequencing; the following are some examples for PIMA isolation sequence: (a) Deng et al. (32) had the positive SSR-containing clones selected and sequenced by Beijing Genomics Institute Co., Ltd (b) Lin et al. (33) choose the strategy to begin with plasmid isolation. Plasmid DNA of positive clones was purified using the Plasmid Miniprep Kit (BioKit, Miaoli, Taiwan). 10 μl of plasmid DNA with a concentration of 100 ng/μl was used in each sequencing reaction. DNA sequencing in both directions of the insert DNA was conducted using an Applied Biosystems 3730 DNA Analyzer with BigDyeR Terminator v3.1 Cycle Sequencing Kit (Applied Biosystems, Foster, California, USA) (c) Lin et al. (36) had both strands of the insert DNA sequencing in both directions conducted with an Applied Biosystems Model 377A automated sequencer (Applied Biosystems) (d) Wu et al. (37) had the positive clones sequenced using the ABI Prism BigDye Terminator Cycle Sequencing Ready Reaction Kit (Applied Biosystems) and ABI 3730 Genetic Analyzer (e) Gu et al. (38) had the positive colonies sequenced in the forward direction using DYEnamic fluorescent cycle sequencing kit (ABI) and run on a Basestation Sequencer (GRI). Sequences containing a microsatellite motif were then sequenced in the reverse direction (f) Harper et al. (35) had the positive clones sequenced using BigDye Terminators (PE-Applied Biosystems) on a Perkin-Elmer ABI 377 automated sequencer (see Note 12)

(continued)

Table 2
(continued)

Steps of PIMA	Materials needed
Primer design	1. PRIMER 3: this widely used software for primer design is developed by Rozen and Skaletsky (39), which is available online at http://primer3.sourceforge.net/ 2. Tandem repeat finder (version 2.02) using by Liu et al. and Wang et al. (40–42). The criteria used in tandem repeat finder to identify microsatellites are as follows: 7 repeats for dinucleotide repeat, 5 repeats for trinucleotide repeat, and 4 repeats for tetranucleotide repeat 3. Bioedit Sequence Alignment Editor software (http://www.mbio.ncsu.edu/BioEdit/BioEdit.html) were used to analyze sequences by Deng et al. (32). The repeat numbers were determined using software SSRHunter1.3 (43). PCR primer pairs were designed using PRIMER 3 software (39) 4. Sabater-Muñoz et al. (30) checked the electropherograms and assembled consensus sequences by using Staden Package software for each clone (44). After comparison, primers were designed in conserved regions of microsatellite loci using Oligo software version v4.0 (45); amplification conditions were set up for each marker 5. Ma et al. (46) screened microsatellite sequences using the software SSRHunter 1.3 (43) with the criteria as follows: the minimum of three repeats for di-, tri-, and tetranucleotide repeats. Primers were designed using the software Primer Premier 5.0 (Palo Alto, Canada)

3. Lipid membrane removal: Removing the lipid membrane of cells could be accessed by adding a detergent or surfactants (e.g., CTAB buffer, 100 ml 1 M Tris–HCl pH 8.0, 280 ml 5 M NaCl, 40 ml of 0.5 M EDTA, 20 g of CTAB (cetyltrimethylammonium bromide), bring total volume to 1 l with ddH_2O).
4. Protein removal: Removing the protein of cells could be accessed by adding a protease (optional step) (e.g., Qiagen Cat. no. 19155).
5. RNA removal: Removing the RNA of cells could be accessed by adding an RNase (often done) (e.g., RNase A (17,500 U) 2.5 ml (100 mg/ml; 7,000 U/ml, solution); Qiagen Cat. no. 19101).
6. DNA precipitation with an alcohol—usually ice-cold ethanol or isopropanol. Since alcohols did not dissolve DNA, DNA will aggregate together as a pellet during centrifugation. This step also removes alcohol-soluble salt.
7. A chelating agent to sequester divalent cations such as Mg^{2+} and Ca^{2+}: They could higher the efficiency of this technique, since it prevents enzymes like DNase from degrading the DNA (e.g., 15 mM $MgCl_2$).
8. Protease: Cellular and histone proteins bound to the DNA can be removed either by adding a protease or by having the proteins precipitated with sodium or ammonium acetate, or extracted with a phenol–chloroform mixture prior to the DNA precipitation (e.g., Qiagen Cat. no. 19155).
9. Alkaline buffer or ultrapure water: The DNA can be resolubilized in a slightly alkaline buffer or in ultrapure water.
10. Several commonly used approaches of DNA isolation were listed (16–24) (Table 1) (see Note 4).
    (a) For CTAB method:
        - Device Preparation: Turn on the water bath and set to 55 °C.
        - Solutions Needed: CTAB buffer, isopropanol, 70 % ethanol, 95 % ethanol, TE buffer.

        CTAB buffer preparation (100 ml 1 M Tris–HCl pH 8.0, 280 ml 5 M NaCl, 40 ml of 0.5 M EDTA, 20 g of CTAB (cetyltrimethylammonium bromide), bring total volume to 1 l with ddH_2O).

        Prior to starting extraction, add 0.8 g polyvinylpyrrolidone (PVP) and 100 μl β-mercaptoethanol into 20 ml CTAB buffer (see Note 5).

        Put the solution in the water bath for 10–20 min to dissolve the PVP (see Note 6).

TE buffer: 1 mM EDTA, ethylenediaminetetraacetic acid (pH 8.0) and 10 mM Tris–HCl (pH 7.5).

2.3 Preparation of Random Amplified Polymorphic DNA Primers

1. RAPD markers: Arbitrary nucleotide sequence to generate DNA fragments from PCR amplification of random segments of genomic DNA. Since RAPD had been used frequently, the arbitrary nucleotide sequence could be obtained easily (e.g., MDBIO, Piscataway, New Jersey, USA).
2. Alternatively, one can generate arbitrary nucleotide sequences by RAPD-primer generator free online (http://www2.uni-jena.de/biologie/mikrobio/tipps/rapd.html; J. Wöstemeyer, Institute of General Microbiology and Microbial Genetics, Germany) and then synthesize the nucleotide sequences to be used.

2.4 PCR-Based Identification of Microsatellite Arrays Experiments

Cloning of small fragments involved four basic steps including fragmentation, ligation, transformation, and screening/selection:

2.4.1 Preparation for Cloning Experiments

1. Fragmentation means to provide fragments of target DNA. In PIMA approach, these fragments were generated by RAPD–PCR, which was listed in Subheading 2.3.
2. RAPD fragments with moderate size, which depends on the desire of the experiments, on agarose gel electrophoresis profile from an agarose gel could be extracted by a technique called gel extraction (or gel isolation). Three major methods could be selected for gel isolation, including spin column extraction, dialysis, and the traditional method. Spin column extraction is popular in these years since they are available as gel extraction kits from several major biotech manufacturers and the only devices needed is an ultracentrifuge (e.g., DF100, Geneaid, Taiwan).
3. Ligation means to glue pieces of DNA together in a desired sequence.
   (a) T4 DNA ligase: For routine ligation, one would need a T4 DNA ligase (25) which can be bought from many different companies. Companies that produce restriction enzymes usually produced T4 DNA ligase (e.g., Cat. No. 15224-041, Invitrogen, USA) (Table 1).
   (b) Vector: A DNA molecule used as a vehicle to transfer foreign genetic material into another cell. There are four major types of vectors, including plasmids, viruses, cosmids, and artificial chromosomes. The following are vectors that had been used in PIMA, such as pGEM-T Vector, pGEM-T Easy Vector (Promega, Madison, Wisconsin, USA), pMD18-T vector (TaKaRa, Japan), PTZ57R TA cloning vector (Fermentas, USA), pT7 Blue T-vector (Novagen, USA), plasmid vectors (Bluescript PBC KS–, Stratagene, UK), etc. (Table 1).
4. Transformation means to insert the newly formed pieces of DNA into cells. Bacteria capable of being transformed, whether

naturally or artificially, are called competent. The bacterial cell is called competent cell in transformation procedure. *Escherichia coli* was reported to show perfect artificially induced competence and allowed to be used as a host for convenient manipulation of DNA and protein expression. Different kinds of *E. coli* competent cell were used to perform the transformation experiment, such as *E. coli* Top10 competent cells, *E. coli* DH5α competent cells (Takara), XL1–blue *E. coli* competent cells (Stratagene, USA), or *E. coli* JM109 competent cells (Promega, USA) (Table 1) (see Note 7).

5. Furthermore, several solutions were used during the transformation experiments:
   (a) LB (Luria–Bertani) medium: 1,000 ml deionized water, 10 g Bactotryptone, 5 g Bacto yeast, 5 g NaCl, 1 ml 5 M NaOH, 1 ml 1 M HCl.
   (b) LB plate: 1,000 ml deionized water, 10 g Bactotryptone, 5 g Bacto yeast, 5 g NaCl, 1 ml 5 M NaOH, 1 ml 1 M HCl, 15 g agar, 1 mg 1,000× ampicillin (for 40 plate).
   (c) Ampicillin.
   (d) IPTG.
   (e) X-gal.
   (f) LB plates with ampicillin/IPTG/X-gal and SOC medium.
   (g) SOC medium (20 g Tryptone, 5 g Yeast Extract, 0.5 g NaCl, 10 ml 250 mM KCl, adjust volume to 1 l with ddH_2O).
6. Screening/selection means to select the cells that were successfully transformed with the new DNA.
   (a) We should select *E. coli* colony containing vector sequences because only a small portion of the cells will actually take up plasmid vectors. In artificial genetic selection, cells that do not contain vector DNA are killed selectively, and only those cells that contain vectors can survive. These cells can actively replicate DNA containing the selectable marker gene encoded by the vector (26, 27). These selectable markers are usually genes when bacterial cells are used as host organisms; it confers resistance to an antibiotic that would otherwise kill the cells, typically ampicillin. When the cells harbor the vector, they will survive when exposed to the antibiotic, while those didn't take up vector sequences will die.
   (b) Blue–white screen experiment: To check each individual colony for the presence of the insert is time-consuming. Blue–white screen is a method for the detection of the insert. It is therefore useful for making this procedure less time and labor intensive. It allows for identification of

successful products of cloning reactions through the color of the bacterial colony. The method is based on the principle of α-complementation of the β-galactosidase gene (see Note 8).

(c) X-gal is an organic compound containing galactose linked to a substituted indole (28). In gene cloning, X-gal is used as a visual indicator of whether a cell expresses a functional β-galactosidase enzyme in blue/white screening. When cells are grown in X-gal-containing plates, the presence of an active β-galactosidase may be detected, since the blue-colored product precipitated within cells resulted in the blue colonies. In reverse, when a gene of interest successfully ligated into the plasmid vector, it therefore disrupts the lacZα gene, and no functional β-galactosidase can form, resulting in white colonies. The cells with ligated insert can then be easily identified by its white color from the unsuccessful blue ones. Example of cloning vectors used for this test are pUC19, pBluescript, and pGEM-T Vectors, and it also requires the use of specific *E. coli* host strains such as DH5α which carries the mutant lacZΔM15 gene (29) (Table 1) (see Note 9).

2.4.2 Preparation for Positive Clone Screening by PCR

1. Colonies with positive reaction means the fact that the vector with small DNA fragment insertion had existed in the *E. coli* cell. Two major DNA templates for PCR screening include the plasmid DNA or colony PCR (30).
2. In order to obtain the small fragment genomic DNA, plasmid DNA would be isolated.
   (a) 1 cm^3 liquid culture of *E. coli* containing DNA fragment insertion, grown overnight at 37 °C. LB (Luria–Bertani) medium: 1,000 ml deionized water, 10 g Bactotryptone, 5 g Bacto yeast, 5 g NaCl, 1 ml 5 M NaOH, 1 ml 1 M HCl.
   (b) For plasmid DNA isolation, kits are available from varying manufacturers to purify plasmid DNA, which are named by size of bacterial culture and corresponding plasmid yield (see Note 10).
   (c) Mini-preparation of plasmid DNA isolation is used in the process of molecular cloning to analyze bacterial clones. It is based on the alkaline lysis method invented by Birnboim and Doly (31). Generally, plasmid DNA yield of mini-preparation is 20–30 μg depending on the cell strain (e.g., Cat. No. A1222, Promega, USA).
3. In addition, one could use the colony DNA as the PCR template without plasmid DNA isolation, which could higher the efficiency of microsatellite screening.

2.4.3 Microsatellite Detection

1. PIMA–PCR: For detecting microsatellite in the plasmid of colony, PCR reaction would be operated by one repeat-specific primer and two vector primers.
2. Repeat-specific primer preparation: Comparing to traditional microsatellite isolation approaches, which have a radioactive hybridization step for detecting microsatellite-containing fragments. In order to skip the dangerous radioactive operation, PIMA use repeat-specific primer for detecting potential microsatellite.
3. The following are some examples for repeat-specific primer design from 2000 to 2012:
   (a) As Lunt et al. (7) report, they chose TG-repeat primer (5′-TGTGGCGG CCGC(TG)8V-3′) as the repeat-specific primer.
   (b) Deng et al. (32) had used microsatellite-specific primer E (VRV $(TTC)_{10}$) or primer F (VRV $(ATC)_{10}$) to isolate microsatellite.
   (c) Lin et al. (33) operated the clone screening using repeat-specific primers including $(AC)_5$, $(AG)_5$, $(AT)_5$, $(CG)_5$, $(CT)_5$, and $(GT)_5$ and two vector primers including forward M13 and reverse M13 primers.
   (d) Sim and Othman (34) used RAPD-based library to screen with three repeat-specific primers, namely, 5′-(GA)7H-3′, 5′-(CA)7D-3′, and 5′-(TG)7V-3′, where H = A/C, D = A/G, and V = A/C/G.
   (e) Harper et al. (35) had designed (CA)n repeats to screen the colonies for using a colony-PCR-based approach (PIMA). The DNA from each colony was amplified using three primers: M13 forward and M13 reverse primers, plus a (CA)*n*-specific oligonucleotide (5′-TGTGGCGGC CGC(TG)8V-3′).
4. Most microsatellite isolation using PIMA strategy followed the dinucleotide repeat-specific primer reported by Lunt et al. (7). Actually, various types of microsatellite were detected including dinucleotide and trinucleotide repeats. According to the microsatellite repeat reported by Deng et al. (32), they choose trinucleotide repeat-specific primer (TTC and ATC) to find trinucleotide repeat microsatellites (see Note 11).

2.4.4 Colony Sequencing

Positive bands were being seen on the PCR profile, which show out one more bands than standard reaction. This extra band indicated the presence of microsatellite loci within the inserted fragments. In order to obtain the microsatellite repeat sequences and its flanking nucleotide pattern, plasmids of positive colony should be isolated and the plasmid DNA could be applied to be sequenced.

These years, many companies provide the services of sequencing; the following are some examples for PIMA isolation sequence:

1. Deng et al. (32) had the positive SSR-containing clones selected and sequenced by Beijing Genomics Institute Co., Ltd.
2. Lin et al. (33) choose the strategy to begin with plasmid isolation. Plasmid DNA of positive clones was purified using the Plasmid Miniprep Kit (BioKit, Miaoli, Taiwan). 10 μl of plasmid DNA with a concentration of 100 ng/μl was used in each sequencing reaction. DNA sequencing in both directions of the insert DNA was conducted using an Applied Biosystems 3730 DNA Analyzer with BigDyeR Terminator v3.1 Cycle Sequencing Kit (Applied Biosystems, Foster, California, USA).
3. Lin et al. (36) had both strands of the insert DNA sequencing in both directions conducted with an Applied Biosystems Model 377A automated sequencer (Applied Biosystems).
4. Wu et al. (37) had the positive clones sequenced using the ABI Prism BigDye Terminator Cycle Sequencing Ready Reaction Kit (Applied Biosystems) and ABI 3730 Genetic Analyzer.
5. Gu et al. (38) had the positive colonies sequenced in the forward direction using DYEnamic fluorescent cycle sequencing kit (ABI) and run on a Basestation Sequencer (GRI). Sequences containing a microsatellite motif were then sequenced in the reverse direction.
6. Harper et al. (35) had the positive clones sequenced using BigDye Terminators (PE-Applied Biosystems) on a Perkin-Elmer ABI 377 automated sequencer (see Note 12).

2.4.5 Primer Design

After sequencing, several softwares could help in finding microsatellite sequences and also some software for designing the sequence-specific primers.

1. PIMER 3: This widely used software for primer design is developed by Rozen and Skaletsky (39), which is available online at http://primer3.sourceforge.net/.
2. Liu et al. (40) had the microsatellite sequences screened using tandem repeat finder (version 2.02) (41). The criteria used in tandem repeat finder to identify microsatellites are as follows: seven repeats for dinucleotide repeat, five repeats for trinucleotide repeat, and four repeats for tetranucleotide repeat.
3. Wang et al. (42) designed specific primers according to the nucleotide sequences upstream and downstream of the repetitive DNA using PRIMER 3 (39).
4. Deng et al. (32) made the sequences analyzed using Bioedit Sequence Alignment Editor software (http://www.mbio.ncsu.edu/BioEdit/BioEdit.html). The repeat numbers were

determined using software SSRHunter1.3 (43). PCR primer pairs were designed using PRIMER 3 software (39).

5. Sabater-Muñoz et al. (30) checked the electropherograms and assembled consensus sequences by using Staden Package software for each clone (44). After comparison, primers were designed in conserved regions of microsatellite loci using Oligo software version v4.0 (45); amplification conditions were set up for each marker.
6. Ma et al. (46) screened microsatellite sequences using the software SSRHunter 1.3 (43) with the criteria as follows: the minimum of three repeats for di-, tri-, and tetranucleotide repeats. Primers were designed using the software Primer Premier 5.0 (Palo Alto, Canada).

3 Methods

3.1 Genomic DNA Preparation

1. Tissues of target species should be prepared for genomic DNA isolation. Tissues are stored in −20 °C for keeping the target DNA from degradation.

The following are several protocols commonly used by laboratories.

3.1.1 Cetyltrimethylammonium Bromide Method (16)

1. Collect the tissue that you need from the −80 °C freezer and keep them on dry ice. Or you could use liquid nitrogen to grind the tissue up.
2. Put 50–60 mg of frozen tissue in an eppendorf tube without direct contact by hand (see Note 13).
3. Then, you could submerge the tube in liquid nitrogen or you could directly use liquid nitrogen to grind the tissue by pestles.
4. Add 500 μl of CTAB buffer and mix the tissue in the tubes (see Note 14).
5. Incubate the mixture in tubes at 55 °C for few hours, mixing once after 30 min.
6. Add 1.5 μl RNase A after incubating for 1 h, and incubate the mixture at 37 °C for 15 min (see Note 15).
7. Add 500 μl of chloroform into the samples and mix gently shaking tubes (see Note 16).
8. Centrifuge for 7 min at 16,000 × *g*. And balance of tubes should be carefully concerned.
9. Transfer the top layer (aqueous layer) and estimate the volume into the new tube (see Note 17).

10. Add 0.08 volumes cold 7.5 M ammonium acetate, and add 0.54 volumes of cold isopropanol into the tube. Mix the mixture by inverting tubes 20–30 times.
11. Incubate the tube with mixture on ice for 30–40 min.
12. Centrifuge the tube for 3 min at 16,000 × *g*.
13. Discard supernatant and add 700 μl 70 % EtOH, invert the tubes for 5–10 times.
14. Centrifuge the tube for 1 min at 16,000 × *g*.
15. Discard supernatant and add 700 μl 95 % EtOH, invert the tubes for 5–10 times.
16. Centrifuge the tube for 1 min at 16,000 × *g*.
17. Discard the supernatant from the tube without dislodging the pellet.
18. Invert the tubes on a clean tissue and allow drying for 10–15 min upside down, or until pellet looks drying (see Note 18).
19. Finally, hydrate pellets with 50 μl TE and store the DNA -20 °C.

3.1.2 DNeasy Plant Mini Kit Method

1. Disrupt samples using a mortar and pestle (see Note 19).
2. Add 400 μl Buffer AP1(disruption buffer) and 4 μl RNase A in the tube containing the disrupted tissues. Vortex and incubate the tubes for 10 min at 65 °C (usually in water bath). Invert the tubes 2–3 times during incubation (see Note 20).
3. Add 130 μl Buffer AP2 (acetic acid) into the tube. Mix and then incubate the tube for 5 min on ice.
4. Centrifuge the lysate for 5 min at 20,000 × *g*.
5. Pipet the lysate into a QIAshredder spin column placed in a 2 ml collection tube. Centrifuge the tube for 2 min at 20,000 × *g*.
6. Transfer the flow-through into a new tube (see Note 21).
7. Then add 1.5× volumes of Buffer AP3/E (guanidine hydrochloride) into the tube, and mix the mixture by pipetting.
8. Transfer 650 μl of the mixture into a DNeasy Mini spin column placed in a 2 ml collection tube. Centrifuge the tube for 1 min at ≥6,000 × *g*.
9. Discard the flow-through out of the tube. Repeat this step until the remaining sample ran out.
10. Place the spin column into a new 2 ml collection tube. Add 500 μl Buffer AW (wash buffer), and centrifuge for 1 min at ≥6,000 × *g*. Discard the flow-through.
11. Add another 500 μl Buffer AW. Centrifuge for 2 min at 20,000 × *g* (see Note 22).

12. Transfer the spin column to a new microcentrifuge tube (1.5 ml or 2 ml).
13. Add 100 μl Buffer AE (10 mM Tris–Cl, 0.5 mM EDTA, pH 9.0.) for elution. Incubate for 5 min at room temperature (15–25 °C). Centrifuge for 1 min at ≥6,000 × *g*.
14. Repeat step 13.

3.2 Random Amplified Polymorphic DNA Fragment Construction

For small DNA fragments, PIMA skips the traditional hard laboring library construction. RAPD–PCR is performed by random primers, genomic DNA, dNTP, $MgCl_2$, Taq buffer (mineral oil, for thermocyclers without a heated lid), ddH_2O, and Taq DNA polymerase.

In our lab we had set the conditions of reactions to run on a MyCycler™ Thermal Cycler using the following conditions:

1. 3 min of denaturation at 94 °C.
2. 45 cycles at 94 °C for 1 min, annealing temperature specific to each primer for 1 min, and extension at 72 °C for 2 min.
3. Final extension at 72 °C for 5 min (33).

Furthermore, several programs for RAPD–PCR from 2000 to 2011were listed in Note 23 (7, 40, 46–51).

3.3 Clone Operation and Screening

Cloning protocols vary with different vectors and competent cells. The following common system produced by Promega Company was used in our laboratory. The content and procedure of protocol could be modified depending on species and demand.

3.3.1 Ligation Using the pGEM-T and pGEM-T Easy Vector (Promega)

1. Ligation Reactions: Use high-efficiency competent cells for transformations. Ligation of fragments with a single-base overhang can be inefficient, so it is essential to use cells with a transformation efficiency of 1 × 108 cfu/μg DNA in order to obtain a reasonable number of colonies. Other host strains may be used, but they should be compatible with blue/white color screening and standard ampicillin selection.
2. If you are using competent cells other than JM109 High-Efficiency Competent Cells purchased from Promega, it is important that the appropriate transformation protocol be followed. Selection for transformants should be on LB/ampicillin/IPTG/X-gal plates. For best results, do not use plates that are more than 1 month old.

3.3.2 Transformation Experiment

1. Materials prepared: LB plates with ampicillin/IPTG/X-gal and SOC medium (see Subheading 2.4.1).
2. Prepare two LB/ampicillin/IPTG/X-gal plates for each ligation reaction, plus two plates for determining transformation efficiency. Equilibrate the plates to room temperature.

3. Centrifuge the tubes containing the ligation reactions to collect the contents at the bottom. Add 2 μl of each ligation reaction to a sterile (17 × 100 mm) polypropylene tube or a 1.5 ml microcentrifuge tube on ice. Set up another tube on ice with 0.1 ng uncut plasmid for determination of the transformation efficiency of the competent cells (standard control set).
4. Remove tube of frozen JM109 competent cells from storage and place it in an ice bath until just thawed (it takes several minutes depending on weather and temperature at that time). Mix the cells by gently flicking the tube. Avoid excessive pipetting, as the competent cells are extremely fragile.
5. Carefully transfer 50 μl of cells into each tube prepared in step 3 (use 100 μl of cells for determination of transformation efficiency).
6. Gently flick the tubes to mix and place them on ice for 20 min.
7. Heat-shock, put the cells for 45–50 s in a water bath at exactly 42 °C (do not shake).
8. Immediately return the tubes to ice for 2 min.
9. Add 950 μl room-temperature SOC medium to the tubes containing cells transformed with ligation reactions and 900 μl to the tube containing cells transformed with uncut plasmid (LB medium may be substituted, but colony number may be lower).
10. Incubate for 1.5 h at 37 °C with shaking softly.
11. Plate 100 μl of each transformation culture onto duplicate LB/ampicillin/IPTG/X-gal plates. For the transformation control, a 1:10 dilution with SOC medium is recommended for plating. If a higher number of colonies are desired, the cells may be pelleted by centrifugation at 1,000 × *g* for 10 min, resuspended in 200 μl of SOC medium, and 100 μl plated on each of two plates.
12. Incubate the plates overnight at 37 °C. If 100 μl is plated, approximately 100 colonies per plate are routinely seen using competent cells that are 1 × 108 cfu/μg DNA. Longer incubations or storage of plates at 4 °C (after 37 °C overnight incubation) may be used to facilitate blue color development. White colonies generally contain inserts; however, inserts may also be present in blue colonies (see Notes 24 and 25).

3.3.3 Transformation Efficiency Calculation

After 100 μl of competent cells are transformed with 0.1 ng of uncut plasmid DNA, the transformation reaction is added to 900 μl of SOC medium (0.1 ng DNA/ml). From that volume, a 1:10 dilution with SOC medium (0.01 ng DNA/ml) is made and 100 μl placed on two plates (0.001 ng DNA/100 μl). If 300 colonies are obtained, the transformation efficiency could be calculated as follows:

$$\frac{300\text{cfu}}{0.001\text{ng}} = 3\times10^{5}\,\text{cfu} / \text{ng} = 3\times10^{8}\,\text{cfu} / \mu\text{g DNA}$$

3.3.4 Screening the Transformants for Inserts

Successful cloning of an insert into the T-vector interrupts the coding sequence of β-galactosidase; recombinant clones can be identified by color screening on indicator plates. However, the characteristics of the PCR products cloned into the vectors can significantly affect the ratio of blue:white colonies obtained. Usually clones containing PCR products produce white colonies, but blue colonies can result from PCR fragments that are cloned in-frame with the lacZ gene. Such fragments are usually a multiple of 3 base pairs long (including the 3′-A overhangs) and do not contain in-frame stop codons. There have been reports of DNA fragments up to 2 kb that have been cloned in-frame and have produced blue colonies.

3.4 Microsatellite Detection

Two vector primers and one repeat-specific primer should be used in PCR (see Subheading 2.4.3). After electrophoresis on agarose gel, one should choose colonies whose PCR shows an additional smaller band in PCR reaction (7). Detailed description of microsatellite selection on PCR profile was shown in Fig. 1 (see Notes 26 and 27).

3.5 Data Analysis

After isolating microsatellite successfully, the primer pairs designed could be used for further analysis in genetic diversity studies, species identification, the inheritance breeding, the population genetics, the physical map construction, the management and security of germplasm, and the marker-assisted breeding.

The following are several examples for application of microsatellites:

1. In our lab, we developed microsatellites from *Setaria italica* and had made several analyses for the microsatellites. The average number of allele (Na) and the average observed (Ho) and expected heterozygosities (He) were calculated using the software CERVUS 3.0 (52). Test of deviation of Hardy–Weinberg equilibrium (HW) and linkage disequilibrium (LD) were performed using the GenePop program (53). The sequences were searched against the GenBank nucleotide collection database using TBLASTX for functional annotation with a threshold of *E*-value $<1.00\text{E}^{-05}$. In addition, cross-species amplification of the SSR primers was applied to six other related species, demonstrating the value of this tool (33). Not only cross-species analyses, microsatellites were also used to analyze the genetic diversity of the *S. italica*, helping us to understand the deep relationship of the crop and the culture (54).

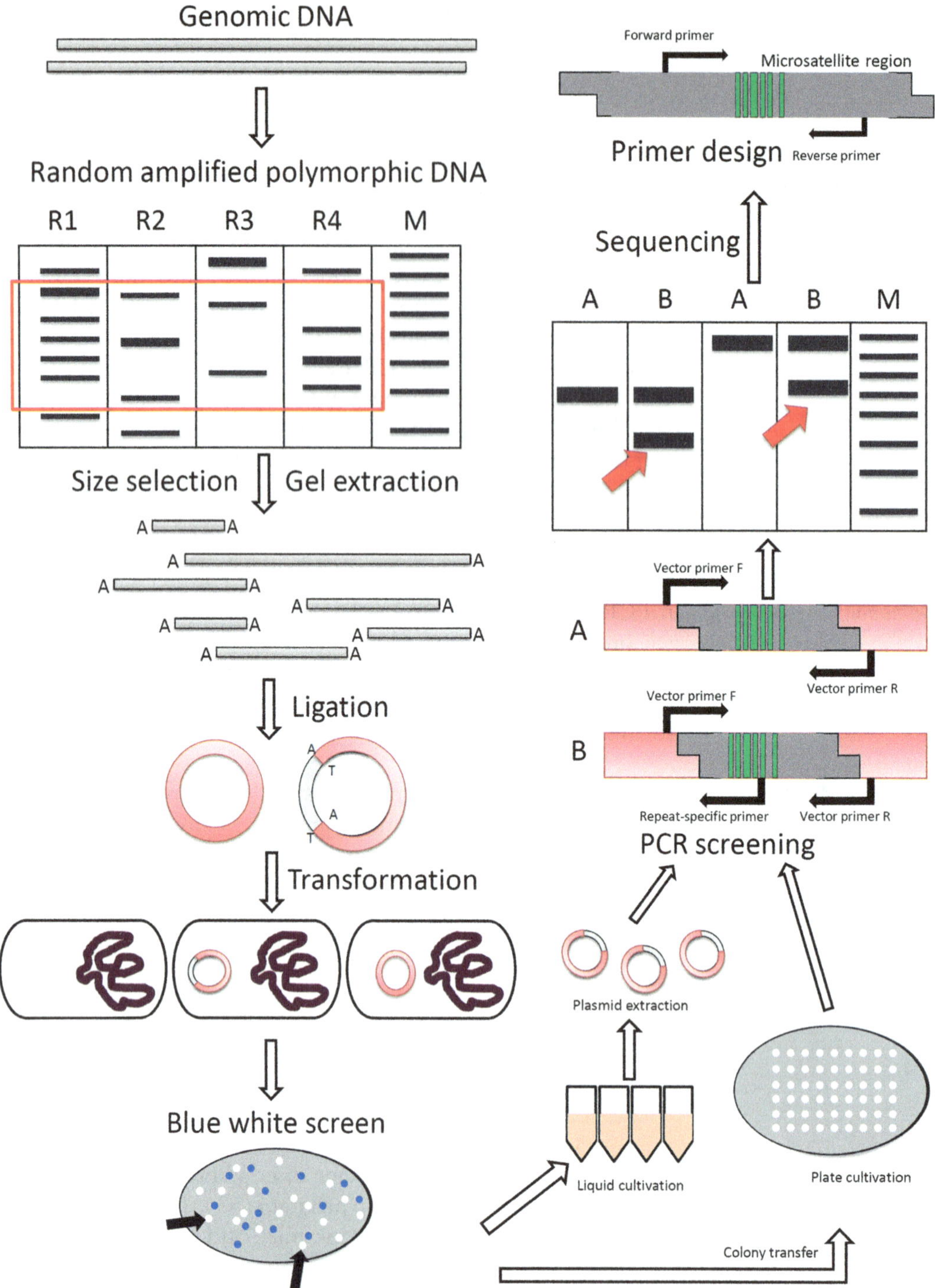

Fig. 1 Schematic flow chart of PCR-based identification of microsatellite arrays (PIMA). These steps include the genomic DNA extraction, RAPD fragments generation, size selection and gel extraction, and cloning: ligation, transformation, blue–white screen (*black arrows* mean white colony are positive clone and should be selected for PCR screening), PCR screening (A: one repeat-specific primer and two vector primers; B: two vector primers), microsatellite detection (*red arrows* on gel profile indicate the existing of microsatellites), and primer design

2. Huang et al. (47) had the allele number, size range, number of bands per individual, and expected (HE) and observed heterozygosities (HO) quantified using the Arlequin version 3.0 (55). GENEPOP (53) online version (http://genepop.curtin.edu.au/) was used to assess Hardy–Weinberg equilibrium (HWE) and linkage disequilibrium (LD).
3. Zhang et al. (49) applied the microsatellite in several analyses including the observed heterozygosity (HO), the unbiased expected heterozygosity (HE), and fixation index (FIS) and were calculated using GDA 1.1 (56). Deviations from Hardy–Weinberg equilibrium (HWE) for each locus and genotypic linkage disequilibrium (LD) between all pairs of loci were tested using FSTAT 2.9.3 (57).
4. Deng et al. (32) calculated the number of alleles, observed heterozygosities (HO), expected heterozygosities (HE), and polymorphic information content (PIC) by using Cervus version 3.0.3 (52). The PIC for each primer was calculated according to Cordeiro et al. (58).
5. Segarra-Moragues et al. (50) obtained the genetic diversity indices and deviations from Hardy–Weinberg equilibrium and linkage disequilibrium between pairs of microsatellite loci using FSTAT software (57).
6. Harper et al. (35) calculated the observed heterozygosity (HO) for each locus using Arlequin (59). HO values were consistently lower than HE. Inbreeding within these small populations may cause this reduction in heterozygosity, and is currently under investigation.

4 Notes

1. The automatic system is controlled by a microcomputer under the flow of 96 and 37 °C through a multi-well sample holder so that the temperature in the samples in the holder varies as required for DNA denaturation, primer annealing, and DNA polymerization. The microcomputer automatically performs multiple thermal cycles and is sufficiently flexible that the temperature profile can be varied from cycle to cycle (13).
2. Therefore, these steps should or should not be taken depending on what tissues you want to isolate DNA at that time.
3. Plant genomic DNAs could be isolated from leaf or any tissue of each individual such as using a DNeasy Plant Mini Kit (Qiagen, Hilden, Germany), or based on a CTAB methodology (16). Animal genomic DNA could be isolated depending on different tissues such as muscle tissue (60); muscle sample of red panda, giant panda, Indian false vampire bat, and Asiatic

black bear using standard phenol–chloroform procedures (17); and blood tissues by salting out procedure (18, 19). According to standard protocol described by Strauss (20), (see bass muscle tissue using the phenol–chloroform extraction methods (21)), genomic DNA was extracted from ethanol-preserved liver tissue by a standard proteinase K-SDS digestion followed by phenol–chloroform extraction (22), bird genomic DNA was extracted using the Genomic DNA Mini Kit (Geneaid, Taipei, Taiwan), or vertebrate genomic DNA could be extracted (24).

4. DNA isolation strategies vary among different laboratories since differences existed in various target species. If the target species has not been accessed of DNA isolation, one could try to follow the isolation strategies operated in related species. The following are several commonly used approaches of DNA isolation:
   (a) Plant genomic DNAs could be isolated from leaf or any tissue of each individual such as using a DNeasy Plant Mini Kit (Qiagen, Hilden, Germany), or based on a CTAB methodology (16).
   (b) Animal genomic DNA could be isolated depending on different tissues such as muscle tissue of red panda, giant panda, Indian false vampire bat, and Asiatic black bear using standard phenol–chloroform procedures (17) and blood tissues by salting out procedure (18, 19); according to standard protocol described by Strauss (20) (see bass muscle tissue using the phenol–chloroform extraction methods (21)), genomic DNA was extracted from ethanol-preserved liver tissue by a standard proteinase K-SDS digestion followed by phenol–chloroform extraction (22), bird's genomic DNA was extracted using the Genomic DNA Mini Kit (Geneaid, Taipei, Taiwan), insect DNA was isolated following the "salting out" protocol (23), or vertebrate genomic DNA could be extracted by protocol of Gemmell and Akiyama (24) (Table 1).
5. Once these have been added the shelf life of the buffer is only 2–3 days.
6. Avoid shaking the solution since the detergent will bubble up too much.
7. Traditionally, plasmids are used for transformation in *E. coli*. In order to be stably maintained in the cell, a plasmid DNA molecule must contain an origin of replication, which allows it to be replicated in the cell independent of the replication of the cell's own chromosome (61).
8. The phenomenon of α-complementation was first demonstrated in the laboratory of François Jacob and Jacques Monod

where the function of an inactive mutant β-galactosidase with deleted sequence was shown to be rescued by a fragment of β-galactosidase where that same sequence, the α-donor peptide, is still intact (62).

9. This method of screening is a convenient way of distinguishing a successful cloning product from other unsuccessful ones. X-gal could be purchased from common chemical companies, such as Invitrogen, Promega, etc.
10. These are the miniprep, midiprep, maxiprep, megaprep, and gigaprep. The plasmid DNA yield will vary depending on the plasmid copy number, type and size, the bacterial strain, the growth conditions, and the kit. Minipreparation of plasmid DNA is a rapid, small-scale isolation of plasmid DNA from bacteria.
11. Otherwise, dinucleotide repeat microsatellites were mostly reported in the research using dinucleotide repeat-specific primer. These above results show out a trend that microsatellite types discovered seem to depend on the type of repeat-specific primer used.
12. Actually, several companies also provide colony-sequencing services, which could only provide the vector primer information and the large volumes of bacterial suspension or you can just provide the colonies you want to sequence.
13. Keeping the tissue frozen during the process of weight measurement.
14. All tissue is entirely in solution with clustering in the tube button.
15. Some laboratories are not including this step when they operate microsatellite extractions. If you want clean, RNA-free extractions, this step should be included (e.g., for cloning, or any RNA work).
16. The following steps are best done in batches of 10–20, depending on how quickly you can work.
17. Be careful to avoid transferring any chloroform.
18. If the pellet dried too long upside down, it will fall out. Continue to dry upright but covered by a clean tissue for 30–45 min.
19. Weight of tissue should be smaller than 100 mg in wet tissue or smaller than 20 mg in lyophilized tissue. Weight of tissue samples depends on different kit or different species.
20. Do not mix Buffer AP1 and RNase A before use. They should be added step by step.
21. If there are pellets that exist, don't disturb it.
22. Remove the spin column from the collection tube carefully so that the column does not come into contact with the flow-through.

23. Example programs for RAPD–PCR from 2000 to 2011:
    (a) Huang et al. (47) performed PCRs in a 10 μl volume containing 10 ng of genomic DNA, 0.2 mM dNTP, 2 mM $MgCl_2$, and 0.12 μM of each primer. PCR took place as follows: 94 °C for 2 min followed by 40 cycles at 94 °C for 20 s, 30 s at primer-specific annealing temperature, 72 °C for 45 s, and a final extension step at 72 °C for 2 min. Electrophoresis was conducted in denaturing 6 % polyacrylamide gels and these were visualized using silver staining (48). Allele sizes were estimated using a 10-bp ladder molecular size standard (Invitrogen).
    (b) Zhang et al. (49) operated the RAPD–PCR amplification in 10 μl reaction mixtures, consisting of approximately 5 ng of template DNA, 50 mM KCl, 20 mM Tris–HCl (pH 8.0), 1.5 mM $MgCl_2$, 0.5 μM of each primer, 0.6 mM of each dNTP, and 0.5 U of Taq DNA polymerase (Takara). The reaction mixture was subjected to PCR amplification in a PTC-100 (MJ) using a PCR program (4 min at 94 °C, followed by 35 cycles of 94 °C for 30 s, 48–64 °C (depending on locus) annealing temperature for 30 s, and 72 °C for 45 s, followed by 10 min at 72 °C). PCR products were then resolved on 6 % denaturing polyacrylamide gels and visualized by silver staining.
    (c) Segarra-Moragues et al. (50) had the 100 μl PCR cocktail containing 50 pmol of "13M" (forward) and 5 pmol of "reverse" primers, 0.2 mM of each dNTP, 2 mM $MgCl_2$, 1× Taq Buffer (Biotools, Madrid, Spain), 3 U of Taq polymerase (Biotools), and 20 μl of the ligation as template. The PCR program consisted of one cycle of 4 min at 94 °C followed by 30 cycles each of 94 °C, 1 min; 50 °C, 1 min; and 72 °C 1 min. PCRs were carried out in a PE GeneAmp9700 thermal-cycler (Applied Biosystems, Madrid, Spain).
    (d) Freitas et al. (51) had the amplification reaction containing 1× buffer, 1.5 mm $MgCl_2$, 0.2 mm of each dNTP, 100 ng of each primer, and 1 U of Taq DNA polymerase in a 20 μl total volume, and was performed through 40 cycles of 40 s at 95 °C, 40 s at 55 °C, and 1 min at 72 °C. PCR products were first resolved in a 1 % agarose gel at 100 V, and the positive clones that presented two bands were sequenced on an ABI PRISM 377 automatic sequencer (Applied Biosystems), using DYEnamic ET Terminator Cycle Sequencing (GE Healthcare).
    (e) Ma et al. (46) performed the PCR on the GeneAmp PCR System 9700 with the following cycles: one cycle of denaturation at 94 °C for 4 min and 45 cycles of 30 s at 94 °C,

50 s at 37 °C, and 50 s at 72 °C. At the last step, products were extended for 7 min at 72 °C. The total 25 μl reaction mixture includes 0.4 mM of each primer, 0.2 mM of each dNTP, 1× PCR buffer, 1.5 mM $MgCl_2$, 0.75 U *Taq* polymerase, and approximately 100 ng template DNA.

(f) Lunt et al. (7) operated the PCR cycles as follows: 5 min preamplification denaturation at 94 °C, 45 cycles of 30 s at 94 °C, 1 min annealing at 37 °C, and 2 min extension at 72 °C. As a final step, products were extended for 5 min at 72 °C. Amplification products were then ligated into a T-vector that was used to transform into competent bacteria.

(g) Liu et al. (40) set PCR cycles as follows: 5 min preamplification denaturation at 94 °C, 35 cycles of 45 s at 94 °C, 40 s at a primer-specific annealing temperature, and 1 min at 72 °C. As a final step, products were extended for 5 min at 72 °C.

24. Colonies containing β-galactosidase activity may grow poorly relative to cells lacking this activity. After overnight growth, the blue colonies may be smaller than the white colonies, which are approximately 1 mm in diameter.

25. Blue color will become darker after the plate has been stored overnight at 4 °C.

26. In positive clones, PCR electrophoresis would show two DNA fragments, of which one PCR product contains microsatellite signal. In contrast, only the whole inserted fragment could be found in negative clones (63).

27. Some lanes marked with a white arrow correspond to clones containing microsatellites, as the two PCR products are formed by combination of universal M13 primers and microsatellite motif PIMA primer (32).

References

1. Jarne P, Lagoda PJL (1996) Microsatellites, from molecules to populations and back. Trends Ecol Evol 11:424–429
2. Tautz D (1989) Hypervariability of simple sequences as a general source for polymorphic markers. Nucleic Acids Res 17:6463–6471
3. Zane L, Bargelloni L, Patarnello T (2002) Strategies for microsatellite isolation: a review. Mol Ecol 11:1–16
4. Ovesna J, Polakova K, Leisova L (2002) DNA analyses and their application in plant breeding. Czech J Genet Plant Breed 38:29–40
5. Kalia RK, Rai MK, Kalia SR, Singh R, Dhawanet AK (2011) Microsatellite markers: an overview of the recent progress in plants. Euphytica 177:309–334
6. Rakoczy-Trojanowska M, Bolibok H (2004) Characteristics and a comparison of three classes of microsatellite-based markers and their application in plants. Cell Mol Biol Lett 9:221–238
7. Lunt DH, Hutchinson WF, Carvalho GR (1999) An efficient method for PCR-based isolation of microsatellite arrays (PIMA). Mol Ecol 8:891–894
8. Ramser J, Weising K, Terauchi R, Kahl G, Lopez-Peralta C, Terhalle W (1997) Molecular marker based taxonomy and phylogeny of

Guinea yam (Dioscorea rotundata—D. cayenensis). Genome 40:903–915

9. Cifarelli RA, Gallitelli M, Cellini F (1995) Random amplified hybridization microsatellites (RAHM): isolation of a new class of microsatellite-containing DNA clones. Nucleic Acids Res 23:3802–3803
10. Ender A, Schwenk K, Stadler T, Streit B, Schierwater B (1996) RAPD identification of microsatellites in Daphnia. Mol Ecol 5: 437–441
11. Grist SA, Firgaira FA, Morley AA (1993) Dinucleotide repeat polymorphisms isolated by the polymerase chain reaction. Biotechniques 15:304–309
12. Cooper SJB, Bull CM, Gardner M (1997) Characterization of microsatellite loci from the socially monogamous lizard Tiliqua rugosa using a PCR-based isolation technique. Mol Ecol 6:793–795
13. Weier HU, Gray JW (1988) A programmable system to perform the polymerase chain reaction. DNA 7:441–447
14. Pavlov AR, Pavlova NV, Kozyavkin SA, Slesarev AI (2004) Recent developments in the optimization of thermostable DNA polymerases for efficient applications. Trends Biotechnol 22:253–260
15. Bartlett JMS, Stirling D (2003) A short history of the polymerase chain reaction. Methods Mol Biol 226:3–6
16. Doyle JJ, Doyle JL (1987) A rapid DNA isolation procedure for small quantities of fresh leaf tissue. Phytochem Bull 19:11–15
17. Sambrook J, Fritsch EF, Maniatis T (1989) Molecular cloning: a laboratory manual, 2nd edn. Cold Spring Harbor Laboratory, New York
18. Miller SA, Dykes DD, Polesky HF (1988) A simple salting out procedure for extracting DNA from human nucleated cells. Nucleic Acids Res 16:1215
19. Shokrollahi B, Amirinia C, Djadid ND, Amirmozaffari N, Kamali MA (2009) Development of polymorphic microsatellite loci for Iranian river buffalo (*Bubalus bubalis*). Afr J Biotechnol 8:6750–6755
20. Strauss WM (1995) Preparation of genomic DNA from mammalian tissue. In: Ausubel FM, Brent R, Kingston RE et al (eds) Current protocols in molecular biology. Wiley, New York, pp 2.2.1–2.2.3
21. Taggart JB, Hynes RA, Prodohl PA, Fergusson A (1992) A simplified protocol for routine total DNA isolation from salmonid fishes. J Fish Biol 40:963–965
22. Blin N, Stafford DW (1976) A general method for isolation of high molecular weight DNA from eukaryotes. Nucleic Acids Res 3: 2303–2308
23. Sunnucks P, Hales DF (1996) Numerous transposed sequences of mitochondrial cytochrome oxidase I–II in aphids of the genus Sitobion (Hemiptera: Aphididae). Mol Biol Evol 13:510–523
24. Gemmell N, Akiyama S (1996) An efficient method for the extraction of DNA from vertebrate tissues. Trends Genet 12:338–339
25. Lehnman IR (1974) DNA ligase: structure, mechanism, and function. Science 186:790–797
26. Brown T (2006) Gene cloning and DNA analysis: an introduction. Blackwell, Cambridge, MA
27. Russell DW, Sambrook J (2001) Molecular cloning: a laboratory manual. Cold Spring Harbor Laboratory, Cold Spring Harbor, NY
28. Horwitz JP, Chua J, Curby RJ, Tomson AJ, Darooge MA, Fisher BE, Mauricio J, Klundt I (1964) Substrates for cytochemical demonstration of enzyme activity. I. Some substituted 3-indolyl-β-D-glycopyranosides. J Med Chem 7:574–575
29. Joung J, Ramm E, Pabo C (2000) A bacterial two-hybrid selection system for studying protein-DNA and protein-protein interactions. Proc Natl Acad Sci U S A 97:7382–7387
30. Sabater-Muñoz B, Legeai F, Rispe C et al (2006) Large-scale gene discovery in the pea aphid *Acyrthosiphon pisum* (Hemiptera). Genome Biol 7:R21
31. Birnboim HC, Doly J (1979) A rapid alkaline extraction procedure for screening recombinant plasmid DNA. Nucleic Acids Res 7:1513–1523
32. Deng X, Long SH, He DF, Li X, Wang YF, Hao DM, Qiu CS, Chen XB (2011) Isolation and characterization of polymorphic microsatellite markers from flax (*Linum usitatissimum* L.). Afr J Biotechnol 10:734–739
33. Lin HS, Chiang CY, Chang SB, Kuoh CS (2011) Development of simple sequence repeats (SSR) markers in *Setaria italica* (Poaceae) and cross-amplification in related species. Int J Mol Sci 12:7835–7845
34. Sim MP, Othman AS (2005) Isolation and characterization of microsatellite DNA loci in sea bass, *Lates calcarifer* Bloch. Mol Ecol Notes 5:873–875
35. Harper GL, Piyapattanakorn S, Goulson D, Maclean N (2000) Isolation of microsatellite markers from the Adonis blue butterfly (*Lysandra bellargus*). Mol Ecol 9:1919–1952
36. Lin C-J, Wang J-P, Lin H-D, Chiang T-Y (2007) Isolation and characterization of

polymorphic microsatellite loci in *Hemibarbus labeo* (Cyprinidae) using PCR-based isolation of microsatellite arrays (PIMA). Mol Ecol Notes 7:788–790

37. Wu H, Zhan X-J, Yan L, Liu S-Y, Li M, Hu J-C, Wei F-W (2008) Isolation and characterization of fourteen microsatellite loci for striped field mouse (*Apodemus agrarius*). Conserv Genet 9:1691–1693
38. Gu X-D, Liu S-Y, Wang Y-Z, Wu H (2009) Development and characterization of eleven polymorphic microsatellite loci from South China field mouse (*Apodemus draco*). Conserv Genet 10:1961–1963
39. Rozen S, Skaletsky HJ (2000) Primer3 on the WWW for general users and for biologist programmers. In: Krawetz S, Misener S (eds) Bioinformatics methods and protocols: methods in molecular biology. Humana, Totowa, NJ, pp 365–386
40. Liu YG, Liu LX, Lei ZW, Gao AY, Li BF (2006) Identification of polymorphic microsatellite markers from RAPD product in turbot (*Scophthalmus maximus*) and a test of cross-species amplification. Mol Ecol Notes 6: 867–869
41. Benson G (1999) Tandem repeats finder: a program to analyze DNA sequences. Nucleic Acids Res 27:573–580
42. Wang K-H, Wu M-J, Chiang T-Y, Chou C-H (2009) Isolation and characterization of polymorphic microsatellite DNA makers for *Euphrasia nankotaizanensis* (Orobanchaceae) and cross amplification in another *Euphrasia* L. Conserv Genet 10:1163–1165
43. Li Q, Wan JM (2005) SSRHunter: development of a local searching software for SSR sites. Yi Chuan 27:808–810
44. Staden R, Judge DP, Bonfield JK (2003) Analysing sequences using the staden package and EMBOSS. In: Krawetz SA, Womble DD (eds) Introduction to bioinformatics: a theoretical and practical approach. Humana, Totawa, NJ, 07512
45. Rychlik W (1992) OLIGO 4.06, primer analysis software. National Biosciences Inc., Plymouth, England
46. Ma H, Ma C, Ma L, Cui H (2010) Novel polymorphic microsatellite markers in *Scylla paramamosain* and cross-species amplification in related crab species. J Crust Biol 30:441–444
47. Huang C-C, Chiang T-Y, Hsu T-W (2008) Isolation and characterization of microsatellite loci in *Taxus sumatrana* (Taxaceae) using PCR-based isolation of microsatellite arrays (PIMA). Conserv Genet 9:471–473
48. Creste S, Neto AT, Figueira A (2001) Detection of single sequence repeat polymorphisms in denaturing polyacrylamide sequencing gels by silver staining. Plant Mol Biol Rep 19:299–306
49. Zhang M, Wang Z-F, Jian S-G, Ye W-H, Cao H-L, Zhu P, Li L (2009) Isolation and characterization of microsatellite markers for *Cycas hainanensis* C. J. Chen (Cycadaceae). Conserv Genet 10:1175–1176
50. Segarra-Moragues JG, Gleiser G, González-Candelas F (2008) Isolation and characterization of microsatellite loci in *Acer opalus* (Aceraceae), a sexually-polymorphic tree, through an enriched genomic library. Conserv Genet 9:1059–1062
51. Freitas PD, Jesus CM, Galetti PM Jr (2007) Isolation and characterization of new microsatellite loci in the Pacific white shrimp *Litopenaeus vannamei* and cross-species amplification in other penaeid species. Mol Ecol Notes 7:324–326
52. Kalinowski ST, Taper ML, Marshall TC (2007) Revising how the computer program CERVUS accommodates genotyping error increases success in paternity assignment. Mol Ecol 16:1099–1106
53. Raymond M, Rousset F (1995) Genepop (Version-1.2)—population genetics software for exact tests and ecumenicism. J Hered 86:248–249
54. Lin HS, Liao GI, Chiang CY, Kuoh CS, Chang SB (2012) Genetic diversity in the foxtail millet (*Setaria italica*) germplasm as determined by agronomic traits and microsatellite markers. Aust J Crop Sci 6:342–349
55. Excoffier L, Laval G, Schneider S (2005) Arlequin ver. 3.0: an integrated software package for population genetics data analysis. Evol Bioinform Online 1:47–50
56. Lewis PO, Zaykin D (2001) Genetic data analysis: computer program for the analysis of allelic data. Version 1.0 (d16c). Free program distributed by the authors over the internet from http://lewis.eeb.uconn.edu/lewishome/software.html
57. Goudet J (1995) FSTAT (version 1.2): a computer program to calculate F-statistics. J Hered 86:485–486
58. Cordeiro GM, Pan YB, Henry RJ (2003) Sugarcane microsatellites for the assessment of genetic diversity in sugarcane germplasm. Plant Sci 165:181–189
59. Schneider S, Roessli D, Excoffier L (2000) Arlequin, Version 2.000: a software for population genetics data analysis. Genetics & Biometry Laboratory, University of Geneva, Geneva, Switzerland
60. Chiang T-Y, Lao S-C, Lee Y-F, Yao C-T (2008) Development of 15 polymorphic microsatellite

loci from *Garrulax morrisonianus* (Timaliidae), an endemic avian species of Taiwan. Conserv Genet 9:1711–1713

61. Hanahan D (1983) Studies on transformation of Escherichia coli with plasmids. J Mol Biol 166:557–580
62. Ullmann A, Jacob F, Monod J (1967) Characterization by in vitro complementation of a peptide corresponding to an operator-proximal segment of the beta-galactosidase structural gene of Escherichia coli. J Mol Biol 24:339–343
63. Huang T-J, Chen Y-Y, Wang K-H, Hsieh T-H, Chou C-H (2008) Isolation and characterization of microsatellite loci in *Ajuga taiwanensis* Nakai ex Murata using PCR-based isolation of microsatellite arrays (PIMA). Bot Stud 50:21–24
64. Hung K-H, Chiang T-Y, Chiu C-T, Hsu T-W, Ho C-W (2009) Isolation and characterization of microsatellite loci from a potential biofuel plant *Miscanthus sinensis* (Poaceae). Conserv Genet 10:1377–1380
65. Hung C-Y, Wang K-H, Huang C-C, Gong X, Ge X-J, Chiang T-Y (2008) Isolation and characterization of 11 microsatellite loci from *Camellia sinensis* in Taiwan using PCR-based isolation of microsatellite arrays (PIMA). Conserv Genet 9:779–781
66. Huang T-J, Chen Y-Y, Li Y-P, Hung C-Y, Chiang T-Y, Chou C-H (2008) Isolation and characterization of microsatellite loci in *Pedicularis verticillata* L. using PCR-based isolation of microsatellite arrays (PIMA). Conserv Genet 9:1389–1391
67. Hung C-Y, Chen Y-Y, Hsu T-W, Huang T-J, Chiang T-Y (2008) Isolation and characterization of 12 microsatellite loci from *Suzukia shikikunensis* (Lamiaceae), a genus endemic to Taiwan and Ryukyus. Conserv Genet 9:1337–1339
68. Huang C-C, Chiang T-Y, Hsu T-W, Hung C-Y, Chiang Y-C, Hung K-H (2009) Isolation and characterization of eight polymorphic microsatellite loci from *Ludwigia polycarpa* (Onagraceae), a threaten herb in North America. Conserv Genet 10:1381–1383
69. Huang C-C, Hung K-H, Hsu T-W, Wang K-H, Lin C-Y, Chiang T-Y (2008) Isolation and characterization of 11 polymorphic microsatellite loci from *Fatsia polycarpa* (Araliaceae), an element of evergreen forests in Taiwan. Conserv Genet 9:1333–1335
70. Huang Y-W, Chiang T-Y, Chiou W-L (2008) Isolation and characterization of microsatellite loci of *Lycopodium fordii* Bak. (Lycopodiaceae, Pteridophyta). Conserv Genet 9:775–777
71. Wu H, Zhan X-J, Zhang Z-J, Zhu L-F, Yan L, Li M, Wei F-W (2009) Thirty-three microsatellite loci for noninvasive genetic studies of the giant panda (*Ailuropoda melanoleuca*). Conserv Genet 10:649–652
72. Wu H, Zhang S-N, Wei F-W (2010) Twelve novel polymorphic microsatellite loci developed from the Asiatic black bear (*Ursus thibetanus*). Conserv Genet 11:1215–1217
73. Wu H, Zhan X-J, Guo Y, Zhang Z, Zhu L, Yan L, Li M, Wei F-W (2009) Isolation and characterization of 12 novel microsatellite loci for the red panda (*Ailurus fulgens*). Conserv Genet 10:523–525
74. Emmanuvel Rajan K, Arul Sundari A, Marimuthu G (2009) Isolation and characterization of microsatellite loci in the Indian false vampire bat *Megaderma lyra*. Conserv Genet 1:369–371
75. Liu YG, Bao BL, Liu LX, Wang L, Lin H (2008) Isolation and characterization of polymorphic microsatellite loci from RAPD product in half-smooth tongue sole (*Cynoglossus semilaevis*) and a test of cross-species amplification. Mol Ecol Resour 8:202–204
76. Lin H-D, Lee T-W, Lin F-J, Lin C-J, Chiang T-Y (2008) Isolation and characterization of microsatellite loci in the endangered freshwater fish *Pararasbora moltrechti* (Cyprinidae) using PCR-based isolation of microsatellite arrays (PIMA). Conserv Genet 9:945–947
77. Chiang T-Y, Lee T-W, Lin F-J, Huang K-H, Lin H-D (2008) Isolation and characterization of microsatellite loci in the endangered freshwater fish *Varicorhinus alticorpus* (Cyprinidae). Conserv Genet 9:1399–1401
78. Yang J-Q, Zhou X-D, Liu D, Liu Z-Z, Tang W-Q (2011) Isolation and characterization of microsatellite loci in the fish *Coilia mystus* (Clupeiformes: Engraulidae) using PCR-based isolation of microsatellite arrays. Genet Mol Res 10:1514–1517
79. Chiang T-Y, Ju Y-M, Fang L-S, Lin C-J (2009) Isolation and characterization of polymorphic microsatellite loci in *Candidia barbata* (Cyprinidae) using PCR-based isolation of microsatellite arrays (PIMA). Conserv Genet 10:503–505
80. Hsu K-C, Wang J-P, Chen X-L, Chiang T-Y (2004) Isolation and characterization of microsatellite loci in *Acrossocheilus paradoxus* (Cyprinidae) using PCR-based isolation of microsatellite arrays (PIMA). Conserv Genet 5:113–115
81. Seyoum S, Tringali MD, Sullivan JG (2005) Isolation and characterization of 27 polymorphic microsatellite loci for the common snook, *Centropomus undecimalis*. Mol Ecol Notes 5:924–927

82. Sun Y, Lin H-D, Tang W-Q, Ju Y-M, Liu Z-Z, Liu D, Yang J-Q (2011) Polymorphic microsatellite loci isolated from the *Squalidus argentatus* using PCR-based isolation of microsatellite arrays (PIMA). Int J Mol Sci 12:5666–5671

83. Bond JM, Porteous R, Hughes S, Mogg RJ, Gardner MG, Reading CJ (2005) Polymorphic microsatellite markers, isolated using a simple enrichment procedure, in the threatened smooth snake (*Coronella austriaca*). Mol Ecol Notes 5:42–44

84. Chiang T-Y, Lin H-D, Chan T-Y, Hung C-Y, Lin F-J (2008) Isolation and characterization of microsatellite loci in the commercially important mudshrimp *Austinogebia edulis* (Upogebiidae) using PCR-based isolation of microsatellite arrays (PIMA). Conserv Genet 9:1653–1655

85. Han C-C, Chang C-S, Chiang T-Y, Chung P-H, Lin H-D (2008) Isolation and characterization of 15 microsatellite loci from *Caridina gracilipes* (Atyidae, Decapoda). Conserv Genet 10:1065–1068

Chapter 4

Fast Isolation by AFLP of Sequences Containing Repeats

Kaisa Rikalainen

Abstract

Fast isolation by AFLP of sequences containing repeats (FIASCO) is a rapid and simple method for separating microsatellite-containing DNA fragments from genomic DNA de novo. The method takes the advantage of the amplified fragment length polymorphism (AFLP) technique that relies on effective digestion–ligation reaction. The repeat-containing fragments are selectively hybridized to biotinylated probes and harvested by streptavidin-coated magnetic beads. The enriched microsatellite-containing fragments can be cloned and sequenced to yield a variety of microsatellite loci for applications in many different fields in molecular genetics.

Key words AFLP, AFLP adaptor, FIASCO, Genomic DNA, Microsatellite, Microsatellite library, *MseI*

1 Introduction

Microsatellite isolation is a crucial step in obtaining powerful genetic markers for applications in various fields of research, from population genetics to forensic DNA analyses. A widely used and effective procedure for enriching microsatellite-containing DNA fragments from genomic DNA is a method called fast isolation by AFLP of sequences containing repeats (FIASCO), introduced by Zane et al. (1). In brief, the method utilizes the amplified fragment length polymorphism (AFLP) technique (2) where genomic DNA is simultaneously fragmented by restriction endonuclease and ligated to specific oligonucleotide adaptors. This DNA–adaptor combination serves as a template for amplification by polymerase chain reaction (PCR). The amplified fragments are hybridized with a biotinylated probe and selectively captured by streptavidin-coated beads. The DNA separated from the beads–probe complex is precipitated and amplified to yield an enriched microsatellite library.

An overview on published peer-reviewed literature shows the popularity of the FIASCO method: it is unquestionably one of the most widely used microsatellite enrichment protocols.

Stella K. Kantartzi (ed.), *Microsatellites: Methods and Protocols*, Methods in Molecular Biology, vol. 1006,
DOI 10.1007/978-1-62703-389-3_4, © Springer Science+Business Media, LLC 2013

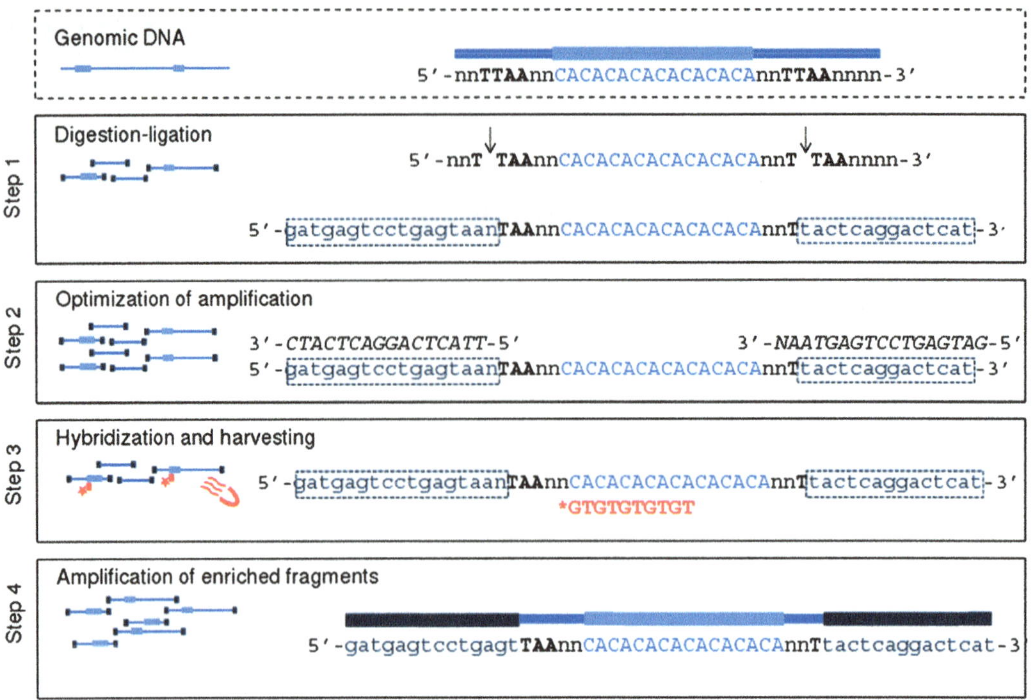

Fig. 1 The FIASCO workflow. In step 1, genomic DNA is simultaneously digested with *Mse*I (restriction sites are shown in *boldface*, *arrows* indicate the cutting sites) and ligated to *Mse*I adaptors (enclosed within *rectangles*). In step 2, DNA is amplified in optimized conditions using *Mse*I primers (shown in *italic*). In step 3, amplified fragments are hybridized to biotinylated probe (here, $(GT)_5$ probe). Hybridized complexes are harvested by streptavidin-coated magnetic particles within a magnetic field. In step 4, recovered DNA is amplified to yield a microsatellite-enriched library. Microsatellite flanking regions carry also adaptor sequences

Since published, the method has been used with various species such as mollusks (3), insects (4), crustacean (5), plants (6), fish (7), birds (8), and mammals (9). For its easiness the method requires neither special laboratory equipment nor supreme laboratory skills. Yet the enrichment is very effective, since the percentage of clones containing dinucleotide repeats was shown to vary between 50 and 95 % (1). The time needed for completing the enrichment can be reduced to three working days, covering the whole procedure from genomic DNA digestion to the cloned products (1). The initial cost of enrichment reagents is low (<1,000€ per 10 samples when the basic laboratory reagents are excluded), the most expensive of those being the magnetic beads.

The method has four major steps (Fig. 1): digestion–ligation (step 1), optimization of amplification (step 2), hybridization and harvesting (step 3), and amplification of enriched fragments (step 4). At the first step the genomic DNA is digested with a "frequent-cutter" AFLP restriction enzyme *Mse*I (restriction sequence T↓TAA) and simultaneously ligated to *Mse*I AFLP adaptor. AFLP adaptors have two important features: they are not phosphorylated which prevents self-ligation, and their ligation with digested DNA knocks out the *Mse*I site of the DNA, thus allowing digestion and ligation to perform simultaneously (1).

The amplification step (step 2) is performed by mixing *Mse*I primers carrying all four possible selective bases (*Mse*I-N) to allow amplification of all fragments flanked by *Mse*I sites. This procedure provides the advantage of going back just one step in case of undesired bands appearing in the PCR amplification. The PCR should result in a visible smear on agarose gel electrophoresis, and one or more distinctive bands probably represent multicopy sequences in the original genome (1). The optimal number of cycles needed in PCR amplification is achieved by stepwise increasing the number of cycles; over-amplification can undesirably increase the average size of amplified fragments.

Amplified DNA is hybridized (step 3) with 5′-biotinylated probe carrying the desired repeat motif ($(GT)_5$ in Fig. 1). Hybridized DNA is captured by streptavidin-coated magnetic beads and magnetic particle processor. Prior to the capture, the beads are coated with unrelated PCR product to minimize nonspecific binding of the DNA. After hybridization, nonspecific DNA is removed by three nonstringent washes and three stringent washes, and after each wash, DNA–bead–probe complex is recovered by a magnetic particle separator. Finally DNA is separated from the bead–probe complex by two denaturation steps.

The DNA is precipitated by a two-step procedure. The precipitated DNA is the best candidate for producing highly enriched microsatellite library and is therefore amplified by PCR using *Mse*I-N primers (step 4). The amplified PCR can be cloned to vector by standard procedures, and the clones containing inserts of desired length are sequenced to yield a panel of microsatellite loci for further analyses.

2 Materials

Prepare all solutions using ultrapure sterile water, analytical grade, and pipettes. Use only sterile equipments, such as pipette tips and tubes. Prepare and store all reagents at room temperature unless otherwise indicated. Follow all waste disposal regulations when disposing waste materials, including pipette tips and tubes.

2.1 Digestion–Ligation Reagents

1. 10× Buffer R, $MgCl_2$ free: Store at +4 °C.
2. Dithiothreitol (DTT), 100 μM: Dissolve in ultrapure sterile water. Store at −20 °C.
3. Adenosine-5′-triphosphate (ATP), 5 mM: Dilute in ultrapure sterile water. Store at −20 °C.

4. *Mse*I restriction endonuclease, 10 U/μl: Store at −20 °C (see Note 1).
5. *MseI* AFLP adaptor, 50 μM: *Mse*I1: 5′-GATGAGTCCTGAGTAAN-3′ (forward), *Mse*I2: 5′-TACTCAGGACTCAT-3′ (reverse). Store at −20 °C.
6. T4-DNA ligase, 1 U/μl: Store at −20 °C.

2.2 PCR Amplification Reagents

1. 10× Reaction Buffer, $MgCl_2$ free (see Note 2).
2. Magnesium chloride, ($MgCl_2$), 50 mM: Store at −20 °C.
3. *Mse*I primer (*Mse*I-N), 7.5 μM, 5′-GATGAGTCCTGAGTAAN-3′, containing all four selective bases adenine, thymine, cytosine, and guanine: Dissolve in ultrapure sterile water. Store at −20 °C.
4. dNTP mixture, 2 mM, containing the following dNTPs: 2′-deoxyadenosine-5′-triphosphate (dATP), 2′-deoxyguanosine-5′-triphosphate (dGTP), 2′-deoxycytidine-5′-triphosphate (dCTP), and 2′-deoxythymidine-5′-triphosphate (dTTP): Dilute in ultrapure sterile water. Store at −20 °C.
5. DNA polymerase, 5 U/μl: Store at −20 °C.

2.3 Hybridization and Washing Reagents

1. Biotinylated repeat probe, 10 μM: a biotinylated oligonucleotide containing the repeat you want to enrich in your sample (e.g., $(CA)_{11}$ probe) (see Note 3). Store at −20 °C.
2. 20× Saline–sodium citrate (SSC), pH 7.5: 3 M sodium chloride (NaCl), 0.3 M trisodium citrate. Adjust to pH 7.5 with hydrochloric acid (HCl) or caustic soda (NaOH).
3. 10 % sodium dodecyl sulfate (SDS).
4. TEN_{100}, pH 7.5: 10 mM tris–hydrochloric acid (tris–HCl), 1 mM ethylenediaminetetraacetic acid (EDTA), 100 mM sodium chloride (NaCl). Adjust to pH 7.5 with hydrochloric acid (HCl) or caustic soda (NaOH).
5. TEN_{1000}, pH 7.5: 10 mM tris–hydrochloric acid (tris–HCl), 1 mM ethylenediaminetetraacetic acid (EDTA), and 1 M sodium chloride (NaCl). Adjust to pH 7.5 with hydrochloric acid (HCl) or caustic soda (NaOH). Preheat to +40 °C before using in washes (see Subheading 3.3, item 1).
6. Streptavidin magnetic particles, 20 mg/2 ml: Store at +4 °C.
7. Unrelated single-stranded DNA (e.g., salmon sperm DNA) for coating: Store at −20 °C.
8. 0.2× Saline–sodium citrate buffer–0.1 % sodium dodecyl sulfate (0.2× SSC, 0.1 % SDS). Preheat to +40 °C before using in washes.
9. Tris–hydrochloric acid–ethylenediaminetetraacetic acid buffer (TE buffer), pH 8.0: 10 mM tris–HCl (pH 8.0), 1 mM EDTA. Adjust to pH 8.0 with hydrochloric acid (HCl) or caustic soda (NaOH).

2.4 Other Reagents and Equipments

1. Tris–acetate–EDTA buffer (TAE buffer), pH 8.0: 40 mM tris–acetate, 1 mM ethylenediaminetetraacetic acid (EDTA). Adjust to pH 8.0 with hydrochloric acid (HCl) or caustic soda (NaOH).
2. 1.5 % Agarose in TAE buffer, gel for electrophoresis: Adjust the mixture volume to your gel tank. For example, mix 4.5 g of agarose D1 low EEO and 300 ml of 1× TAE buffer in Erlenmeyer bottle and heat in microwave oven, gently stirring a few times until completely dissolved (for about 2 min). Cool down to approximately 55 °C and add an appropriate amount of DNA gel stain to be able to visualize the DNA with blue light or UV. Mix well but try to avoid air bubbles. Assembly the gel tank and pour the gel.
3. 3 M Sodium acetate (NaAc).
4. 0.16 M acetic acid (HAc).
5. Isopropanol.
6. 70 % ethanol, not denatured, use ice-cold. Store at −20 °C.
7. PCR product purification reagents. It is recommended to use a commercial kit for purification to yield high-quality DNA.
8. Magnetic particle separator.
9. Heat blocks and water bath.

3 Methods

Carry out all procedures at room temperature unless otherwise specified. Mix carefully all reagents by vortex before using (NB: Do not mix enzymes by vortex to retain their activity). All PCR amplifications should be performed in microcentrifuge tubes (e.g., volume of 200 μl). Use sterile (filtered) disposable pipette tips, centrifuge tubes, and reagents (when possible), and wear gloves to avoid contamination. Set up PCR amplifications in a laminar hood and close the tube caps before leaving the hood. Keep the PCR reagents and enzymes on ice/in cold block to make sure that the DNA polymerase and T4-ligase do not activate too early. It is recommended to carefully examine the methods before starting up, because some reagents need to be preheated before use and some reactions need to be performed in warm conditions (see text for details).

3.1 Digestion–Ligation

1. In microcentrifuge tube (volume of, e.g., 1.5 ml), perform the digestion–ligation reaction of the extracted genomic DNA (see Notes 4–6) as follows: mix 2.5 μl 10× Buffer R, 1 μl DTT (100 μM), 1 μl ATP (5 mM), 0.625 μl *Mse*I restriction endonuclease (10 U/μl), 0.5 μl *Mse*I adaptor (50 μM), 1 μl T4-DNA ligase (1 U/μl), and 5 μl genomic DNA (25–250 ng). Adjust the volume to 25 μl by adding ultrapure sterile water.

2. Incubate the digestion–ligation mix at +37 °C for 3 h.
3. After incubation dilute the mixture 1:10 by adding 225 μl of ultrapure sterile water.

3.2 Optimization of Amplification

1. Optimize the number of cycles needed in PCR amplification by making five separate PCR reactions in distinct microcentrifuge tubes. Each of the reactions is terminated after different number of cycles (12, 15, 18, 21, and 24, see Note 7). For one reaction, mix 2 μl 10× Reaction Buffer, 0.6 μl $MgCl_2$ (50 mM), 4 μl *Mse*I primers, 2 μl dNTPs (2 mM), 0.08 μl DNA polymerase (5 U/μl), and 5 μl restricted–ligated DNA. Adjust the volume to 20 μl by adding ultrapure sterile water. Make one reaction for each termination point, i.e., altogether five reactions.
2. Perform the PCR amplification. The amplification program is as follows: +94 °C 2 min (initial denaturation); 24× (+94 °C 45 s, +53 °C 45 s, +72 °C, 1 min); and +72 °C, 7 min (terminal elongation). Remember to take out the corresponding tube after 12, 15, 18, and 21 cycles and keep it at +4 °C until all the reactions are completed.
3. Check your PCR products by agarose gel electrophoresis (see Note 8); PCR conditions (i.e., corresponding number of cycles) producing the lighter visible smear is considered optimal and selected for further use.
4. Replicate PCR amplification under optimal conditions as described previously (Subheading 3.2, steps 1 and 2). Purify your selected PCR products and proceed to hybridization.

3.3 Hybridization and Harvesting

1. Preheat a water bath to +100 °C, and two blocks, one to +95 °C (later used at +98 °C) and the other to +40 °C. Preheat the washing solutions (TEN_{1000} and 0.2× SSC, 0.1 % SDS) to +40 °C during hybridization (Subheading 3.3, step 5).
2. Wash 100 μl of streptavidin-coated beads with 100 μl of TEN_{100} and resuspend the beads in 40 μl of TEN_{100} in a centrifuge tube. Add about 1 μg (in a volume of 10 μl) of unrelated-single-stranded DNA (e.g., salmon sperm single-stranded DNA) to minimize nonspecific binding of DNA. The total volume of suspension is now 50 μl.
3. In a microcentrifuge tube, mix 45 μl of the purified PCR product, 5 μl biotinylated probe (10 μM), and 30 μl of 20× SSC and 1 μl 10 % SDS. Add 19 μl of ultrapure sterile water to reach the volume of 100 μl.
4. Carefully close the lid and denature the DNA in the water bath at +100 °C for 5 min and then put it into the heat block preheated at +95 °C.

5. Add 300 μl of TEN_{100} and resuspend the magnetic beads (50 μl). Incubate the hybridization mixture at +40 °C for 30 min, stirring gently a few times during incubation. Preheat the heat block to +98 °C for denaturation (Subheading 3.3, step 8).
6. Separate the hybridization buffer from the beads in magnetic particle separator, discarding the buffer.
7. Wash the beads in a tube to remove any nonspecific DNA: Wash the beads with 400 μl of TEN_{1000} (preheated to +40 °C) three times (nonstringent washes) and then with 400 μl of SSC 0.2× SDS 0.1 % (preheated to +40 °C) three times (stringent washes). After each wash, carefully recover the DNA–beads–probe complex by magnetic field separation (see Note 9).
8. For separating the DNA from the beads–probe complex, elute the washed beads first with 50 μl of TE buffer and incubate it at +98 °C for 5 min (the first denaturation step).
9. Quickly separate and save the supernatant (50 μl) containing the target DNA fragments (see Note 10), and elute the beads again, now with 14 μl of 0.1 M NaOH to denature remaining target fragments (the second denaturation step).
10. Separate the supernatant containing the target DNA fragments and neutralize it before storing by adding 6.5 μl 0.16 M acetic acid. Add TE buffer to reach the final volume of 50 μl (see Note 10).

3.4 Amplification of the Enriched Fragments

1. Precool the centrifuge to +9 °C for DNA precipitation. Freeze some 70 % ethanol on ice/in freezer.
2. Precipitate the DNA from both of the eluted samples (eluted in TE at the first denaturation step and in NaOH at the second denaturation step) by adding 5 μl of 3 M sodium acetate and 55 μl isopropanol. Mix well and centrifuge the samples at +9 °C with 16,060 rcf (=16,060 × *g*, or at maximum speed) for 15 min. Discard the supernatant and wash the pellets containing target DNA with 110 μl of ice-cold 70 % ethanol. Centrifuge again to discard the supernatant and let the pellets dry out with tube caps open in laminar hood for a few minutes to evaporate the remaining ethanol. Dissolve the DNAs in 50 μl of ultra-pure sterile water.
3. Amplify 2 μl of the eluted DNAs by PCR as described in "Subheading 3.2, step 1," but this time using only 2 μl of DNA and adjusting total volume to 20 μl. Perform 30 cycles of the amplification program and check the PCR product on 1.5 % agarose gel electrophoresis. The visualization should display a smear above 200 bp (1). Based on the visualization, select the best candidate for cloning and sequencing (see Notes 11 and 12).

4. Clone the PCR products to the vector that is convenient to you and transform the cloned vectors to competent bacteria (e.g., *Escherichia coli*). Clones containing inserts can be detected, e.g., by blue–white selection, depending on your vector type. The inserted DNA from the selected clones is extracted and amplified, and the amplified product is checked on agarose gel electrophoresis. The colonies containing an insert of desired length (e.g., PCR products of 500 base pair and over) can then be sequenced.

4 Notes

1. You can also use other AFLP restriction endonucleases, e.g., *Taq*I restriction enzyme, with the restriction site T↓CGA. In this case, remember to use adaptors with compatible cohesive ends and suitable primers (see, e.g., ref. 2). The digestion should yield fragments that are of suitable length to PCR. The size of the restricted fragment length can be derived from the equation $f=(1/4)^N$, where f is the frequency of the restriction sites and N is the length of the restriction sequence (10).
2. You can use the PCR chemistry that is the most convenient for you. For most of the primers, the PCR conditions, such as $MgCl_2$ concentration, have to be optimized. Therefore, it is better to use $MgCl_2$-free buffer and add the required amount of $MgCl_2$ when preparing the master mix (see Note 6).
3. You can also make a multi-probe hybridization, where multiple probes carrying different repeat motifs are mixed. In this case, make a mixture of equal amounts of probes and use this mixture as a probe in hybridization.
4. To increase the amount of genomic DNA or to maximize the likelihood of polymorphism, DNA from several individuals can be pooled prior to the first step (see, e.g., ref. 11).
5. To avoid and track contamination, it is recommended to prepare a negative sample for each sample set simultaneously with DNA extraction; negative sample can then be further processed with the samples throughout the whole procedure. It is also recommended to prepare a negative control for each PCR set. Both the negative sample and control are prepared as the original sample/PCR with the exception of using ultrapure sterile water instead of DNA.
6. Multiple PCRs are easily performed at the same time in automatized PCR machine either in tube strips or in PCR plates. For this, prepare "a master mix" for the reactions by mixing the buffer, $MgCl_2$, dNTPs, the primers, and water, and use vortex to achieve complete mixture. Then add ice-cold DNA polymerase and mix by tapping with finger; avoid mixing the

polymerase by vortex or by pipetting. Pipette the master mix onto plate/into strip and add the individual samples (i.e., the target DNA). Close the plate/tubes and centrifuge shortly to avoid any drops on the tube walls. Then perform the PCR amplification.

7. You can try different sets of terminations. Originally, Zane et al. (1) proposed the termination of reactions to be at 14, 17, 20, 23, and 26 cycles. In our laboratory the set described previously (Subheading 3.2, step 1) has been found to be the best.
8. Different agarose gel equipments require different settings for running time and voltage. When running the electrophoresis for the first time, watch the time to avoid running the samples for too long. Remember to run also a suitable DNA ladder into the gel.
9. The efficiency of the washes can be detected by amplifying 2 μl of each recovered fraction by 30 cycles of PCR under the conditions described in Subheading 3.4.
10. DNA can be stored at +4 °C for a couple of days. For a longer storage time, −20 °C is recommended.
11. You should amplify both of the eluted DNAs (i.e., elutes from the two elution steps) to see which one gives the best visualization on the agarose gel electrophoresis. In my experience, it is the sample from the last elution step (sodium acetate) that is the most suitable for cloning.
12. Also the last nonstringent wash and the last stringent wash should have a proportion of DNA fragments containing the selected repeat and should carry the *Mse*I-N primer site at each end (1).

Acknowledgments

The work was supported by the Centre of Excellence in Evolutionary Research of the Academy of Finland and the Finnish Cultural Foundation. Alessandro Grapputo, Elina Virtanen, and Sari Viinikainen provided valuable comments on the method and on the manuscript.

References

1. Zane L, Bargelloni L, Patarnello T (2002) Strategies for microsatellite isolation: a review. Mol Ecol 11:1–16
2. Vos P et al (1995) AFLP: a new technique for DNA fingerprinting. Nucl Acids Res 23:4407–4414
3. Chen L et al (2011) Isolation and characterization of sixteen polymorphic microsatellite loci in the golden apple snail *Pomacea canaliculata*. Int J Mol Sci 12:5993–5998
4. Grapputo A (2006) Development and characterization of microsatellite markers in the colorado potato beetle, *Leptinotarsa decemlineata*. Mol Ecol Notes 6:1177–1179
5. Xu XJ et al (2009) Isolation and characterization of ten new microsatellite loci in the mud

crab, *Scylla paramamosain*. Conserv Genet 10:1877–1878

6. Li Y, Liang L, Ge XJ (2010) Development of microsatellite loci for *Pinus koraiensis* (Pinaceae). Am J Bot. doi:10.3732/ajb.1000098
7. Guo S, Zou G, Yang G (2009) Development of microsatellite DNA markers of grass carp (*Ctenopharyngodon idella*) and their cross-species application in black carp (*Mylopharyngodon piceus*). Conserv Genet 10: 1515–1519
8. Satio DS et al (2006) Isolation and characterization of microsatellite markers in red-flanked bushrobin, *Tarsiger cyanurus* (aves: Turdidae). Mol Ecol Notes 6:425–427
9. Rikalainen K et al (2008) A large panel of novel microsatellite markers for the bank vole (*Myodes glareolus*). Mol Ecol Res 8:1164–1168
10. Bastié-Sigeac F, Lucotte G (1983) Optimal use of restriction enzymes in the analysis of human DNA polymorphism. Hum Genet 63:162–165
11. He Y, Wang J (2010) Temporal variation in genetic structure of the Chinese rare minnow (*Gobiocypris rarus*) in its type locality revealed by microsatellite markers. Biochem Genet 48:312–325

Chapter 5

Microsatellite DNA Capture from Enriched Libraries

Elena G. Gonzalez and Rafael Zardoya

Abstract

Microsatellites are DNA sequences of tandem repeats of one to six nucleotides, which are highly polymorphic, and thus the molecular markers of choice in many kinship, population genetic, and conservation studies. There have been significant technical improvements since the early methods for microsatellite isolation were developed, and today the most common procedures take advantage of the hybrid capture methods of enriched-targeted microsatellite DNA. Furthermore, recent advents in sequencing technologies (i.e., next-generation sequencing, NGS) have fostered the mining of microsatellite markers in non-model organisms, affording a cost-effective way of obtaining a large amount of sequence data potentially useful for loci characterization. The rapid improvements of NGS platforms together with the increase in available microsatellite information open new avenues to the understanding of the evolutionary forces that shape genetic structuring in wild populations. Here, we provide detailed methodological procedures for microsatellite isolation based on the screening of GT microsatellite-enriched libraries, either by cloning and Sanger sequencing of positive clones or by direct NGS. Guides for designing new species-specific primers and basic genotyping are also given.

Key words Microsatellite enrichment, Next-generation sequencing, Primer design, Genotyping

1 Introduction

Microsatellites or SSR (simple-sequence repeats) are ubiquitous, codominant (i.e., allow the discrimination of homozygotes and heterozygotes) genetic markers that show high levels of length polymorphism due to their higher rates of mutation with respect to the rest of the genome (1–3). They are defined by noncoding DNA motifs of one to six base pairs (bp) repeated in tandem, and can be classified into perfect, imperfect, interrupted, or composite based on the repeat composition (4). Their distribution and frequency varies among eukaryotic (5) and prokaryotic (6) genomes, as well as between coding and noncoding regions (5). Their high variability, ease of genotyping, and high reproducibility make them powerful genetic markers, broadly employed in many evolutionary and population genetic studies (2, 3, 7), and still preferable to other

Stella K. Kantartzi (ed.), *Microsatellites: Methods and Protocols*, Methods in Molecular Biology, vol. 1006, DOI 10.1007/978-1-62703-389-3_5,

markers (such as e.g., single nucleotide polymorphisms, SNPs) in detecting population structure in specific studies where a high per-marker variability is needed (8).

While microsatellite applications in population genetic studies are well established and considered to render robust results, there are still difficulties associated with the de novo development of microsatellites in species for which no genomic resources are available (9). Apart from the species-specific characteristics of the genome (i.e., complexity and microsatellite frequency), additional technical factors associated with microsatellite isolation methods (i.e., cost and time investments) could potentially impact the effectiveness of microsatellite loci mining.

In this chapter, we present detailed methodological procedures for an efficient and rapid way for screening short inserts from GT microsatellite-enriched libraries in a broad variety of taxa, taking advantage of different methods developed for DNA enrichment and sequencing. Moreover, the process of primers design and basic genotyping of the newly developed microsatellite loci is also provided.

2 Materials

2.1 Microsatellite Isolation

2.1.1 Microsatellite Enrichment and Library Construction

1. 10× TE buffer: 100 mM Tris–Cl, pH 7.5, and 10 mM EDTA disodium salt. Dissolve all reagents in distilled water and sterilize by autoclaving. A 1× working solution is obtained by adding one part of concentrated 10× TAE to nine parts of distilled water. Store at room temperature.
2. DEPC-treated water (RNase free).
3. Agarose (low melting).
4. 10× TAE buffer: 400 mM Tris–acetate, 10 mM EDTA (pH 8). Dissolve all reagents in distilled water and sterilize by autoclaving. A 1× working solution is obtained by adding one part of concentrated TAE to nine parts of distilled water. Store at room temperature.
5. SYBR safe DNA gel stain (Invitrogen) (see Note 1).
6. 1 kb and 100 bp DNA ladders.
7. 10,000 U/ml *Rsa*I or *Nhe*I restriction enzymes (e.g., New England Biolabs).
8. 20,000 U/ml *Xmn*I restriction enzyme (e.g., New England Biolabs).
9. 10× NEBuffer 4 (supplied with *Rsa*I enzyme): 500 mM potassium acetate, 200 mM Tris–acetate, 100 mM magnesium acetate, 10 mM dithiothreitol (pH 7.9).
10. 100 μg/ml bovine serum albumin (BSA, supplied with *Xmn*I restriction enzyme).

11. 400,000 U/ml (high concentration) T4 DNA ligase.
12. SNX linkers (10): Forward SNX linker: 5′-CTAAGGCCTTG CTAGCAGAAGC-3′; phosphorylated reverse SNX linker: 5′-pGCTTCTGCTAGCAAGGCCTTAGAAAA-3′. It is recommended that primers be desalted (or purified with e.g., HPLC). Dissolve with DEPC-treated water to obtain a 100 μM stock solution. Store in aliquots at −20 °C.

    To prepare double-stranded SNX linkers, combine 10 μl of each 100 μM forward and reverse linker with 80 μl of TE buffer (1×) and mix well. Heat at 95 °C for 15 min and let the solution cool at room temperature for a half hour. Store at −20 °C and slowly defrost on ice before use in the ligation step.
13. 100 mM deoxynucleotide triphosphates (dNTPs). A 10 mM working solution is obtained by adding one part of the mix (100 mM) to nine parts of DEPC-treated water. Store in aliquots at −20 °C.
14. 100 mM deoxyadenosine triphosphate (dATP). A 10 mM working solution is obtained by adding one part of the mix (100 mM) to nine parts of DEPC-treated water. Store in aliquots at −20 °C.
15. 5,000 U/ml Taq DNA polymerase.
16. 3′-Biotinylated CA_{12} oligonucleotide. Dehydrate with DEPC-treated water and store away from natural daylight in aliquots at −20 °C.
17. DynaMag-2 magnetic holder (Invitrogen).
18. 20× SSC buffer: 3 M NaCl and 0.3 M sodium acetate (pH 7.0). Dissolve all reagents in distilled water and sterilize by autoclaving. Store at room temperature.
19. 1× Hybridization buffer: 6× SSC, 0.1 % sodium dodecyl sulfate (SDS). Store at room temperature.
20. TBS buffer: 100 mM Tris, pH 7.5, 150 mM NaCl. Dissolve all reagents in distilled water and sterilize by autoclaving. Store at room temperature.
21. TBST buffer: 100 mM Tris, pH 7.5, 150 mM NaCl, 0.1 % Tween-20. Dissolve all reagents in distilled water and sterilize by autoclaving. Store at room temperature.
22. 1× Wash Buffer (low stringency): 2× SSC, 0.1 % SDS. Dissolve all reagents in distilled water and sterilize by autoclaving. Store at room temperature.
23. 1× Wash Buffer (high stringency): 0.5× SSC, 0.1 % SDS. Dissolve all reagents in distilled water and sterilize by autoclaving. Store at room temperature.
24. 1 mg/ml Streptavidin MagneSphere Paramagnetic beads (Promega).

25. 10× TLE buffer: 10 mM Tris, 0.1 mM EDTA disodium salt. Dissolve all reagents in distilled water and sterilize by autoclaving. A 1× working solution is obtained by adding one part of concentrated 10× TAE to nine parts of distilled water. Store at room temperature.
26. 3 M sodium acetate.
27. 100 and 70 % ethanol.
28. pGEM-T Easy vector (Promega). The kit comes with 50 ng/μl pGEM-T Easy vector, 3 Weiss units/μl T4 DNA ligase, and 2× ligation buffer: 60 mM Tris–HCl, pH 7.8, 20 mM $MgCl_2$, 20 mM DTT, 2 mM ATP, 10 % polyethylene glycol (MW8000, ACS Grade).

2.1.2 Screening and Isolation of Positive Clones Containing Microsatellites

1. SOC medium: Mix 2 % bacto-tryptone, 0.5 % yeast extract, 10 mM NaCl, 2.5 mM KCl, 10 mM $MgCl_2$, and 10 mM $MgSO_4$. Sterilize in autoclave. Add glucose (sterilized by passage through a 0.22-μm filter) to a final concentration of 20 mM. Store in the dark at room temperature.
2. 50 mg/ml ampicillin. Dissolve the powder in distilled water and sterilize by passage through a 0.22-μm filter. Store in aliquots at −20 °C.
3. 2 % X-Gal. Dissolve it in dimethylformamide and store in aliquots at −20 °C.
4. 0.1 M IPTG. Dissolve the powder in distilled water and sterilize by passage through a 0.22-μm filter. Store in the dark in aliquots at −20 °C.
5. XL10-Gold® bacteria (Stratagene).
6. LB: 1 % Bacto-tryptone, 0.5 % yeast extract, 10 mM NaCl, and 1.5 % agar. Dissolve all reagents in distilled water and sterilize by autoclaving. Store at room temperature.
7. LB/agar: Add 1.5 % agar to the above medium. Dissolve all reagents in distilled water and sterilize by autoclaving. Store at room temperature.
8. LB/ampicillin: Prepare and autoclave the LB medium as described above. Ampicillin (and other antibiotics and medium supplements) is heat sensitive, so ensure that the LB has cooled to below 50 °C. Add 50 mg/ml ampicillin to the liquid medium, just before use, to a final concentration of 100 μg/ml. Work under sterile conditions.
9. LB/agar/ampicillin/X-Gal/IPTG plates: Prepare and autoclave the LB/agar as described above and let it cool to below 50 °C. Add the 50 mg/ml ampicillin to a final concentration of 50 μg/ml. While still liquid, pour the medium into the Petri dishes (approximately 25 ml). Allow the plates to harden at room temperature. Prior to adding the bacterial culture, spread

40 μl of 2 % X-Gal (see item 3) and 10 μl of 0.1 M IPTG (see item 4) with a wire loop onto each plate. Work under sterile conditions.

10. Glycerol. Sterilize in autoclave and store at room temperature.
11. CA_{12} nonbiotin-labeled oligonucleotide. Dehydrate with DEPC-treated water to obtain a 100 μM stock solution and store at −20 °C.
12. QIAprep Spin Miniprep kit (Qiagen).

2.2 Next-Generation Sequencing Methods for Microsatellite Capture

1. GS FLX Titanium Rapid Library Preparation kit (Roche).
2. 70 % ethanol.

2.3 PCR Primer Design, Testing, and Basic Genotyping

1. Species-specific primers. Dissolve with DEPC-treated water to obtain a 100 μM stock solution. The 10 μM working stocks are prepared by combining one part of the 100 μM stock solution with nine parts of DEPC-treated water. Store in aliquots at −20 °C.
2. Labeled species-specific forward primers of interest (e.g., Applied Biosystems, ABI). The primer can be labeled at the 5′ end with different third-generation dyes (e.g., 6-FAM, PET, NED or VIC). VIC (green), NED (yellow), and PET (red) are ABI proprietary dyes and must be ordered through the company, but 6-FAM (blue) and HEX (the replacement for VIC from the second-generation dyes) are nonproprietary dyes available from most primer synthesis companies. Dissolve with DEPC-treated water to obtain a 100 mM stock solution. The 10 μM working stocks are prepared by combining one part of the 100 μM stock solution with nine parts of DEPC-treated water. Store away from natural daylight in aliquots at −20 °C.
3. Type-it Microsatellite PCR kit (Qiagen).

3 Methods

3.1 Microsatellite Isolation

Since microsatellites have become the marker of choice for kinship, population genetic, and evolutionary biology studies (2, 3, 7) numerous microsatellite-isolation methods have been developed (9, 11). The early methods based on a cloning–screening process of complete libraries (12) were time-consuming and cumbersome, with a rather low efficiency. At present, they have been replaced by different methods based on enrichment and hybrid capture of the repeat sequences that efficiently isolate specific DNA microsatellites from a wide variety of genomes. Essentially, these methods

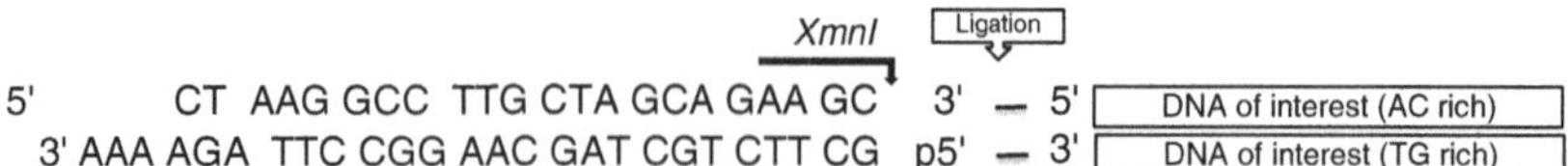

Fig. 1 Scheme of the double-sequence (ds) SNX linker ligated to DNA fragments. The recognition site for *Xmn*I is indicated in the figure. *p* indicates the phosphorylation of the 5′-ends of the reverse oligo. The dsSNX linker has a 3′ overhang (AAAA) that decreases the formation of linker dimers during the restriction/ligation process. Additionally, it contains the recognition sequence for the enzyme *Xmn*I (a 10-base cutting enzyme) that cleaves the end of the linker, therefore helping to keep the end of their sequences free for ligation to the genomic fragments

require the construction of a genomic library enriched for repeat motifs and posterior screening, isolation, and sequencing of microsatellite-containing clones (9, 10, 13).

More recently, next-generation sequencing (NGS) technologies have been also applied successfully to the isolation of microsatellites. In this case, microsatellite-enriched genomic fragments are ligated to platform-specific adapters and directly sequenced, producing tens of thousands of reads. Although the cost is still relatively high, this high-throughput sequencing approach generates a larger number of microsatellite markers in comparison with Sanger sequencing. Hence, it is particularly useful when the sequencing of enriched libraries fails to isolate enough quality microsatellite loci (e.g., when a lower quantity of repetitive DNA is present for the identification of microsatellites (14) or even in extinct species (15), for which the initial sample of DNA is of low quality). The selection of one or the other approach will depend on available resources, species intrinsic limitations, and the tackled questions, which define the minimum number of polymorphic loci needed.

3.1.1 Microsatellite Enrichment and Library Construction

The methods described under this and next subheadings are based on (10, 11, 16) and rely on the construction of a genomic library of blunt-ended DNA fragments enriched for GT repeat sequences ligated to SNX linkers. Briefly, genomic DNA is completely digested with one or two restriction enzymes (e.g., *Rsa*I and/or *Nhe*I), and then ligated to the double-stranded linkers. The fragment size selection step of extracting/purifying the DNA from an electrophoresis agarose gel has been eliminated since the use of *Rsa*I along or in combination with other 4-, 5-, or 6-cutter restriction enzymes already generates a desirable size range for cloning. *Xmn*I (an enzyme that has a restriction site at the end of the linker; see Fig. 1) is also included in the ligation to avoid the dimerization of the linkers. Pooled aliquots of different digests are enriched for GT microsatellites by hybridization and magnetic capture with a CA biotinylated probe. Enriched fragments are made double-stranded by PCR using the SNX forward linker as primer, and TA cloned into pGEM-T Easy vectors.

The CA dinucleotide repeat probes are chosen for the selective hybridization step due to the higher levels of polymorphism reported for CA/GT in comparison to other dinucleotide motifs (17) and the higher frequency of this motif in many organisms (18). Nevertheless, when attempting to isolate microsatellites for the first time in a species, it is advisable to perform an exploratory search on available genomic sequences in data bases (e.g., Genbank) and on published literature (e.g., the Molecular Ecology Resources database; http://tomato.bio.trinity.edu/) for the target species, as well as for closely related species, and, if needed, to adapt this protocol to the use of other possible microsatellite probes.

A standard method of DNA purification by phenol/chloroform extraction (19), followed by ethanol precipitation and elution in TE buffer, typically renders high molecular weight in a broad variety of taxa and tissue sources. However, if preferred, good DNA yields are also obtained with available commercial DNA extraction kits, such as DNeasy Tissue kit (Qiagen).

1. Extract high molecular weight genomic DNA following the method of purification of preference, elute it in TE buffer, and preserve at –20 °C. For best results prepare and manage samples at 4 °C before and after extraction. It is preferable to work with the highest quality of genomic DNA possible (approximately 10 μg) with little fragmentation of its structure. Check quantity and purity (absorbance 260/280 ratio) of 1 μl in, e.g., Thermo Scientific NanoDrop 2000. Run an aliquot of 5 μl on a 0.8 % TAE agarose gel with SYBR safe for checking quality and concentration of DNA by comparison against a size marker (e.g., 1 kb DNA; see Fig. 2).
2. We recommend digesting 0.5–1.5 μg of genomic DNA with a two- to tenfold excess of enzyme. Digest the DNA with *Rsa*I in a final volume of 50 μl (see Note 2). An example of a mix would contain 10–20 μl of genomic DNA, 40 U of the enzyme, 5 μl of 10× NEBuffer 4 (supplied with *Rsa*I), and 23–33 μl of DEPC-treated water. Incubate at 37 °C overnight followed by heating to 65 °C for 20 min to denature the remaining restriction enzyme. A successful digestion should show most of the DNA fragments within a 200–1,500 bp size range (see Note 3). Run an aliquot of 5 μl on a 2 % TAE agarose gel with SYBR safe to check for the completeness of digestion (see Fig. 2).
3. Use 13 μl of the digested DNA and add 400 U of high concentration T4 DNA ligase, 1.5 μl of 10 mM dATP, 20 U of *Xmn*I restriction enzyme, 1 μl of 100× BSA (provided with *Xmn*I), 14 μl of 10 μM dsSNX (Fig. 1) linker mix (see Subheading 2.1.1 on how to anneal the linkers together), 3.5 μl of 10× NEBuffer 2 (supplied with the *Nhe*I), and DEPC-treated water to a final volume of 35 μl. Incubate at room temperature for 4h to overnight. Reaction can be stored at 4 °C.

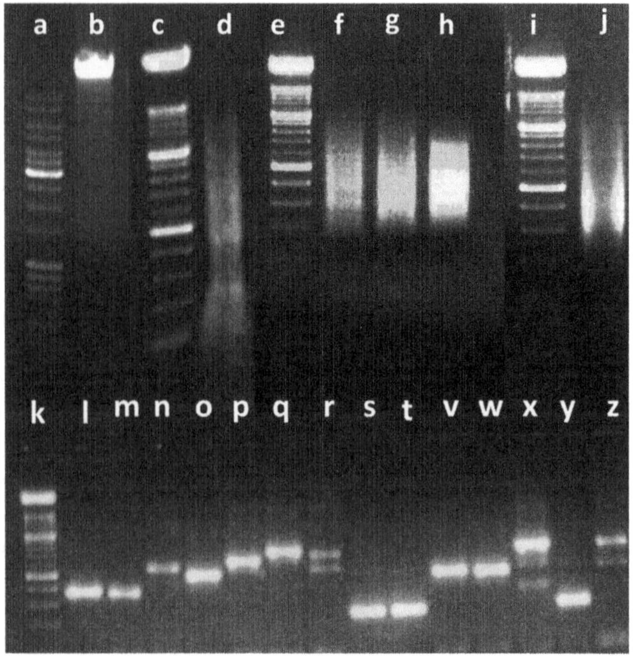

Fig. 2 Agarose gels stained with SYBR safe to visualize the DNA product at different stages of a microsatellite-isolation process performed with the European sardine (*Sardina pilchardus*) (39). *Lane b*: genomic extraction. *Lane d*: restriction analysis of genomic DNA with *Rsa*I. *Lanes f–h*: PCR test of restriction–ligation success at 20–25 and 30 cycles (lanes *f*, *g*, and *h*, respectively). *Lane j*: PCR amplification of enriched capture fragments. *Lanes l–z*: PCR screening of positive clones containing GT microsatellites. *Lanes r*, *x*, and *z* contained putative microsatellite loci. *Lanes a*, *c*, *e*, *i*, and *k* correspond to 1 kb (Invitrogen) and 100 bp (Roche) molecular size markers

4. To test for restriction ligation success, prepare a PCR reaction containing 5 μl of 10× standard PCR buffer, 2 μl of 25 mM $MgCl_2$, 1.5 μl of 10 mM dNTPs, 3 μl of 10 μM forward SNX linker, 2 U of Taq DNA polymerase, 4 μl of the cut-ligated DNA from the previous step, and DEPC-treated water to a final volume of 50 μl. Run three PCR reactions simultaneously, with the following conditions: 1 cycle of 2 min at 72 °C (that serves as an elongation step, which fills any gap in ligated products), followed by 1 cycle of denaturing at 95 °C for 2 min; 20, 25, and 30 cycles (corresponding to the three different tubes) of denaturing at 95 °C for 20 s, annealing for 20 s at 60 °C and extension at 72 °C for 90 s; followed by 1 cycle of a 10 min extension at 72 °C. A successful PCR of linked-ligated DNA should show a clean "smear" in the target nucleotide range (i.e., 400–1,000 bp; see Note 4). Run an aliquot of 5 μl on a 2 % TAE agarose gel with SYBR safe to check for amplification success (see Fig. 2).

5. To prepare the linker-ligated-probe mixture, combine 80 μl of the 1× hybridization buffer, 5 μl of 100 μM biotinylated microsatellite probe, and 15 μl of the previous PCR reaction (see Note 5). Incubate in a thermocycler at 95 °C for 15 min followed by a cycle of 15 min at 50 °C.
6. In the meantime, prepare the streptavidin-coated magnetic beads. Vortex the magnetic beads to a uniform suspension. Add 100 μl of the beads to a tube and place it in the magnetic holder for approximately 30 s to capture the particles. Carefully remove the storage buffer. Remove the tube from the magnet.
7. Add 400 μl of TBST buffer and pipette up and down to mix. Place the tube in the magnetic holder for approximately 2 min and carefully remove the supernatant buffer. Repeat this step three times. Meantime, set two thermo-blocks to 50 and 65 °C, and keep aliquots of TBT and TBST at 50 °C, and Wash Buffer at 65 °C.
8. Add 300 μl of TBT buffer, resuspend the beads by gently vortexing the tube, and keep the beads at 50 °C while shaking (e.g., 70 rpm, just enough to maintain beads suspended in solution) until step 5 is complete.
9. Add the enrichment product from step 5 to the magnetic beads and mix. Again place the tube in the thermo-block, and incubate for 1–2 h at 50 °C while shaking.
10. Place the tube in the magnetic holder for 2 min and remove the supernatant. Add 400 μl of TBST buffer at 50 °C and pipette to mix. Repeat this step two times.
11. Place the tube in the magnetic holder for 2 min and remove the supernatant. Add 300 μl of Wash Buffer (low stringency) at 45 °C and pipette to mix. Repeat this step three times. In the meantime, set a thermo-block to 65 °C.
12. Place the tube in the magnetic holder for 2 min and remove the supernatant. Add 300 μl of Wash Buffer (high stringency) at 65 °C and pipette to mix. Repeat this step three times, and at the end of the last wash, incubate at 65 °C for 5 min (see Note 6). In the meantime, set a thermocycler to 95 °C.
13. To release the enriched product from the biotinylated oligos, place the tube in the magnetic holder for 2 min and remove the supernatant. Add 50 μl of TLE and incubate for 5 min at 95 °C in the thermocycler. After incubation, place again the tube in the magnetic holder for 1–2 min. Quickly recover the supernatant (that contains the enrichment product) and transfer it to an Eppendorf tube. Some beads may be transferred with the pipette, but they will not substantially affect subsequent PCR amplifications. This supernatant can be stored at −20 °C.

14. To perform a PCR of the single-stranded DNA product enriched for GT microsatellites, prepare reactions containing 5 μl of 10× standard PCR buffer, 2 μl of 25 mM $MgCl_2$, 1.5 μl of 2.5 mM dNTPs, 0.5 μl of 10 μM forward SNX linker, 2.5 U of Taq DNA polymerase, 4 μl of the washed elution product (enriched DNA) from the previous step, and DEPC-treated water to a final volume of 50 μl. Run the PCR, with the following conditions: 1 cycle of denaturing at 95 °C for 2 min; 30 cycles of denaturing at 94 °C for 30 s, annealing for 30 s at 60 °C and extension at 72 °C for 90 s; followed by 1 cycle of 15-min extension at 72 °C. Run an aliquot of 5 μl on a 2 % TAE agarose gel with SYBR safe to check for amplification success (see Fig. 2). A successful PCR of enriched DNA should show a clean "smear" in the target nucleotide range (i.e., 400–1,000 bp).
15. For the purification of the enriched PCR product, increase the volume to 100 μl with DEPC-treated water. Add 1/10 volume of 3 M sodium acetate (pH 6.8) and two volumes of cold 100 % ethanol to each tube. Incubate 5 min at −70 °C and spin down at 10,000 × *g* for 10 min. Carefully discard the ethanol and add two volumes of cold 70 % ethanol. Centrifuge at 10,000 × *g* for 2 min at room temperature and discard the ethanol, being very careful not to disturb the pellet when removing the ethanol. Dry under vacuum, and resuspend it in 25 μl of TLE buffer.
16. Prepare the PCR-enriched product for the TA ligation. Prepare a mix of 0.3 μl of 10× standard PCR buffer, 0.4 μl of 25 mM $MgCl_2$, 1 μl of 2.5 mM dATP, 2.5 U of Taq DNA polymerase, and 3 μl of DEPC-treated water and add it directly to each 25 μl enriched precipitated PCR product from the above step. Incubate for 30 min at 72 °C in the thermocycler.
17. Mix 3 μl of the A-tailed enriched product from above with 5 μl of 2× ligation buffer (provided with the T4 DNA ligase), 3 Weiss U of T4 DNA ligase, and 1 μl of 50 ng/μl pGEM-T Easy vector. Incubate overnight at 4 °C.

3.1.2 Screening, Isolation, and Sequencing of Clones Containing Microsatellites

Once the (GT microsatellite rich) recombinant plasmids are prepared, they can be transformed into XL10-Gold® *Escherichia coli* cells and plated onto LB/agar/ampicillin/X-Gal/IPTG plates. Recombinant clones are then selected based on blue-white screening. The time-consuming colony hybridization step has been eliminated, making the process equally efficient but much quicker. Instead, a PCR amplification screening process has been incorporated (11). The positive colonies (most containing microsatellite motifs) are then regrown in fresh LB/ampicillin medium; plasmids are purified by using mini preps and sequenced with Sanger technology.

1. Transform the ligation reaction into XL10-Gold® *E. coli* or any other available ampicillin-sensitive bacteria (e.g., DH5-α competent cells). Do so by thawing 50 μl of XL10-Gold® cells on ice and adding 5 μl of the ligated product to each tube. Incubate on ice for 30 min and mix from time to time by gently swirling the tubes every 10 min. Work under sterile conditions.
2. Heat shock at 42 °C for exactly 45 s and transfer the tubes back onto ice for 2 min. Do not shake the bacteria in this step.
3. Add 900 μl of SOC medium (preheated to 42 °C) and incubate at 37 °C for 1 h with gentle shaking (200 rpm). In the meantime prepare the plates (see Subheading 2.1.2).
4. Pellet the bacteria by centrifugation at 6,000 × *g* for 5–8 min. Discard 500 μl of the medium and resuspend by gently pipetting. The volume of transformed bacteria to be spread onto LB/agar/ampicillin/X-Gal/IPTG plates needs to be estimated empirically. With fresh bacteria and medium, 50–80 μl of the transformed bacteria should generate about 1,400 white colonies per plate and this number is a good indicator of a successful ligation–transformation. Incubate the plates upside down overnight at 37 °C. The remaining (unplated) bacteria can be stored at −80 °C (mixed with glycerol).
5. Prepare as many 96-well plates as required for the growing and screening of microsatellite-containing clones (1–2 plates are typically a good starting point). Assemble one flat-bottom 96-well plate for growing bacteria with 200 μl of LB/ampicillin (see Subheading 2.1.2) in each well, and another 96-well PCR replica-plate (with the same number of LB-containing wells) with 20 μl of PCR cocktail in each well. For the screening PCR, combine 2 μl of 10× standard PCR buffer, 1.5 μl of 25 mM $MgCl_2$, 0.4 μl of 2.5 mM dNTPs, 0.5 μl of 10 μM of M13 forward and reverse primers, 0.3 μl of 10 μM of the CA probe (nonbiotin-labeled), 0.5 U of Taq DNA polymerase, and DEPC-treated water to a final volume of 20 μl. Keep the plates at 4 °C throughout the process.
6. Using a sterile toothpick or a 10 μl sterile pipette, touch the center of a white (positive) colony and shake briefly in the PCR mix. With the same toothpick, apply the colony pick to the backup LB/ampicillin well plate in the same well position as in the PCR well plate. Twist the toothpick to dislodge the remaining colony. Work under sterile conditions.
7. Once finished, grow the bacteria at 37 °C for 2 h and store at 4 °C until the PCR amplification (step 8) has been checked on an agarose gel.
8. Place the PCR in the thermocycler and run with the following conditions: 1 cycle of denaturing at 95 °C for 10 min and 30 cycles of denaturing at 95 °C for 30 s, annealing for 45 s at

60 °C and extension at 72 °C for 2 min followed by 1 cycle of 10 min extension at 72 °C. Run an aliquot of 7–10 μl on a 2 % TAE agarose gel with SYBR safe to check for amplification success. The PCR product will be discarded anyway, so load as much as possible on the agarose gel to ensure that bands are clearly visible. The lanes that show two bands or a smear (see Note 7) indicate the presence of a microsatellite with a GT motif in their sequence (Fig. 2).

9. Prepare as many 15 ml falcon tubes as positive colonies recovered from step 8. Fill each with 7–10 ml of LB/ampicillin (see Subheading 2.1.2). Pipette 20–50 μl of the corresponding growing positive colonies from step 7 and inoculate the fresh LB medium in the falcon tubes. Incubate overnight at 37 °C with vigorous shaking. Work under sterile conditions.
10. Centrifuge for 10 min at 6,000 × *g* to pellet the bacteria and discard the remaining LB medium. Proceed following the manufacturer's protocol for the QIAprep Spin Miniprep kit (Qiagen). Elute with 30–50 μl of DEPC-treated water, instead of using the Buffer EB provided with the kit. The DNA products are ready for direct sequencing with M13 forward and reverse primers using the BigDye Terminator Sequencing kit (Applied Biosystems) following manufacturer's instructions on an automated sequencer (e.g., ABI 3100 or ABI 3739 Genetic Analyzers). Other universal primers for sequencing can be used (e.g., SP6 and T7 primers).
11. For visualizing, BLAST searching (20), and editing of the sequences, 4Peaks (available at http://www.mekentosj.com/science/4peaks), FinchTV (available at http://www.geospiza.com/Products/finchtv.shtml), or other suitable alignment software can be used. These programs can also be utilized to identify and remove low-quality or complex microsatellite sequences.
12. Sequence assembling and trimming of vector and linker sequences can be performed with Sequencher 5.0 (GeneCode), CodonCode Aligner, or Clustal X (21). For Sequencher, and in order to avoid the assembling of contigs based on the repeat themselves, reduce the strictness of sequencing matching and overlapping when doing the "Sequence Assembly" (e.g., by adjusting the "Minimum Match Percentage" and the "Minimum Overlap" values to 60 and 10, respectively. See Note 8). Export sequences of interest into a FASTA format.

3.2 Next-Generation Sequencing Methods for Microsatellite Capture

In contrast to conventional Sanger sequencing, high-throughput next-generation DNA sequencing technologies (e.g., HiSeq 2000 of Solexa/Illumina or the GS FLX/454 of Roche) are providing access to whole-genome information (including microsatellites and other types of markers) of many non-model organisms and thereby present new and exciting biological applications in population

Table 1
List of some external services (provided by academic institutions or private companies) for developing microsatellite markers[a]

Name, country	Web access
Bioprofiles, England	http://www.bioprofiles.co.uk/
Cornell University, USA	http://www.brc.cornell.edu/brcinfo/index.php?p=microsatellite
Ecogenics, Switzerland	http://www.ecogenics.ch/
Genetic Identification Services, USA	http://www.genetic-id-services.com/index.htm
Genetic Markers Services, England	http://www.geneticmarkerservices.com
GenoScreen, France	http://www.genoscreen.com/
Genterprise, Germany	http://www.genterprise.de/
Savannah River Ecology Laboratory, USA	http://www.srel.edu/microsat/Microsat_DNA_Development.html
The University of Arizona, USA	http://uagc.arl.arizona.edu/index.php/microsatellite-capture.html

[a]Other companies might accept samples for NGS, but not specifically for developing microsatellite markers

genetic and conservation studies (22). The relatively short-length read feature of initial NGS technology in comparison with Sanger sequencing lessens in the last years the chance of obtaining enough flanking sequence for the design of primers and successive amplification, and was occasionally problematic in species with highly repetitive or complex genomes (23). Among all currently available NGS methods, the GS FLX Titanium is the choice of preference for searching for microsatellites (see Table 1 for an example of microsatellite-developing services), since it generates reads up to 1 kb, and offers a competitive cost-effective output, generating huge amounts of data (>400.000 sequence reads) quickly vs. other sequencing methods (14). Although Illumina (24) is already about ten times cheaper, the problems associated to the assembly of small sequences (average 100 bp length) prevent for the time being the general use of this technology for microsatellite development (although see (25) and (26) for promising advances).

For the purpose of developing microsatellites, the starting material could be a microsatellite-enriched library (27) or total genomic DNA (14), with the latter eliminating the benchwork time for library preparation. In general, the success in obtaining a fair amount (>40) of potentially useable microsatellite loci (i.e., a unique sequence with a flanking region that do not match other recover loci and with enough space for designing primers) is related to the frequency of microsatellites in the genome of the species of interest, which highly varies among plants, vertebrates, and invertebrates (9, 24). For instance, the Roche GS FLX Titanium sequencing of a 96-well plate usually yields on average lower usable microsatellite sequences in invertebrates (3.4 %) compared to vertebrates (12 %) and plants (14 %) (24). Thus, depending on the taxonomic group, taking into account that many projects only require the development of small amounts of microsatellite markers and given that most of the 454 sequencing services usually offer partitions of a full plate in half, 1/4, 1/8, or 1/16 sections, using an enriched genomic library as DNA source and running a small plate section may be the option of choice. Moreover, by using coded Multiplex Identifier (MID) adapters (28), it is possible to sequence different microsatellite-enriched libraries in multiplex pools in a very efficient way (25), decreasing sequencing cost and also enabling the preferential selection of a specific microsatellite motif, if desired.

Briefly, the protocol used for the enriched library construction (based on Subheading 3.1, but without cloning and transforming the microsatellite-containing DNA product) would continue with the addition of MID adaptors to the PCR-enriched products followed by their immobilization on DNA capture beads and clonal amplification via PCR emulsion on the bead, following the recommended Titanium Rapid Library Preparation and Titanium emPCR protocols (October 2009, Roche). If the starting material is genomic DNA, it should be fragmented into 300–800 bp (generally by nebulization) prior to the ligation of the adapters (but see Note 9).

1. Start working with 16 μl of the PCR-enriched product (≥30 ng/μl) of step 15 (Subheading 3.1.1) and add to it successively 2.5 μl of 10× RL PNK buffer, 2.5 μl of dATP 1 μl of dNTPs, 1 μl of RL T4 DNA polymerase, 1 μl of RL T4 PNK, and 1 μl of RL Taq DNA polymerase (all reagents supplied with the GS FLX Titanium Rapid Library Preparation kit). Incubate for 20 min at 25 °C followed by 20 min at 72 °C in the thermocycler.
2. In the meantime, prepare the AMPure beads. To do so, vortex the magnetic beads to a uniform suspension. Add 125 μl of the beads to a microcentrifuge tube and place it in the magnetic holder to capture the particles, and carefully remove the storage buffer. Remove the tube from the magnet and add 500 μl of Sizing Solution (supplied with the kit). Resuspend the beads by gently vortexing and keep the tube on ice until it is used.
3. Add 1 μl of RL MID adaptor and 1 μl of RL ligase (supplied with the kit) to the A-tailed PCR mix from step 1. Incubate at 25 °C for 10 min.
4. Add MID-ligated DNA to the beads, gently mix by vortexing, and incubate at room temperature for 5 min.

5. Place the tube in the magnetic holder until the beads are fully pelleted, and remove the supernatant. Add 100 μl of TE buffer (supplied with the kit) and 500 μl of Sizing Solution, gently mix by vortexing, and incubate at room temperature for 5 min. Repeat this step two times.
6. Place again the tube in the magnetic holder and wash the beads with 1 ml of 70 % ethanol. Once the pellet is air-dry completely, add 53 μl of TE buffer and pipette to mix. With the tube placed again in the magnetic holder, transfer 50 μl of the supernatant to a new tube, being very carefully not to carry over any beads with the pipetting.
7. Quantitate the DNA library (50 μl in duplicate) by fluorometry (e.g., by using the TBS 380 fluorometer). For that, prepare a standard curve with a serial dilution of 2.5×10^8 molecule/μl of the RL Standard mix (supplied with the kit) on TE buffer. To calculate the library sample concentration (in molecules/μl), use the Rapid Library Quantitation Calculator (available at www.545.com/my454). Based on that information, prepare the DNA library to a working stock of 1×10^7 molecules/μl in TE buffer. To assess the quality of the library, run it on a High-Sensitivity chip on the BioAnalyzer 2100 to validate that the fragment size is between 600 and 900 bp and less of the 10 % of the fragments are below 350 bp.
8. The DNA is then ready to be sent to the NGS facility, where it will be titrated, emulsified, and sequenced on the Genome Analyzer FLX following manufacturer's instructions. Sequences will be processed for quality, length of sequence reading, and MID adaptor trimming using the shotgun signal pipeline, using different stringency filters on the amplicon signal.
9. Finally, sequence data provided to the user by the sequencing facility are screened for microsatellite motifs by using the software Msatcommander ((29); available at https://github.com/brantfaircloth/msatcommander). This software allows for the use of different filtering criteria to select unique sequences with specific microsatellite length motifs and numbers of repeats, after discarding those that are redundant ones or lack sufficient sequence for subsequent primer design.

3.3 PCR Primer Design, Testing, and Basic Genotyping

Primer pairs should be thoughtfully designed taking into account that the sizes of the amplicons and the dye labels of each pair of primers must differ sufficiently in order to be distinguished from one another. Using a four-color dye set (a popular group is 6-FAM, VIC, NED, and PET, with the orange dye LIZ used for the size standard) allows for a four-PCR product (or multiplex) combination in the same lane. If there is nonoverlapping size distribution of alleles at pairs of loci, this combined number of loci could be increased and can include loci labeled with the same dye. Moreover, some of the characteristics of the dyes can be considered, for example, combining the most intense (6-FAM) with the least intense (NED) dyes to label the weakest and the most intensely amplified products, respectively.

1. There are several open source software packages available for primer design. One of the most common is Primer3 ((30); available at http://primer3.sourceforge.net/), which is integrated into other software (e.g., Msatcommander (29), BatchPrimer3 (31), available at http://probes.pw.usda.gov/batchprimer3/index.html, or QDD (32), available at http://gsite.univ-provence.fr/gsite/Local/egee/dir/meglecz/QDD.html) for high-throughput processing allowing batch input of a large number of sequences in FASTA format. As a general rule, the primer design strategy should take into account the following parameters: (1) the number of repetitions in the microsatellite motif should be no less than six; (2) the final PCR amplicons should be a total length of around 400 bp; (3) approximately ten or more bp on both sides of the microsatellite sequence should be maintained; and (4) the reverse primer should begin (5′-end) with a guanine to increase the proportion of PCR products with an A-tail (called "PIG-tailing," see also Note 9). The primer selection conditions, although flexible, should address some of the following parameters: (a) primer length of 18–22 bp; (b) melting temperature (T_m) of 58–62 °C; (c) T_m difference between primers (ΔT_m) of <1–2 °C; and (d) G+C content (% GC) of 45–60 %. The other criteria for the stability of primer secondary structures are usually left as default parameters in Primer3. Once designed (see Note 10), order them from the company of preference.
2. Extract DNA from 8 to 16 individual samples from the target species following the preferred method of extraction and elute in TE buffer. Quantify the DNA concentration on a spectrophotometer and adjust it to a minimum of 10 ng/μl for future reactions (see Note 11).
3. To test the primers in a PCR amplification, prepare reactions containing 1 μl of 10× standard PCR buffer, 1.5 μl of 25 mM $MgCl_2$, 1 μl of 2.5 mM dNTPs, 0.16 μl of 10 μM forward and reverse primers, 0.5 U of Taq DNA polymerase, 1 μl of genomic DNA (aprox. 10 ng), and DEPC-treated water to a final volume of 15 μl. The mixture is then subjected to gradient thermocycler with the following conditions: 1 cycle of denaturing at 95 °C for 2 min; 30 cycles of denaturing at 94 °C for 30 s, annealing for 30 s at 52–60 °C and extension at 72 °C for 90 s; followed by 1 cycle of 15 min extension at 72 °C. Run an aliquot of 5 μl on a 2 % TAE agarose gel with SYBR safe to check for amplification success. The forward primers yielding

PCR products showing clear and distinct bands can be synthesized again with a 5′ fluorescent dye (see Note 12).

4. Run the PCR using the reaction mix as indicated above using the labeled primers and with the following two-step amplification profile: 1 cycle of 2 min at 95 °C, 30 cycles of denaturing at 95 °C for 30 s, primer-specific annealing temperature for 45 s, and extension at 72 °C for 45 s, followed by 1 complete cycle of denaturing at 95 °C for 30 s, annealing for 45 s at 53 °C, and extension at 72 °C for 45 s. PCR finish with a 15-min extension at 72 °C cycle.
5. The dilution needed to run the samples on a capillary sequencer needs to be estimated empirically, but preparing a dilution series of 1/10 and 1/50 is generally a good starting point. The products can then be visualized by separating the samples on a capillary sequencer (e.g., using LIZ as a size standard (Amersham Biosciences), according to the manufacturer's instructions).
6. The loci showing a clear allele pattern and allele variability can be used for further population genetic analyses. In many cases, using the Type-it Microsatellite PCR to perform a multiplex PCR and the Multiplex Manager software (33) to design the PCR multiplex combination of the variable loci is highly advisable to reduce costs. Run the PCR using the PCR conditions as recommended by the kit manufacturers but reducing the reaction mix volume to 10 μl as follows: 2.5 μl master mix, 0.2 μl of Q buffer (both of them supplied with the kit), 5 μl of primer mix, 0.4 μl DEPC-treated water, and 10–40 ng of template DNA. The primers should be previously mixed in an equimolar concentration.
7. Several software packages are available to perform allele scoring of microsatellite electropherograms (e.g., GeneMapper from ABI or the open source STRand software, available at http://www.vgl.ucdavis.edu/informatics/strand.php.).
8. Once the allele score is completed and results are retrieved into a text or table format, use Convert (34) or MSA (35) to convert spreadsheet data into input files for a large number of downstream applications (e.g., population genetic analyses).

4 Notes

1. Ethidium bromide (EtBr) can be also used for DNA staining; however it has been progressively replaced by SYBR Safe in the laboratories, which has been documented as less harmful as the EtBr, and it does not require UV for visualization.
2. Alternatively, other species-specific restriction enzymes can be used (e.g., *Bst*UI, *Alu*I, *Bsa*AI, *Hinc*II), although it is advisable

not to use those that have a recognition site in the SNX linker. To prevent overcutting of the DNA or the ligation of the DNA fragments into chimeras, conduct steps 2 and 3 in separate reactions for each enzyme combination. Digested DNA can be pooled afterwards (before step 4) and volume can be concentrated using a MinElute Purification kit (Qiagen). It is often wise to do a previous digestion test with the selected enzymes (using approx. 1 μg of genomic DNA) to determine that the digestion yields fragments of about 200–1,500 bp in length.

3. If the digestion was not complete, add more units of enzyme and incubate at 37 °C for 2–4 additional hours or until complete digestion.

4. The differences in the number of cycles will help to choose the PCR product with the cycle combination that produces an intense smear around the desirable DNA-fragment range. Take into account that the increase in the number of cycles will be biased towards the smaller DNA fragments. Also, other PCR-induced bias could occur when the polymerase over-amplifies specific regions of the genome. Discard the PCR product if some spurious bands appear in the amplification.

5. Alternatively, it is possible to perform the enrichment directly from the cut-ligated DNA of step 3 using the same volume.

6. The temperature used in this step (65 °C) corresponds to the probe-specific hybridization temperature (T_{hyb}) for CA_{12}. If other probes (or a mix of probes with similar annealing temperatures, T_m) are used for the enrichment process, their specific T_m should be calculated first. Use OligoCalc (36) (accessible at http://www.basic.northwestern.edu/biotools), which is a Web-based tool that calculates this value taking into account the salt concentration of the stringency washes. The salt solution that is more often used in hybridization experiments is SSC. For the protocol used here, the SSC content (2×) of the buffer solution rendered a final Na^+ concentration of 0.33 M that should be considered when using the software. Once calculated, the T_{hyb} recommended for the oligo mix is usually T_m − (10–12 °C), reducing the stringency and ensuring that the small microsatellite motif sequences are also retained during elution.

7. In case there is no microsatellite motif in the insert, only one band will be amplified that corresponds to the amplification of the entire cloned fragment. However, in inserts that contain a GT motif, the CA probe (here functioning as a primer) will anneal at various positions within the microsatellite yielding an extra band or a smear of variable length.

8. Several "contigs" might contain the same (or similar) flanking sequence regions with different microsatellite sequence lengths. This could correspond to a minisatellite region sequence.

Minisatellites are another class of tandem repeats defined by a short (6–100 bp) motif spanning 0.5 to several kilobases. However, because they differ from microsatellites in size, mutation processes, and chromosomal distribution (37), they are not suitable for genotyping along with microsatellite loci.

9. Special attention should be paid to the average size length of the enriched products, since larger fragments (>500 bp) will not be completely sequenced in both directions, with a test for insert size on an agarose gel being advisable before proceeding with the next steps. The size range of PCR products suitable for the GS FLX Titanium method (500–800 bp) can be extracted for an agarose gel using a QiaQuick Gel Extraction kit (Qiagen), following the manufacturer's instructions.
10. Adding a guanine to the 5′ end of the reverse primers promotes adenylation by Taq DNA polymerase (i.e., the addition of an adenine at the end of the PCR products), reducing PCR stutter and making the peak calling easier. For the same purpose, it is possible also to add a "GTTTCTT" tail to the 5′ end of the reverse primers to facilitate genotyping (38).
11. It is also advised to check the amplification and polymorphism of the new primers in closely related species. Do so by extracting genomic DNA from a minimum number of 8 to 16 individual organisms from other species. Then, proceed from step 3 of Subheading 3.3.
12. A two-stage touchdown amplification profile could be applied when specific PCR products are not found. In that case, PCR profile should include 1 °C reduction of the T_m at each cycles (e.g., from 65 to 55 °C) followed by 20–25 cycles at 55 °C.

Acknowledgments

EGG is sponsored by a postdoctoral fellowship of the Ministerio de Educación Cultura y Deporte (MECD). The research of R.Z. is funded by the Ministerio de Economía y Competitividad (CGL2010-18216).

References

1. Ellegren H (2004) Microsatellites: simple sequences with complex evolution. Nat Rev Genet 5:435–445
2. Jarne P, Lagoda PJL (1996) Microsatellites, from molecules to populations and back. Trends Ecol Evol 11:424–429
3. Schlotterer C (2004) The evolution of molecular markers—just a matter of fashion? Nat Rev Genet 5:63–69
4. Weber JL (1990) Informativeness of human (dC–dA)n.(dG–dT)n polymorphisms. Genomics 7:524–530
5. Toth G, Gaspari Z, Jurka J (2000) Microsatellites in different eukaryotic genomes: survey and analysis. Genome Res 10:967–981
6. Kassai-Jager E, Ortutay C, Toth G et al (2008) Distribution and evolution of short tandem

repeats in closely related bacterial genomes. Gene 410:18–25

7. Selkoe KA, Toonen RJ (2006) Microsatellites for ecologists: a practical guide to using and evaluating microsatellite markers. Ecol Lett 9:615–629
8. Haasl RJ, Payseur BA (2011) Multi-locus inference of population structure: a comparison between single nucleotide polymorphisms and microsatellites. Heredity 106:158–171
9. Zane L, Bargelloni L, Patarnello T (2002) Strategies for microsatellite isolation: a review. Mol Ecol 11:1–16
10. Hamilton MB, Pincus EL, Fleischer RC (1999) Universal linker and ligation procedures for construction of genomic DNA libraries enriched for microsatellites. Biotechniques 27:500–515
11. Gardner MG, Cooper SJB, Bull CM et al (1999) Isolation of microsatellite loci from a social lizard, *Egernia stojesii*, using a modified enrichment procedure. J Hered 90:301–304
12. Rassmann K, Schlotterer C, Tautz D (1991) Isolation of simple-sequence loci for use in polymerase chain reaction-based DNA fingerprinting. Electrophoresis 12:113–118
13. Prochazka M (1996) Microsatellite hybrid capture technique for simultaneous isolation of various STR markers. Genome Res 6:646–649
14. Abdelkrim J, Robertson BC, Stanton JL et al (2009) Fast, cost-effective development of species-specific microsatellite markers by genomic sequencing. Biotechniques 46: 185–192
15. Allentoft ME, Schuster SC, Holdaway RN et al (2009) Identification of microsatellites from an extinct moa species using high-throughput (454) sequence data. Biotechniques 46:195–200
16. Glenn TC, Schable NA (2005) Isolating microsatellite DNA loci. Methods Enzymol 395:202–222
17. Bachtrog D, Agis M, Imhof M et al (2000) Microsatellite variability differs between dinucleotide repeat motifs—evidence from *Drosophila melanogaster*. Mol Biol Evol 17:1277–1285
18. Li YC, Korol AB, Fahima T et al (2002) Microsatellites: genomic distribution, putative functions and mutational mechanisms: a review. Mol Ecol 11:2453–2465
19. Sambrook J, Fritsch EF, Maniatis T (1989) Molecular cloning: a laboratory manual. Cold Spring Harbor Laboratory, New York
20. Altschul SF, Gish W, Miller W et al (1990) Basic local alignment search tool. J Mol Biol 215:403–410
21. Thompson JD, Gibson TJ, Plewniak F et al (1997) The CLUSTAL_X windows interface: flexible strategies for multiple sequence alignment aided by quality analysis tools. Nucleic Acids Res 25:4876–4882
22. Ellegren H (2008) Sequencing goes 454 and takes large-scale genomics into the wild. Mol Ecol 17:1629–1631
23. Lerner HRL, Fleischer RC (2010) Prospects for the use of next-generation sequencing methods in ornithology. Auk 127:4–15
24. Gardner MG, Fitch AJ, Bertozzi T et al (2011) Rise of the machines—recommendations for ecologists when using next generation sequencing for microsatellite development. Mol Ecol Resour 11:1093–1101
25. Jennings TN, Knaus BJ, Mullins TD et al (2011) Multiplexed microsatellite recovery using massively parallel sequencing. Mol Ecol Resour 11:1060–1067
26. Castoe TA, Poole AW, Gu W et al (2010) Rapid identification of thousands of copperhead snake (*Agkistrodon contortrix*) microsatellite loci from modest amounts of 454 shotgun genome sequence. Mol Ecol Resour 10:341–347
27. Santana QC, Coetzee MPA, Steenkamp ET et al (2009) Microsatellite discovery by deep sequencing of enriched genomic libraries. Biotechniques 46:217–223
28. Binladen J, Gilbert MTP, Bollback JP et al (2007) The use of coded PCR primers enables high-throughput sequencing of multiple homolog amplification products by 454 parallel sequencing. Plos One 2:e197
29. Faircloth BC (2008) MSATCOMMANDER: detection of microsatellite repeat arrays and automated, locus-specific primer design. Mol Ecol Resour 8:92–94
30. Rozen S, Skaletsky HJ (2000) Primer3 on the WWW for general users and for biologist programmers. In: Krawertz S, Misener S (eds) Bioinformatics methods and protocols: methods in molecular biology. Humana, Totowa, NJ, pp 365–386
31. You FM, Huo NX, Gu YQ et al (2008) BatchPrimer3: a high throughput web application for PCR and sequencing primer design. BMC Bioinformatics 9:253
32. Meglecz E, Costedoat C, Dubut V et al (2010) QDD: a user-friendly program to select microsatellite markers and design primers from large sequencing projects. Bioinformatics 26:403–404
33. Holleley CE, Geerts PG (2009) Multiplex Manager 1.0: a cross-platform computer program that plans and optimizes multiplex PCR. Biotechniques 46:511–517

34. Glaubitz JC (2004) CONVERT: a user-friendly program to reformat diploid genotypic data for commonly used population genetic software packages. Mol Ecol Notes 4:309–310

35. Dieringer D, Schlotterer C (2003) Microsatellite analyser (MSA): a platform independent analysis tool for large microsatellite data sets. Mol Ecol Notes 3:167–169

36. Kibbe WA (2007) OligoCalc: an online oligonucleotide properties calculator. Nucleic Acids Res 35:W43–W46

37. Vergnaud G, Denoeud F (2000) Minisatellites: mutability and genome architecture. Genome Res 10:899–907

38. Brownstein MJ, Carpten JD, Smith JR (1996) Modulation of non-templated nucleotide addition by tag DNA polymerase: primer modifications that facilitate genotyping. Biotechniques 20:1004–1006

39. Gonzalez EG, Zardoya R (2007) Isolation and characterization of polymorphic microsatellites for the sardine *Sardina pilchardus* (Clupeiformes: Clupeidae). Mol Ecol Notes 7:519–521

Chapter 6

Next-Generation Sequencing for High-Throughput Molecular Ecology: A Step-by-Step Protocol for Targeted Multilocus Genotyping by Pyrosequencing

Jonathan B. Puritz and Robert J. Toonen

Abstract

Next-generation sequencing technology can now provide population biologists and phylogeographers with information at the genomic scale; however, many pertinent questions in population genetics and phylogeography can be answered effectively with modest levels of genomic information. For the past two decades, most population-level studies have lacked nuclear DNA (nDNA) sequence data due to the complications and cost of amplifying and sequencing diploid loci. However, pyrosequencing of emulsion PCR reactions, amplifying from only one molecule at a time, can generate megabases of clonally amplified loci at high coverage, thereby greatly simplifying allelic sequence determination. Here, we present a step-by-step methodology for utilizing the 454 GS FLX Titanium pyrosequencing platform to simultaneously sequence 16 populations (at 20 individuals per population) at 10 different nDNA loci (3,200 loci in total) in one plate of sequencing for less than the cost of traditional Sanger sequencing.

Key words 454, NGS, nDNA, Population genetics, Phylogeography, Sequencing cost, Sanger

1 Introduction

For 20 years, mitochondrial DNA (mtDNA) has been the standard molecular sequence marker for population geneticists and phylogeographers (1, 2). The mitochondrial genome, with a rapid mutation rate and small effective population size, contains a wealth of information for recent population histories. Moreover, mtDNA loci are easy to prime, amplify, and sequence due to uniparental inheritance and a lack of recombination. However, the mtDNA genome is essentially a single marker and cannot alone represent all of the evolutionary processes acting upon a population (3, 4). Thus, the need to incorporate nDNA sequence data into analyses has been widely recognized by the field over the last 10 years (5–8).

The use of nDNA sequence loci in phylogeography and population genetics has been limited by laboratory and analytical difficulties (reviewed in 7, 9, 10). nDNA loci that are polymorphic at the

Stella K. Kantartzi (ed.), *Microsatellites: Methods and Protocols*, Methods in Molecular Biology, vol. 1006, DOI 10.1007/978-1-62703-389-3_6, © Springer Science+Business Media, LLC 2013

population level often include large amounts of genetic polymorphism, complete with high levels of heterozygosity and insertion-deletion (INDEL) mutation events (9, 10). Sanger sequencing of a heterozygous locus with an INDEL polymorphism results in double or multiple chromatogram peaks (11); this complex polymorphism requires either laboratory techniques to physically separate the two sequences (i.e., cloning, single-strand conformation, etc.) (9, 10) or the use of computational algorithms to decode heterozygous genotypes (10, 12). The laboratory techniques are expensive and time-consuming, and the computational methods work best in data with lower levels of polymorphism, usually excluding multiple INDELS (9, 12, 13). In short, the extra cost, in both labor time and money, of analyzing polymorphic nDNA loci has largely kept nuclear sequences out of phylogeography.

Next-generation sequencing technology has the capacity to overcome the primary obstacles of including nDNA loci in phylogeography by providing a rapid and inexpensive way to produce large amounts of targeted sequence data (14, 15). The 454 GS FLX Titanium pyrosequencing technology is particularly useful for phylogeography with the ability to generate over 1 million reads of around 400 bp per read (16). Additionally, these reads are obtained from the pyrosequencing of individual emPCR reactions (17) which amplify DNA from only one molecule that is the effective equivalent of bacterial cloning. In short, pyrosequencing technology, when used in combination with barcoded PCR primers and a gasketed 454 sequencing plate, enables the rapid and simultaneous sequencing of thousands of targeted loci at high coverage and low comparative cost (18). Additionally, our previous work demonstrated that the high coverage of each locus generated from this approach greatly simplifies haplotype determination, even in a very polymorphic sea star species (18). Here, we present a more generalized step-by-step protocol to sequence 10 different nDNA loci (approximately 400 bp each) across 16 different populations with 20 individuals per population using the 454 GS FLX Titanium pyrosequencing platform for less than the cost of Sanger sequencing.

Targeted 454 sequencing is one of many platforms in a vast array of next-generation sequencing technology, and each has different possible applications to population and conservation genetics (15, 18). However, only one other platform has been applied in phylogeographic studies; Emerson et al. (19) used restriction-site-associated DNA tags (RAD tags) (20, 21) to determine the evolutionary relationship between recently diverged populations of pitcher plant mosquitoes. RAD tag-based Illumina sequencing can genotype multiple populations at thousands of SNP loci simultaneously but has more limited ability to survey large sample sizes within populations because of the cost. For example, Emerson et al. (20) genotyped 21 different populations at ~3,741 different SNPs but with only six individuals per population. Our targeted 454 sequencing methodology offers a compromise with the ability to sequence a reasonable sample size (20 individuals) from one population for ten different nDNA loci in 1/16th of a plate of 454 sequencing. For perspective, a single lane of Illumina sequencing is more expensive than 1/16th plate of 454 sequencing at current market prices, and only ~16 individuals can be RAD tagged in one lane, albeit with thousands of SNPs per individual (21, 22). The protocol we outline here generates high-quality genetic data with larger sample sizes, and we believe that it is a good compromise between cost and depth of genomic sampling for phylogeographic analyses.

2 Materials

2.1 Experimental Design and Primer Barcoding

This protocol assumes that working nDNA PCR primers are in hand. Products should be between 300 and 400 base pairs for best results. There are several good reviews on nDNA marker development (23, 24), and for those who lack nDNA markers, we have found the universal primer set from Jarman et al. has been particularly useful (26). Alternatively, our implementation of this methodology used primers that were developed directly from EST libraries, degenerate primers from interspecific alignments, and modification of universal intron primers (25).

2.2 Locus Amplification and Purification Materials

1. 3,200 barcoded PCR primers with attached Fusion primers (see Subheading 3.1).
2. AccuSure 2× PCR mix.
3. Additional $MgCl_2$.
4. Nanopure water.
5. 96-well PCR plates.
6. SPRIPlate Super Magnet Plate.
7. 60 ml of AMPure XP (stored at 4 °C for 12 months).
8. 70 % ethanol.
9. TE buffer.

2.3 Quantification and Pooling Materials

1. Black 96-well optical plates.
2. PicoGreen dsDNA Assay Kit, including DNA standard (stored at 4 °C).
3. M2 microplate reader with fluorescence detection.
4. Microcentrifuge tubes.
5. Nanopure water.
6. SoftMax Pro Software and PC to run M2 microplate reader. Make sure that read mode is set to "fluorescence (top read)"; excitation wavelength = 490 nm; emission wavelength = 525 nm;

cutoff wavelength = 515 nm; readings = 10 (*different from the default*); PMT = auto; and plate type is set to 96-well standard opaque.

2.4 Data Analysis Materials

1. Software such as Geneious 5.4 (26) capable of handling next-generation sequencing data. However, data can also be successfully manipulated in a Unix terminal (see Chapter 7).

3 Methods

3.1 Experimental Design and Primer Barcoding

1. The emulsion PCR (emPCR, *16*) of 454 sequencing amplifies a mixture of samples within one reaction; in other words, individual loci need to be tagged within the product (serial tagging), or they need to be physically separated by a gasket on the 454 sequencing plate. For this protocol, each population is a separate library prepared for a 1/16 gasketed region of a 454 picotiter plate, and each individual ($n = 20$) is given a single serial barcode within each population (Fig. 1; see Note 1).
2. 454 forward primers are designed in the following format: *5′-CGTATCGCCTCCCTCGCGCCATCAG-*[*Serial Barcode*]-[*Template Specific Forward Primer*]*-3′*. 454 reverse primers follow the format *CTATGCGCCTTGCCAGCCCGCTCAG-*[*Barcode*]-[*Template Specific Reverse Primer*]*-3′* (see Note 2).
3. For each locus, we first generate 22 unique forward and reverse primer pairs using matching serial barcodes. See Table 1 for the first 22 codes recommended by the manufacturer, otherwise known as multiplex identifiers (MIDs) (see Note 3). The two extra primer sets per locus are included in case of any particular primer-template incompatibilities.
4. Primers should be synthesized with the highest level of purification possible (we used PAGE purification), and all 440 primers should be ordered in bulk to minimize the per primer cost. If possible, primers should also be ordered in 96-well plates (with forward and reverse primers in adjacent rows) to facilitate multichannel pipetting.
5. All sample DNA extractions should be aliquoted (1–5 ng of DNA per μL) into 96-well PCR plates by population in a format that matches the primer plates for easier PCR reaction setup. Eight rows of ten samples each allow for extra space in plates during later steps of the protocol, especially DNA quantification which uses 15 standards. This layout allows four loci from one population to be stored in a single plate. The rest of the protocol is written in a plate-by-plate format assuming a single worker; depending on equipment and personnel, more than one plate may be processed at a single time.

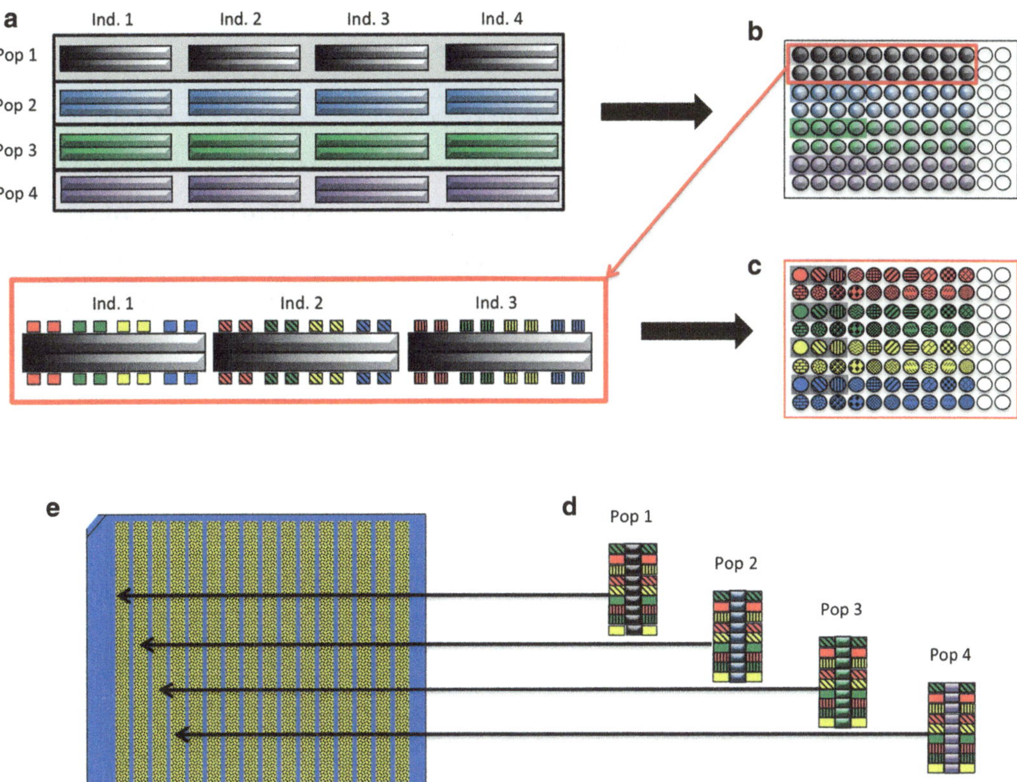

Fig. 1 Graphical representation of the overall experimental protocol. (**a**) Graphical representation of DNA extractions of individuals from populations which are then pipetted into (**b**) 10 by 2 populations in 96-well PCR plates. From one population (*red inset*), multiple loci are amplified with individual barcodes, (**c**) keeping to the same format of the original population extraction. (**d**) Reactions are then cleaned, quantified, and pooled by populations. (**e**) Each population becomes a library and is placed into a unique gasketted section of a picotiter plate for 454 sequencing

3.2 Locus Amplification and Purification

1. Because PCR performance often changes with the longer MID-labeled primers, the first step is to re-optimize PCR conditions for the new longer primers. Look to minimize nonspecific binding, cycle number, and annealing temperature (see Note 4).
2. Set up PCR reactions with 12.5 μL of Accusure 2× PCR mix, 0.6 μL of each 10 μM primer, 0.25–0.5 μL of 50 mM $MgCl_2$, and 1.5 μL of sample template, bringing the final volume to 25 μL with sterile water.
3. Thermocycling profiles should follow the general form of 95 °C denaturation for 10 min (Hot Start), then 30–35 cycles of 95 °C for 30 s, 55–60 °C for 45 s, and 1.5 min at 68 °C, followed by a final extension of 20 min at 68 °C (see Note 5).
4. Check all PCR reactions with UV imaging on a 1 % agarose gel. Any reactions with strong nonspecific bands (i.e., large primer dimer or secondary bands) should be redone.

Table 1
First 22 MIDs from Roche (see Note 3)

MID01	ACGAGTGCGT	MID14	CGAGAGATAC
MID02	ACGCTCGACA	MID15	ATACGACGTA
MID03	AGACGCACTC	MID16	TCACGTACTA
MID04	AGCACTGTAG	MID17	CGTCTAGTAC
MID05	ATCAGACACG	MID18	TCTACGTAGC
MID06	ATATCGCGAG	MID19	TGTACTACTC
MID07	CGTGTCTCTA	MID20	ACGACTACAG
MID08	CTCGCGTGTC	MID21	CGTAGACTAG
MID10	TCTCTATGCG	MID22	TACGAGTATG
MID11	TGATACGTCT	MID23	TACTCTCGTG
MID13	CATAGTAGTG	MID24	TAGAGACGAG

5. Repeat until all loci have been successfully amplified.
6. Aliquot 20 μL of PCR product into a new plate.
7. Follow the AMPure XP protocol with the exception of using 15 μL of AMPure XP bead per reaction instead of 36 μL (see Note 6 for full protocol).
   (a) Add 15 μL AMPure XP beads to each PCR reaction.
   (b) Mix by pipetting ten times and let sit for 5 min.
   (c) Place plate into SPRIPlate 96 Super Magnet Plate for 2 min.
   (d) Aspirate cleared solution from reaction plate and discard.
   (e) Pipette 200 μL of 70 % ethanol into each well of plate and incubate for 30 s.
   (f) Aspirate out the ethanol and discard.
   (g) Repeat for a second wash.
   (h) Remove the plate from the Super Magnet, add 50 μL of elution buffer (TE) to each well, and pipette mix ten times.
   (i) Plate reaction plate into Super Magnet for 1 min to separate beads.
   (j) Transfer the eluant to a new plate.
8. Optional, but recommended—visualize all reactions with UV imaging on a 1 % agarose gel. There should be only single clean bands present in every reaction. Any primer dimer will negatively affect the quality of your libraries.
9. Store plates in freezer until all loci are successfully amplified and cleaned.

3.3 Quantification and Pooling

1. Remove all the plates of cleaned product for a single population from freezer, defrost, and spin down.
2. Follow the "Quantitating Double-Stranded DNA with Quant-it PicoGreen dsDNA Reagent and SpectraMax Fluorecence Microplate Readers" (see Note 7 for full protocol):
   (a) Dilute concentrated TE buffer from kit by 20-fold with nanopure water.
   (b) Create a 200-fold dilution of PicoGreen reagent (using TE Buffer) for a working solution (see Note 8).
   (c) Prepare a standard range of lambda DNA concentrations in TE. 1 μg/mL = 1,000 μL of Lambda stock; 500 ng/mL = 500 μL of Lambda stock plus 500 μL of TE; 100 ng/mL = 100 μL of Lambda stock plus 900 μL of TE; 10 ng/mL = 10 μL of Lambda stock plus 990 μL TE; and a blank of 1,000 μL TE. Note: This standard curve differs from SpectraMax Protocol.
   (d) Pipette standards in triplicate into wells of black microplate that correspond to empty wells of the PCR plate; we used columns 11 and 12.
   (e) For each sample well, pipette 95 μL of PicoGreen working solution.
   (f) Quickly add 5 μL of PCR product from cleaned PCR product plate.
   (g) Incubate for 2–5 min in the dark with mild agitation. A plate shaker at very low RPM works well.
   (h) Place plate into microplate reader and click "Read."
3. Optional: Repeat step 2 and average readings. Quantification and pooling is the most critical aspect for the successful application of this protocol. This protocol has been successfully completed without this optional step, but it is strongly recommended if time and supplies permit.
4. Export data in plate format from software, so it can be readily imported into a spreadsheet program.
5. When all loci from one population are quantified, use a spreadsheet program to calculate the volume needed to pipette 0.5 ng of total DNA from each sample locus. This is done by dividing 0.5 by the concentration reading from the microplate reader. (Readings from software are in μg/ml which is equivalent to ng/μL.)
6. Carefully pipette the specified amount from each sample (from one population) into a single microcentrifuge tube (see Note 9).
7. Store the prepared library at least −20 °C (preferably −80 °C).
8. Repeat steps 1–7 for the remaining 15 populations.

3.4 Data Analysis

1. With several commercial, freeware, GUI, and command line-based systems for data analysis, this section will follow a general format without specific software instructions.
2. The sequencing facility or service should return 16 unique library files, each consisting of all sequences recovered from a single gasket pool—representing a single population.
3. Import the library file and remove all reads that are less than 150 bp.
4. Sort all reads by barcode with 454 Fusion primer trimming.
5. Within each barcode set, create contigs by one of two methods (see Note 10):
   (a) Use a reference sequence for each locus.
   (b) Create contigs using fairly high level of sensitivity (i.e., 20 % maximum gaps, 80 % identity overlap).
6. Within each locus contig:
   (a) Discard any reads where the whole primer sequence is not present and any reads where both serial barcodes do not match.
   (b) Discard or trim any reads with an average quality score below 30.
   (c) Score any basepair below 75 % consensus for all reads "heterozygous."
      - If polymorphism is an INDEL, especially a homopolymer, use both majority consensus and average quality score for the repeats of different lengths to make the call. If less than 10× coverage, use quality score exclusively (see Note 11).
      - If non-INDEL heterozygous bp were determined in a locus, sort that contig by the heterozygous bp. The two most common haplotypes in the contig are the two alleles for the heterozygous locus; however, see Note 12.
7. Save allele(s) from contigs for each individual. These can later be aligned and used in standard population genetic and phylogeographic analyses.

4 Notes

1. This is a generalized experimental design chosen to minimize two factors: (a) the cost of primer synthesis and (b) the impact of errant quantification and pooling of a single sample on overall library quality. However, different levels of gasketing can be used in conjunction with more or less serial barcoding.

More information on barcoding methods can be found in these papers (27, 28).

2. More information on 454 Fusion primer design can be found in Technical Bulletin No. 013-2009 (Roche).
3. More information on 454 MIDs can be found in Technical Bulletin No. 005-2009 (Roche).
4. Keep in mind that these primers are significantly longer than normal PCR primers, with the majority of the primer not designed to bind to the template. A higher concentration of $MgCl_2$ and elimination of BSA from reactions were critical for successful PCR reactions using the full primers, especially for avoiding long primer dimers. Lower cycle number and annealing temperatures help to reduce PCR-induced chimeras and recombination (29).
5. Notice the long hot start and final extension periods. This is critical to ensure that full products are amplified.
6. The ratio of AMPure bead to PCR product allows for size-selective purification. A ratio of 0.75 greatly reduces products under 300 bp in size while still providing a large yield of purified product. In our original protocol, we used a 0.8 ratio, but later quality control procedures at the sequencing facility purified our libraries a second time at a 0.7 ratio to further eliminate smaller products. The protocol can be found at https://www.beckmancoulter.com/wsrportal/bibliography?docname=Protocol_000387v001.pdf.
7. http://www.moleculardevices.com/Documents/general-documents/mkt-appnotes/microplate-appnotes/SpectraMax%20AppNote%2022%20(PicoGreen)%20rev%20B.pdf
8. Plastic containers should be used instead of glass because the reagent may adsorb to glass surfaces. This solution should be protected from light and used within a few hours of prep.
9. Pipetting accuracy is critical at this stage. Make sure to use calibrated pipettes and to familiarize yourself with the proper pipetting procedure of your pipette brand and the level of accuracy. Any samples with calculated pipetting volumes less than 0.5 μL should be diluted tenfold to ensure accurate volumes. Be sure to also multiply the calculated sample volume by 10 as well.
10. Using a reference sequence will help eliminate nonspecific reads from your contig and increase the speed of data processing. However, this will discard a high percentage of reads, and it will be worthwhile for a few individuals to view all possible contigs to look for any inconsistencies or missed variation.
11. Homopolymers are the most consistent source of sequencing errors for 454 sequencing (30). Final alignments of alleles

from populations should be checked for any singletons that differ only by a homopolymer INDEL; this is most likely a sequencing artifact.

12. There may be a small number of "recombinant" reads in the contig. They will be small in coverage number and be a clear mix of the two most common haplotypes.

Acknowledgments

The authors would like to thank Maria Byrne, Sergio Barbosa, Carson Keever, Jason Addison, Michael Hart, and Richard Grosberg for their extensive collaborative and supportive efforts with this project. We also thank Clarissa Murch for extensive help with project organization and lab work. Lastly, we would like to thank Scott Hunicke-Smith of University of Texas for his vast assistance with implementing this unique 454 sequencing project and the Hawai'i Institute of Marine Biology EPSCoR core genetics facility. This project was funded a grant from the National Science Foundation (Bio-OCE 0623699). This is contribution #1520 from the Hawai'i Institute of Marine Biology and 8754 from the School of Ocean and Earth Sciences and Technology (SOEST).

References

1. Avise JC (1998) The history and purview of phylogeography: a personal reflection. Mol Ecol 7:371–379
2. Avise JC, Arnold J, Ball RM et al (1987) Intraspecific phylogeography: the mitochondrial DNA bridge between population genetics and systematics. Annu Rev Ecol Systemat 18:489–522
3. Avise JC (2004) Molecular markers, natural history, and evolution. Sinauer Associates, Sunderland, MA
4. Hoelzer GA (1997) Inferring phylogenies from mtDNA variation: mitochondrial-gene trees versus nuclear-gene trees revisited. Evolution 51:622–626
5. Karl SA, Avise JC (1992) Balancing selection at allozyme loci in oysters: implications from nuclear RFLPs. Science 256:100–102
6. Karl SA, Avise JC (1993) PCR-based assays of mendelian polymorphisms from anonymous single-copy nuclear DNA: techniques and applications for population genetics. Mol Biol Evol 10:342–361
7. Hare M (2001) Prospects for nuclear gene phylogeography. Trends Ecol Evol 16: 700–706
8. Bowen BW, Bass AL, Soares L et al (2005) Conservation implications of complex population structure: lessons from the loggerhead turtle (*Caretta caretta*). Mol Ecol 14: 2389–2402
9. Zhang D-X, Hewitt GM (2003) Nuclear DNA analyses in genetic studies of populations: practice, problems and prospects. Mol Ecol 12: 563–584
10. Creer S (2007) Choosing and using introns in molecular phylogenetics. Bioinformatics 3: 99–108
11. Mallarino R, Bermingham E, Willmott KR et al (2005) Molecular systematics of the butterfly genus *Ithomia* (Lepidoptera: Ithomiinae): a composite phylogenetic hypothesis based on seven genes. Mol Phylogenet Evol 34:625–644
12. Huang Z-S, Ji Y-J, Zhang D-X (2008) Haplotype reconstruction for scnp DNA: a consensus vote approach with extensive sequence data from populations of the migratory locust (*Locusta migratoria*). Mol Ecol 17:1930–1947
13. Salem RM, Wessel J, Schork NJ (2005) A comprehensive literature review of haplotyping

software and methods for use with unrelated individuals. Hum Genom 2:39–66

14. Metzker ML (2009) Sequencing technologies—the next generation. Nat Rev Genet 11:31–46
15. Allendorf FW, Hohenlohe PA, Luikart G (2010) Genomics and the future of conservation genetics. Nat Rev Genet 11:697–709
16. Margulies M, Egholm M, Altman WE et al (2005) Genome sequencing in open microfabricated high density picoliter reactors. Nature 437:376
17. Leamon JH, Lee WL, Tartaro KR et al (2003) A massively parallel PicoTiterPlate based platform for discrete picoliter-scale polymerase chain reactions. Electrophoresis 24:3769–3777
18. Ekblom R, Galindo J (2010) Applications of next generation sequencing in molecular ecology of non-model organisms. Heredity 107: 1–15
19. Emerson KJ, Merz CR, Catchen JM et al (2010) Resolving postglacial phylogeography using high-throughput sequencing. Proc Natl Acad Sci 107:1–5
20. Baird NA, Etter PD, Atwood TS et al (2008) Rapid SNP discovery and genetic mapping using sequenced RAD markers. PLoS One 3:e3376
21. Miller MR, Dunham JP, Amores A et al (2007) Rapid and cost-effective polymorphism identification and genotyping using restriction site associated DNA (RAD) markers. Genome Res 17:240–248
22. Hohenlohe PA, Bassham S, Etter PD et al (2010) Population genomics of parallel adaptation in threespine stickleback using sequenced RAD tags. PLoS Genet 6:e1000862
23. Thomson RC, Wang IJ, Johnson JR (2010) Genome-enabled development of DNA markers for ecology, evolution and conservation. Mol Ecol 19:2184–2195
24. Friesen VL (2000) Introns. In: Baker AJ (ed) Molecular methods in ecology. Blackwell Science Ltd., Oxford, pp 274–294
25. Puritz JB, Addison JA, Toonen RJ (2012) Next-generation phylogeography: a targeted approach for multilocus sequencing of non-model organisms. PLoS One 7(3):e34241
26. Drummond AJ, Ashton B, Buxton S, Cheung M, Cooper A, Duran C, Field M, Heled J, Kearse M, Markowitz S, Moir R, Stones-Havas S, Sturrock S, Thierer T (2011) Geneious 5.4.
27. Meyer M, Stenzel U, Hofreiter M (2008) Parallel tagged sequencing on the 454 platform. Nat Protoc 3:267–278
28. Binladen J, Gilbert MTP, Bollback JP et al (2007) The use of coded PCR primers enables high-throughput sequencing of multiple homolog amplification products by 454 parallel sequencing. PLoS One 2:e197
29. Lahr DJG, Katz LA (2009) Reducing the impact of PCR-mediated recombination in molecular evolution and environmental studies using a new-generation high-fidelity DNA polymerase. Biotechniques 47:857–866
30. Gilles A, Meglecz E, Pech N et al (2011) Accuracy and quality assessment of 454 GS-FLX Titanium pyrosequencing. BMC Genom 12:245

Chapter 7

Optimizing Selection of Microsatellite Loci from 454 Pyrosequencing via Post-sequencing Bioinformatic Analyses

Iria Fernandez-Silva and Robert J. Toonen

Abstract

The comparatively low cost of massive parallel sequencing technology, also known as next-generation sequencing (NGS), has transformed the isolation of microsatellite loci. The most common NGS approach consists of obtaining large amounts of sequence data from genomic DNA or enriched microsatellite libraries, which is then mined for the discovery of microsatellite repeats using bioinformatics analyses. Here, we describe a bioinformatics approach to isolate microsatellite loci, starting from the raw sequence data through a subset of microsatellite primer pairs. The primary difference to previously published approaches includes analyses to select the most accurate sequence data and to eliminate repetitive elements prior to the design of primers. These analyses aim to minimize the testing of primer pairs by identifying the most promising microsatellite loci.

Key words Molecular markers, Next-generation sequencing, Microsatellite marker development

1 Introduction

Microsatellite loci remain one of the most popular genetic markers for a variety of applications from pedigrees to molecular ecology (1). Until recently, the most common procedure to isolate microsatellites was using labeled probes to identify microsatellite-containing sequences from either bulk genomic DNA or libraries of genomic DNA enriched for microsatellite motives (2). Despite extensive protocol optimization (e.g., 3–7), the development of microsatellites using this approach remains labor intensive and costly. Therefore, the development of microsatellite markers in new species has been a bottleneck to the application of microsatellite markers to non-model systems, especially for taxa whose genomes have a low frequency of microsatellite occurrence such as lepidopterans, birds, and bats (8).

A more recent strategy for the isolation of microsatellite loci involves mining large sequence databases for the discovery of

Stella K. Kantartzi (ed.), *Microsatellites: Methods and Protocols*, Methods in Molecular Biology, vol. 1006, DOI 10.1007/978-1-62703-389-3_7, © Springer Science+Business Media, LLC 2013

microsatellite repeats using bioinformatics tools. Although this approach is fast and efficient compared to traditional methods, until recently its application was limited to species for which genomic resources were available. With the decreased cost of generating genomic data by massively parallel "next-generation" sequencing (NGS), this strategy is now broadly available for non-model organisms. As a consequence, the discovery of hundreds of microsatellite loci has become a reality for virtually any species (e.g., 9–14).

Among NGS technologies, the 454 pyrosequencing is the most commonly used to date in microsatellite projects. The longer read lengths of the 454 platform facilitate the design of PCR primers targeting amplicons corresponding to the size range typically used for microsatellite genotyping (~90–400 bp). Among published studies using the 454 platform, approaches that sequence both enriched and non-enriched libraries have been used. Enriched libraries are constructed using capture probes to retain only genomic DNA fragments containing microsatellite repeats, as described in a number of published microsatellite isolation methods (e.g., 5), which are then used as the template for 454 sequencing. In contrast, non-enriched, often referred to as *shotgun*, libraries consist of purified genomic DNA, which is fragmented and used directly as template for sequencing. Details on the preparation of template can be found in (12, *enriched libraries*), (10, *shotgun* method] and (2).

Typically, microsatellite discovery from a previously unstudied species requires only a modest sequencing effort, unless microsatellites are very rare in the genome. Massively parallel sequencing allows templates from different species to be pooled onto a single Picotiter plate region, thereby further reducing the cost of sequencing. This pooling is facilitated by the ligation of multiplex identifier (MID) adaptors, which function as barcodes to sort sequences from different libraries bioinformatically during the post-sequencing processing. Detailed protocols for the ligation of the MID adaptors and template preparation for sequencing are available in the manufacturer's technical bulletins (15).

Here, we present our preferred method to design microsatellite markers from next-generation sequencing data. We assume the reader of this protocol is a novice in bioinformatics and describe the steps to move from the raw 454 sequence data to selecting the most promising microsatellite-flanking primer pairs for laboratory testing (see Fig. 1). This method is simply a series of bioinformatics analysis. Some are basic necessary steps (e.g., sorting out the sequence by MID adaptors, detection of microsatellite-containing sequences, and the design of flanking primers), whereas others are aimed at increasing the likelihood of designing successful primer pairs. Among the laboratory issues with microsatellite loci, failed or inconsistent PCR amplification, which can result in null-alleles, is perhaps the most common. Errors in the source sequences from

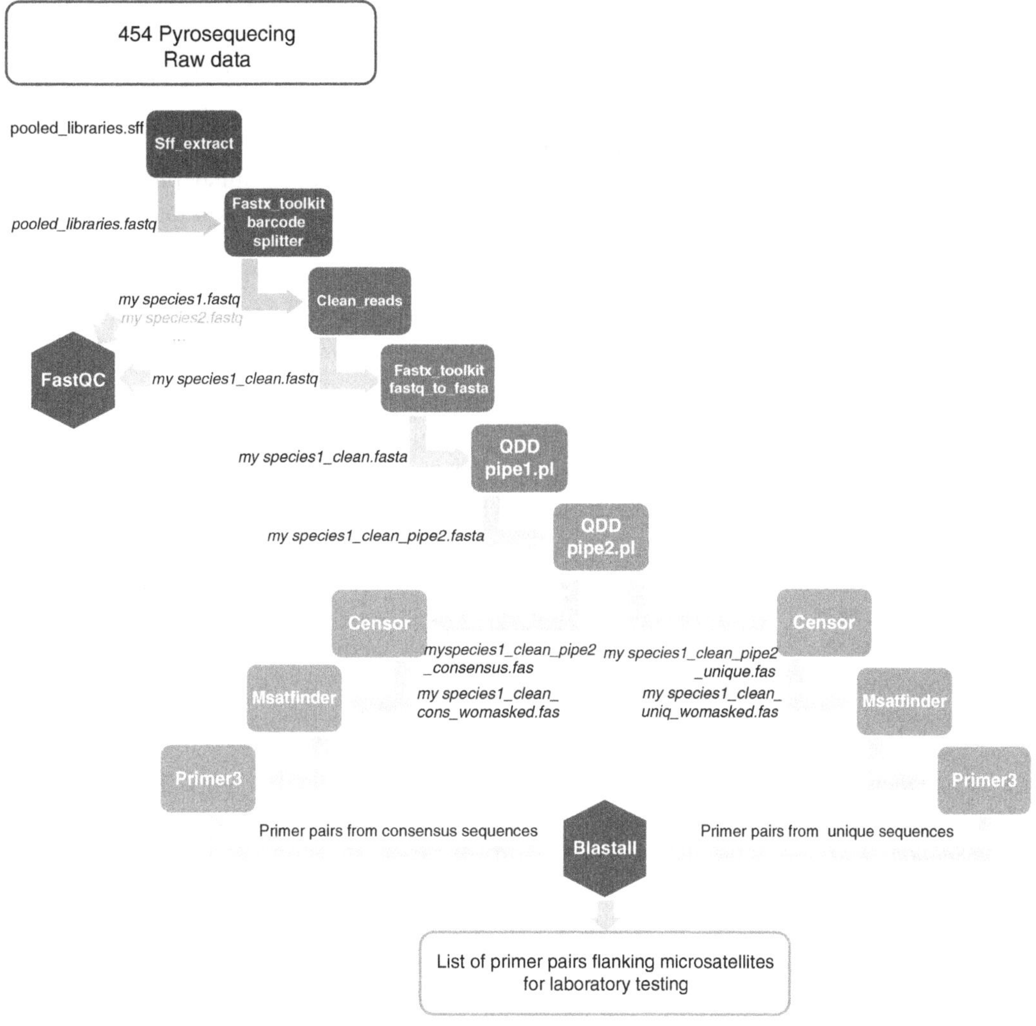

Fig. 1 Flowchart. This method is a pipeline of bioinformatics analysis, from the raw sequencing results to a list of primers for laboratory testing. Each individual analyses is indicated in a *box* or *hexagon*. The names of the input and output files used in the examples throughout the text are indicated (in *black*)

which primers are designed will exacerbate this issue. (For a more detailed discussion see (1)). Another very common issue is the amplification of multiple targets (multiband patterns in electrophoresis), which can result when microsatellite loci fall within repetitive elements (e.g., minisatellites, transposable elements). These issues are why we recommend steps 3 and 7 of our protocol, which are aimed at enhancing sequence accuracy, and steps 5 and 6, which are aimed to eliminate repetitive elements. Although it is obviously possible to develop microsatellites without using these steps, we have found that they increase the success rate of primer pairs tested in the lab and save time and money as a result. Although we present these analyses in what seems like a logical sequence to us, they could also be performed in a different order. Likewise, as faster and better algorithms for the analyses of NGS data continue to being developed, we encourage researchers to explore alternative software for any of the steps of this pipeline. Finally, although we focus on 454 pyrosequencing data, a similar pipeline and the same rationale can be applied for the isolation of microsatellite markers from any large sequence dataset with minor adjustments of steps 1.1–1.3.

Below we provide a step-by-step overview of the method, describe the rationale of the analyses, and point out some decisions that the readers need to make.

1.1 File Conversion from SFF to FASTQ with SFF_EXTRACT (16)

454 sequence reads are usually stored as binary SFF files but need to be converted to the human-readable FASTQ format. See Box 1 for an explanation of what a FASTQ file is.

1.2 Library Splitting by Barcode Matching Using the BARCODE_SPLITTER Script from the FASTX_TOOLKIT (17)

The pooled libraries run in a single Picotiter plate region are all returned in a single file. This pooled sample needs to be split into smaller files, based on barcode matching, each of which contains only the sequences from a single library (e.g., a single species for which microsatellites are being developed).

1.3 Evaluation of the Quality of the Sequences in the Raw Data (Before Quality Control) with FASTQC (18)

In order to see the effect of your quality control (QC), it is necessary to evaluate the quality of the sequences in the datasets before and after QC. The software FASTQC allows you to create a quality report of the data and visualize a variety of QC metrics, providing a fast and intuitive way of evaluating the quality of the data and monitoring the QC process. The FASTQC analyses that are most informative for this pipeline include:

- The *Basis Statistics* module, which indicates the number of sequences in the dataset, the length range of the reads, and the overall % GC in all the bases of all the sequences.
- The *Per Base Sequence Quality* module, which shows an overview of the range of quality values across all bases at each position in the FASTQ file (see Box 1). The *y*-axis represents the Phred quality scores, with values above 30 being very good, values between 20 and 28 being acceptable, and values below 20 indicating bad sequence quality (18).
- The *Per Base Sequence Content* module, which plots out the proportion of each base position for which each of the four normal DNA bases has been called. In a random library, the proportion of bases calls along the sequences should be similar across all positions along the read length; significant changes in GC content likely indicate that the adaptor sequence or MID tags have not been completely removed.

Box 1 Understanding a FASTQ file

In most sequencing platforms, including Roche 454, each base call has an associated base call quality score, which estimates the probability that the base call is incorrect. The quality information is calculated using the Phred scale (*Q*),

$$Q = -10 \log 10\, p$$

where *p* is the probability of error.

For example, a Phred score of $Q = 20$ indicates $p < 0.05$, $Q = 25$ indicates $p < 0.003$, and $Q = 30$ indicates $p < 0.001$.

FASTQ is a text-based format for storing both nucleotide sequences and its corresponding quality (Phred) scores. This is an example of a minimal FASTQ file:

```
@SEQ_ID
GATTTGGGGTTCAAAGCAGTATCGATCAAATAGTAAATCCATTTGTTCAACTCACAGTTT
+SEQ_ID
!''*((((***+))%%%++)(%%%%).1***-+*''))**55CCF>>>>>>CCCCCCC65
```

A FASTQ file uses four lines per sequence read. Line 1 has the symbol @ followed by the sequence identifier and optional sequence descriptors, line 2 corresponds to the nucleotide sequence, line 3 has the symbol + and optionally the same sequence identifier or descriptors found in line 1, and line 4 has a string of the same number of characters as bases in the nucleotide sequence, each of which corresponds to a quality score encoded as an ASCII character by adding 33 to the Phred value (in 454, Sanger, and Illumina 1.8+) or 64 (in Illumina 1.3+, Illumina 1.5+, and Solexa). In the ASCII code, the Phred score of each base call is represented by a one-digit symbol. For instance, in a Roche 454 FASTQ file, a Phred score value of $Q = 30$ is represented by the symbol ?, which is the ASCII code for 63 (Phred + 33 = 30 + 33 = 63).

In 454/Sanger, the range of Phred scores is 0–40 (worst to best quality), which are represented by the ASCII characters indicated below:

```
ASCII        !"#$%&'()*+,-./0123456789:;<=>?@ABCDEFGHIJ
             |                   |    |    |         |
Phred+33     33                  53   58   63        73
Phred        0                   20   25   30        40
```

Different sequencing technologies have different error profiles. For instance, in Roche 454 sequences, most errors are associated with erroneous length calling of homopolymer stretches. Another peculiarity of 454 is that sequences tend to be most reliable near the beginning of each read.

1.4 Quality Control, Elimination of Short Reads, and Barcode Clipping with CLEAN_READS (16)

The software CLEAN_READS performs three important tasks. First, it trims the 5′-ends of the sequences of the primer, linkers, and adapters. Some prefer to be conservative and clip a few extra bp to account for possible sequencing errors such as insertions in the adaptor sequences. Second, low-quality regions of the sequences, as indicated by Phred scores below the cutoff, are removed by using a program called LUCY based on a sliding window algorithm (LUCY is implemented in CLEAN_READS). The sliding window uses three steps. First, low-quality regions at either end of the sequence are removed using the *lucy_bracket* option. The second step flags all regions of a specified length within the sequence whose average quality falls below the threshold quality. The third step trims each remaining sequence based on two quality criteria specified by the *–lucy_error* option: (1) the maximum average probability of error over the final clean range (*max_avg_error*) and (2) the allowable probability of error for the final two bases at each end of the sequence (*max_error_at_ends*). This final step returns the largest region of each sequence that meets all specified criteria as the clean read and discards reads that are too short (generally reads under 90 bp are not useful). These three options, each with two parameters, control how low-quality regions of the sequences are trimmed for the final dataset (see Note 1).

1.5 Evaluation of the Quality of the Sequences in the Clean Data (After Quality Control) with FASTQC

After performing the quality control, it is important to monitor the quality of the clean dataset before moving forward. If the result is not satisfactory, it is advisable to repeat CLEAN_READS with different sets of parameters and evaluate the results with FASTQC as many times as necessary.

1.6 Detection of Microsatellite-Containing Sequences Using the PIPE1.PL Script of QDD (19)

Although this step would not be strictly necessary at this point, because the microsatellite search will be performed again at the end of the pipeline, eliminating the sequences without microsatellites from the pipeline allows all of the subsequent analyses running much faster.

1.7 Similarity Analysis Using the PIPE2.PL Script of QDD

The PIPE2.PL script of QDD compares all the sequences in your dataset (ALL-AGAINST-ALL BLAST) in order to sort the reads in three main categories, for each of which a new data file is created. The first file includes all the singletons, i.e., reads for which no similar sequence is detected, and a *unique* file is written. The second file includes the reads for which at least one other sequence with very high sequence identity exists, typically above 95 % similarity. These redundant reads of the same sequence are then used to build contigs, and from each contig, a single consensus sequence is written to the *consensus* file. The third file includes the reads that are similar but not identical to other sequences in the dataset (i.e., sequences with positive BLAST hits and sequence similarity below the threshold, therefore excluded from the contigs), which are then written to the *grouped* data file. Because similar but non-identical sequences in the genome can cause problems with primer specificity for your microsatellite loci, we recommend against using *grouped* sequences to design microsatellite markers.

1.8 Avoiding Sequences with Homology to Repetitive Elements Using CENSOR (20)

Along the same vein as above, we try to avoid using sequences similar to known repetitive elements for the design of microsatellite markers. The software CENSOR is used to compare the sequences of our dataset against a reference collection of repetitive elements and to mask homologous portions (i.e., substitute the homologous portion of the sequence by a string of "X" of the same length). Sequences containing masking symbols are subsequently eliminated from the pipeline. We use the libraries of repetitive elements published by *RepBase Reports*, each of which contains known sequences representing repetitive DNA from different eukaryotic species. In some cases, it is meaningful to select a subset of libraries (e.g., repetitive elements from vertebrate genomes).

1.9 Design of Primers Flanking Microsatellite Repeats

The software MSATFINDER (21) is used to find and create a list of microsatellite repeats classified by repetitive motive, number of repetitive units, and other features. Using this list, you can then go back to the data files, find the microsatellite-containing sequences, and use the software PRIMER3 (22) to design primers flanking the microsatellite repeats. Although many programs (including MSATFINDER or QDD) can simultaneously find the microsatellite repeats and create a list of primers, in our experience, the design of primers merits careful attention and we recommend visually inspecting the primers or preferably designing them one by one. An added advantage of the latter is that it allows selecting primers on the consensus stretches of the contigs.

In our experience, using accurate sequences for primer design tends to reduce PCR amplification failure rate, as discussed above. However, there is a trade-off between the stringency of the quality control and the number of template sequences for microsatellite design (a very stringent QC will eliminate most sequences from the pipeline). To circumvent this issue, we suggest two different strategies to designing microsatellite primers from available large sequence datasets. The first strategy is to perform relatively low-stringency quality control (e.g., use default parameters, see Note 1) and use only the contigs (*consensus* sequence data files) to find microsatellite repeats and design primers. Another possible strategy includes performing a rigorous quality control with stringent parameters (e.g., lucy_bracket=10.0,0.003; lucy_window=10.0,0.003; lucy_error=0.01,0.01) and then using both the remaining *unique* and *consensus* sequence data files to design primers.

1.10 Similarity Analysis with BLASTALL to Avoid Using Duplicated Sequences for the Design of Microsatellites

Oftentimes, it is meaningful to design primers from different batches of sequences, or after testing a set of priers in the laboratory, we will decide to design additional primers. To avoid designing primers for the same locus twice, it is a good idea to maintain a database of sequences used as template for designing microsatellites and every time we create a new list, cross-compare them to avoid the use of duplicated sequences. The best way to do this is using a sequences similarity analysis, e.g., with BLASTALL.

2 Software

The following list of software is used in this pipeline and has to be installed before getting started. The installation packages are available from the listed websites, where there is also information on dependencies (other programs and libraries) that need to be installed as well.

- SFF_EXTRACT and CLEAN_READS from (http://bioinf.comav.upv.es/)
- FASTX_TOOLKIT (http://hannonlab.cshl.edu/fastx_toolkit/)
- FASTQC (http://www.bioinformatics.bbsrc.ac.uk/projects/fastqc/)
- QDD2.1_BETA (http://gsite.univ-provence.fr/gsite/Local/egee/dir/meglecz/QDD.html)
- CENSOR and REPBASE (http://www.girinst.org/repbase/index.html)
- MSATFINDER (http://www.genomics.ceh.ac.uk/msatfinder/)

In this protocol, we describe how to run these programs using a command line interface in the terminal of Linux systems. If you have a Windows PC or a Mac you have a few options. The first is to install *Linux* using virtual machine software such as the freely available *Virtual Box*. The second is to create a bootable partition in your PC or Mac and boot *Linux* from it. Another interesting option is to use a *cloud computing* environment such as *Amazon Cloud E2*. This latter option has the added advantage of providing as much computing memory as you need without having to give up use of your computer for the hours needed by the most computationally intensive steps of this pipeline. *CloudBioLinux* is a community project that provides an *Amazon Image* with preinstalled bioinformatics software and libraries. This is probably the best alternative for a newcomer to the field because the most commonly used bioinformatics tools are preinstalled and you need only install the few specific packages listed above to run this pipeline. Additionally, *CloudBioLinux* is designed for biologists and includes good documentation for new users (e.g., Getting Started with CloudBioLinux at http://cloudbiolinux.org/).

In order to follow this protocol, you do need a bit of familiarity with the *shell* and *Unix/Linux*, but the computing skills you need are very basic. Still, despite our attempt to give simple and clear guidelines, it is impossible to cover the range of issues that a novice to bioinformatics may encounter and encourage to explore troubleshooting options. Resources on how to use each program are usually included in the *README.txt* files provided with the installation packages, the software help (see Note 2), and the websites from the software developers. There are also a wide range of resources on the web to assist newcomers to the area of bioinformatics, including specialized bioinformatics forums such as http://seqanswers.com/.

A few additional examples include:

- Practical Computing for Biologists (23)*:* Excellent book, which also has a website (http://practicalcomputing.org/), where you can learn how to install *VirtualBox*, how to install software in *Linux*, how to interact with your computer using the *shell*, and many other useful computing skills.

- http://software-carpentry.org/4_0/shell/: Online course covering the basics of using the *Unix* command *shell*.

These are links to websites of the various resources discussed above, where you can also find support documentation and discussion groups:

- Cloudbiolinux (http://cloudbiolinux.org/)
- Virtualbox (https://www.virtualbox.org/)
- Amazon EC2 (http://aws.amazon.com/ec2/)

3 Methods

In this section, we provide examples of command line instructions for each of the software packages to perform the steps we described in Subheading 1 above. Each command line instruction is indicated by a "$" symbol which is not part of the command and must not be typed into the command line in order for it to work. Throughout the text body (not in the command line examples), the names of the programs and scripts are written in SMALL CAPS and the names of the files are written in *italics*.

3.1 File Conversion from SFF to FASTQ with SFF_EXTRACT

Transfer the SFF file to your working directory. Tell SFF_EXTRACT to convert your SFF file into a FASTQ file (using the option –Q) and indicate the name of the new file (option –o). Also indicate that all bases should be capitalized (with the option –u). Inspect the new file using the programs MORE, HEAD, and/or TAIL (see Note 3).

```
$ mkdir working
$ sff_extract –u -Q pooledlibraries.sff -o pooledlibraries.fastq
$ head pooledlibraries.fastq
$ tail pooledlibraries.fastq
```

3.2 Library Splitting by Barcode Matching Using BARCODE_SPLITTER from the FASTX_TOOLKIT

With the help of a text editor (e.g., NANO), create a *barcode* file with two tab separated columns indicating the library identifiers in the first column and barcode sequences in the second column.

Example of *barcode* file:

```
myspecies1	GACTACGAGTAGACT
myspecies2	GACTACGCGTCTAGT
myspecies3	GACTACGTACACACT
myspecies4	GACTACGTACTGTGT
```

Use FASTX_BARCODE_SPLITTER to create individual FASTQ files for each library. Indicate the barcodes file name with the option –bcfile. The --prefix option indicates where the output files should be placed and allows adding a prefix to the file names. The option –mismatches allows indicating a maximum number of mismatches allowed in the barcode to account for sequencing errors; if not specified, the mismatch is 1 by default. You need to add the –bol option to indicate the program to look for the barcodes at the 3′-ends.

$ cat pooled.fastq | fastx_barcode_splitter.pl --bol --bcfile barcode --prefix / working/ --suffix ".fastq"

In this example, six library-specific files will be created called *myspecies1.fastq*, *myspecies2.fastq*, and so on.

3.3 Evaluation of the Quality of the Sequences in the Raw Data (Before Quality Control) with FastQC

Create a new directory and move the FASTQ files into it. Invoke the program FastQC in the command line.

$ mkdir fastqc

$ cp myspecies1.fastq / working/fastqc

$ fastqc

FastQC will open as a graphical user interface. Select the files to analyze (File > Open). Newly opened files will immediately appear in the set of tabs at the top of the screen (it can take a few minutes depending on the size of the file). On the left side of the interactive report, a series of tabs indicate the analyses that were performed and allow you to access the reports by clicking the tabs. Notice the number of sequences in your library and the length range of the reads (shown in the Basis Statistics report). Have a look at the *Per Base Sequence Quality* plots and notice how the sequence quality decays towards the 3′-ends, possibly being unacceptable beyond a certain position (although this depends on how the sequencing run was performed). Examine the *Per Base Sequence Content* and look for strong biases in sequence composition in the initial positions of the sequences due to the presence of barcodes, linkers, and primers at the 5′-ends. It is worth taking the time to look at the results of each module to see if anything stands out. A thorough explanation of each analysis and its meaning is available in the *help* file of FASTQC (see Note 4).

3.4 Quality Control, Elimination of Short Reads, and Barcode Clipping with CLEAN_READS

Use CLEAN_READS to create a new file with only the clean ranges of all sequences. You need to indicate the name of the input file in FASTQ format (-i), a name for the newly created clean data file in FASTQ format (–o), and the sequencing platform used to generate the sequences (-p 454). Specify the number of nucleotides to be clipped from the beginning and end of each sequence as two integers separated by a comma (–e option). Then use default parameters for trimming low-quality regions or specify the –lucy_bracket, --lucy_window, and –lucy_error parameters to use for custom QC thresholds. Define the minimum read length of clean sequence to retain, e.g., 90 bp (-m option).

```
$ clean_reads -i myspecies1.fastq -o myspecies1_clean.fastq -p 454
    -f fastq -e 24,0 --lucy_error=0.003,0.003 --lucy_window=10.0,0.003 --lucy_bracket=10.0,0.003 -m 90.
```

3.5 Evaluation of the Quality of the Sequences in the Clean Data (After Quality Control) with FastQC

Repeat the analysis explained in Subheading 3.3 and compare the quality metrics of the clean set to those of the raw set. You should notice a reduction in both the number of reads and length of the sequences. Have a look at the *Per Base Sequence Quality* plots and observe the overall improvement of the quality along the sequence length. Verify that the low-quality tails at the 3′-ends have been removed.

```
$ cp myspecies1_clean.fastq ./fastqc
$ fastqc
```

If the results of the quality trimming step are not convincing, try different combinations of parameters in Subheading 3.4 and reevaluate the results with FastQC (see Note 5).

3.6 Detection of Microsatellite-Containing Sequences Using the pipe1.pl Script of QDD

Before running QDD, you need to perform a number of preparatory steps. First, change the format of the output file from Subheading 3.5 (the FASTQ file containing the clean ranges of the reads) from FASTQ to FASTA using the fastq_to_fasta script of the Fastx_toolkit (see Note 6). Second, create the scheme of directories required by QDD: a "datain" directory within the "QDD2.1_beta" directory (or the directory where pipe2.pl is found) and a "myproject1" directory within "datain." Finally, move the FASTA file to the "myproject1" directory, and verify that the FASTA file looks right by using the programs head, more, or tail.

```
$ fastq_to_fasta -Q 33 -i myspecies1_clean.fastq -o myspecies1_clean.fasta
$ mkdir /working/QDD2.1_beta/datain
$ mkdir /working/QDD2.1_beta/datain/myproject1
$ mv myspecies1_clean.fasta /working/QDD2.1_beta/datain/my project1
```

Now that you have your input file in FASTA format in your project directory, you can run the pipe1.pl script. Be sure that you are in the same directory where the pipe1.pl script is located, which is generally in the QDD program directory:

```
$ cd /working/QDD2.1_beta
$ perl pipe1.pl
```

```
ubuntu@ip-10-205-13-16:/working2/QDD2.1_beta$ perl pipe1.pl
**********************************************************"
QDD version2.1(beta) 28 June 2011
Emese Meglecz, Aix-Marseille University, Marseille, France
emese.meglecz@univ-provence.fr
Plesae, read the Documentation_QDD2.pdf
**********************************************************"

1 : Operating system (win/linux): linux
2 : Project folder: /working2/QDD2.1_beta/datain/trash
3 : Sort sequences by tag (YES=1/NO=0): 0
4 : Remove adapters from sequences (YES=1/NO=0): 0
5 : Minimum sequence length: 90
6 : Pathway to BLAST executables: /working/software/ncbi-blast-2.2.25+/bin/
7 : Delete intermediate files (YES=1/NO=0): 1

Press enter if all of the settings are correct, or the number of the parameter if you whish to change the settings!
2
Project folder:
/working/QDD2.1_beta/datain/myproject

1 : Operating system (win/linux): linux
2 : Project folder: /working/QDD2.1_beta/datain/myproject
3 : Sort sequences by tag (YES=1/NO=0): 0
4 : Remove adapters from sequences (YES=1/NO=0): 0
5 : Minimum sequence length: 90
6 : Pathway to BLAST executables: /working/software/ncbi-blast-2.2.25+/bin/
7 : Delete intermediate files (YES=1/NO=0): 1

Press enter if all of the settings are correct, or the number of the parameter if you whish to change the settings!
```

Fig. 2 Screenshot of QDD's PIPE1.PL options dialog

Upon launching QDD, a menu will open up with a list of options. Indicate the path to the project directory (e.g., /working/QDD2.1_beta/datain/myproject1) and make sure that the paths to BLAST+ and CLUSTALW are correct. Do not remove adaptors or sort the sequences by tag, as you have already done this earlier in the pipeline (see QDD screenshot in Fig. 2). If the options look correct, accept and run QDD.

This will create a new directory in the project directory named "pipe1_xxx." This should contain a log file with a summary of the results (that can be read with MORE) and a new FASTA file with only the microsatellite-containing sequences, whose name has the ending *_pipe2.fas*.

3.7 Similarity Analysis Using the PIPE2.PL Script of QDD

Now that you have an input file in FASTA format with the suffix *_pipe2.fas* in a directory called "pipe2_xxx" located in your project directory, you can run the PIPE2.PL script (see Note 7).

$ perl pipe2.pl

Upon launching the PIPE2.PL script, a menu will open with a list of options. Verify that the paths that point to the project directory (e.g., /working/QDD2.1_beta/datain) and to the BLAST+ and CLUSTALW executables are correct. Use QDD to build consensus sequences, but there is no reason to keep intermediate files. Specify the minimum percentage of identity between sequences of a contig (95 % is recommended) and the proportion of sequences that must have the same base on the aligned site to accept it as a consensus (default is 66 %) (see Fig. 3).

```
**********************************************************
QDD version2.1(beta) 28 June 2011
Emese Meglecz, Aix-Marseille University, Marseille, France
emese.meglecz@univ-provence.fr
Plesae, read the Documentation_QDD2.pdf
**********************************************************

1 : Operating system (win/linux): linux
2 : Project folder: /working2/QDD2.1_beta/datain/plxi/
3 : Make consensus sequences (YES=1/NO=0): 1
4 : Minimum % of identity between sequences of a contig (80-100): 95
5 : Proportion of sequences that must have the same base at a site to accept it as a consensus (0.5-1): 0.66
6 : Pathway to CLUSTALW2 executables: /working/software/clustalw-2.0.10-linux-i386-libcppstatic/
7 : Pathway to BLAST executables: /working/software/ncbi-blast-2.2.25+/bin/
8 : Delete intermediate files (YES=1/NO=0): 1

Press enter if all of the settings are correct, or the number of the parameter if you whish to change the settings!
2
Project folder:
/working/QDD2.1_beta/datain/myproject1/

1 : Operating system (win/linux): linux
2 : Project folder: /working/QDD2.1_beta/datain/myproject1/
3 : Make consensus sequences (YES=1/NO=0): 1
4 : Minimum % of identity between sequences of a contig (80-100): 95
5 : Proportion of sequences that must have the same base at a site to accept it as a consensus (0.5-1): 0.66
6 : Pathway to CLUSTALW2 executables: /working/software/clustalw-2.0.10-linux-i386-libcppstatic/
7 : Pathway to BLAST executables: /working/software/ncbi-blast-2.2.25+/bin/
8 : Delete intermediate files (YES=1/NO=0): 1

Press enter if all of the settings are correct, or the number of the parameter if you whish to change the settings!
█
```

Fig. 3 Screenshot of QDD's PIPE2.PL options dialog

The analysis will create a directory with the results ("pipe3_xxx") in the project directory. Use MORE to read the log file, which gives a report of the number of sequence reads assigned to each output file. It will also give a warning if errors were encountered during the analysis.

3.8 Avoiding Sequences with Homology to Repetitive Elements Using CENSOR

Download the most recent release of *Repbase* (a set of files with the extension .ref containing sequences of known repetitive elements) to a new directory (e.g., "download_libraries"). These files can be downloaded from http://www.girinst.org/repbase/update/index.html, where a description of each is also available.

Before running the software CENSOR, select which *Repbase* libraries will be used to perform the comparison by placing the libraries in the "biolib" directory. Because *Repbase* is continuously updated and in order to avoid using older versions of the same libraries, it is a good idea to first eliminate any libraries present in the "biolib" directory and then add the new ones. You can narrow the search to only libraries with repetitive elements from certain taxa (e.g., vertebrate genomes), if appropriate libraries are available. Do not include the *simple.ref* library, which contains macrosatellite repeats, as this would mask your target microsatellites.

$ sudo rm /usr/local/share/censor-4.2.27/biolib/*.ref

[This is the location where the library files were saved during the installation]

$ cd /working/download_libraries

```
$ sudo mv fugapp.ref, fugrep.ref, humrep.ref, humsub.ref, mamrep.ref, mamsub.ref, mousub.ref, prirep.ref, prisub.ref, pseudo.ref, ratsub.ref, rodrep.ref, rodsub.ref, synrep.ref, tmpxen.ref, vrtrep.ref, zebapp.ref, zebrep.ref/ usr/local/share/censor-4.2.27/biolib/
```

Next, move the query files (the output files from PIPE2.PL) to the folder where the CENSOR script is and run CENSOR. Do this separately for each the *unique* and *consensus* files.

```
$ mv/working/QDD2.1_beta/datain/pipe2_xxx/myspecies1_clean_pipe2_unique.fas/working/censor-4.2.27
$ censor.ncbi myspecies1_clean_pipe2_unique.fas
$ mv / working/QDD2.1_beta/datain/pipe2_xxx/myspecies1_clean_pipe2_consensus.fas / working/censor-4.2.27
$ censor.ncbi myspecies1_clean_pipe2_consensus.fas
```

Using SED, remove any sequences with homology to repetitive elements (by removing any sequence-containing masking elements or strings of XXX). Do this for both the *consensus* and the *unique* sets. (You can copy the string of characters below and replace the file names to match yours).

```
$ sed -e ‘H;$!d;x;s/\nA/|A/g’ myspecies1_clean_pipe2_consensus.fas.masked | sed -e ‘H;$!d;x;s/\nC/|C/g’ | sed -e ‘H;$!d;x;s/\nT/|T/g’ | sed -e ‘H;$!d;x;s/\nG/|G/g’ | sed -e ‘H;$!d;x;s/\nX/|X/g’ | grep -v “X” | sed -e ‘H;$!d;x;s/|A/\nA/g’ | sed -e ‘H;$!d;x;s/|C/\nC/g’ | sed -e ‘H;$!d;x;s/|T/\nT/g’ | sed -e ‘H;$!d;x;s/|G/\nG/g’ | sed -e ‘H;$!d;x;s/|X/\nX/g’ | grep “[>ACTGX]” > myspecies1_clean_cons_womasked.fas
$ sed -e ‘H;$!d;x;s/\nA/|A/g’ myspecies1_clean_pipe2_unique.fas.masked | sed -e ‘H;$!d;x;s/\nC/|C/g’ | sed -e ‘H;$!d;x;s/\nT/|T/g’ | sed -e ‘H;$!d;x;s/\nG/|G/g’ | sed -e ‘H;$!d;x;s/\nX/|X/g’ | grep -v “X” | sed -e ‘H;$!d;x;s/|A/\nA/g’ | sed -e ‘H;$!d;x;s/|C/\nC/g’ | sed -e ‘H;$!d;x;s/|T/\nT/g’ | sed -e ‘H;$!d;x;s/|G/\nG/g’ | sed -e ‘H;$!d;x;s/|X/\nX/g’ | grep “[>ACTGX]” > myspecies1_clean_uniq_womasked.fas
```

3.9 Design of Primers Flanking Microsatellite Repeats

Move the one of the output files from previous step (FASTA files with the masked sequences removed) to the MSATFINDER directory. Run MSATFINDER as outlined below:

```
$ mv myspecies1_clean_cons_womasked.fas / working/msatfinder-2.0.9
$ cd / working/msatfinder-2.0.9
$ ./msatfinder myspecies1_clean_cons_womasked.fas
```

This will create a series of directories containing results files in the MSATFINDER directory and also a summary results file called

```
>cons_gr151_2
ATTCCATGTGCATCACTAGATTCCACTCAgctgctgctgctgctgctGCAGTTGGCGCAGAGCAAGTCGCAATTATGCGGCAGGGNTGAAGTTTTGTTGGCGAGTCTGGC
TTACAGCCTTAATTTGAGGCTCCATTAAGCCGTGGGCGTGGCCACGTCACGGCCACCCACCCTCCCTTTCCCCCCCTCAGAGAAAACGCGGACGCCCCCCGCGCCGCCGGCGC
GCTGCGCCACGATGGACTCCGCGTCAGCCTCGCTCCGCTCTGCGCGGCTCTGCTG
>GSDTG6M04I0NHF
ATTCCATGTGCATCACTAGATTCCACTCAgctgctgctgctgctgctGCAGTTGGCGCAGAGCAAGTCGCAATTATGCGGCAGGGGTGAAGTTTTGTTGGCGAGTCTGGC
TTACAGCCTTAATTTGAGGCTCCATTAAGCCGTGGGCGTGGCCACGTCACGGCCACCCACCCTCCCTTTCCCCCCCTCAGAGAAAACGCGGACGCCCCCCGCGCCGCCGGCGC
GCTGCGCCACGATGGACTCCGCGTCAGCCTCGCTCCGCTCTGCGCGGCTCTGCTG
>GSDTG6M04ID70N
ATTCCATGTGCATCACTAGATTCCACTCAgctgctgctgctgctgctGCAGTTGGCGCAGAGCAAGTCGCAATTATGCGGCAGGG-TGAAGTTTTGTTGGCGAGTCTGGC
TTACAGCCTTAATTTGAGGCTCCATTAAGCCGTGGGCGTGGCCACGTCACGGCCACCCAC-----------------------------------------------------
-------------------------------------------------------
```

Fig. 4 Example of contig in the *_cons_grouped.fas* file. The *first* and *second lines* correspond to the consensus sequence; the *third* to *sixth lines* correspond to the two sequences grouped in the contig. The number of sequences in the contig is indicated in the last digit of contig name

results.html. You can inspect the *results.html* file using a web browser (e.g., MOZILLA FIREFOX) for a quick overview. Next, go to the "Repeats" directory; here you will find a series of files that report the number of microsatellites found, classified by repeat motif, repetition length, and other features. It is worth looking at these files to understand the microsatellite composition of your libraries (you can find a description of the column headers looking in the MSATFINDER manual, available at (http://www.genomics.ceh.ac.uk/msatfinder/msatfinder_manual.html#files). The most important result file for our purpose here is the *msatfinder.repeats* file. Open the *msatfinder.repeats* file using a spreadsheet application such as GNUMERIC.

```
$ cd Repeats
$ gnumeric msatfinder.results
```

This command will open a GNUMERIC spreadsheet. Go to the Data > Sort menu to classify the sequences by repeat motif, repeat length, number of repeats, etc. Pick a list of microsatellite repeats for which you want to try to find primers. Next, use the sequence identifiers in column 1 to go back to the data files, find the individual sequences, and use these to design primer pairs with PRIMER3. This process is slightly different for the *unique* and *consensus* files, as outlined below.

You can start with the contigs (*consensus* file), from which primers are designed on the consensus stretches of the assembled sequences. In this case, you need to find the alignment that corresponds to each sequence identifier. The alignments are written in the result file with the suffix *cons_subs.fas* in the "pipe2_xxx" directory from QDD. See the example of an alignment in Fig. 4; it consists of a consensus sequence followed by each of the individual sequences used to build the contig. The last digit of the sequence identifier indicates the number of sequences that were used to build the alignment; in this example, "cons_3_2" indicates that this contig is the consensus of two sequences. The total number of text lines in the alignment is twice the number of sequences that were used to build the alignment plus two, in our example six.

To find the alignment in the *cons_subs.fas* file you can use the program GREP. GREP finds a query and outputs the line that contains the query followed by a specified number of lines. In the example, the query is "cons_3_2" and the total number of lines to output is six, that is, five additional lines after the line matching the query.

```
$ cd/working/QDD2.1_beta/datain/ myproject1/pipe2_xxx
$ grep -w "cons_gr2_2" -A 5 myspecies1_clean_pipe2_cons_
  sub.fas
```

Copy the consensus stretch of the contig (*consensus* file), open PRIMER3 online (http://frodo.wi.mit.edu/), and paste the selected *consensus* in the source sequence box. Use the symbols [ ] at both sides of the microsatellite repeat to indicate the target region for which primers will be designed. Use the symbols < > to mark any Ns or repetitive regions of sequence you want to avoid using as priming sites. Finally, indicate what the product size should be (typically 90–400 bp). You can either set the parameters for your own preferences in the design of primers or use the program default values (Fig. 5).

PRIMER3 will output a number of possible primer pairs, and you need to select one of them. This is the first of your list of primers pairs to test at the bench. Repeat this process for each sequence you decide to include until you design as many primer pairs as you wish to test.

With the singletons (*unique* file), the process is very similar. The primary difference is that you need to find the query sequences for PRIMER3 in the data file that has the suffix *unique.fas* in the "pipe2_xxx" directory from QDD. As the query is a singleton, you need to only output one line after the query.

```
$ cd/working/QDD2.1_beta/datain/ myproject1/pipe2_xxx
$ grep -w "cons_gr2_2" -A 1 myspecies1_clean_pipe2_unique.fas
```

You can also allow QDD to select primers for you, but by pulling individual sequences, you can examine each one and control the design of your primers to a greater degree than is possible for an automated pipeline. Our experience is that this extra effort at the primer design stage ultimately saves time and money in the primer testing and optimization stage.

3.10 Similarity Analysis with BLASTALL to Avoid Using Duplicated Sequences for the Design of Microsatellites

Once you have a list of sequences from which you have designed microsatellite primers, you want to verify that there are no duplicate sequences among your selection. Additionally, this step is particularly important if you want to add additional primer pairs to your list, for example, if you did not obtain enough useful primers in your first attempt. To accomplish this task, you take your database of existing sequences and compare your new list of sequences that you wish to add.

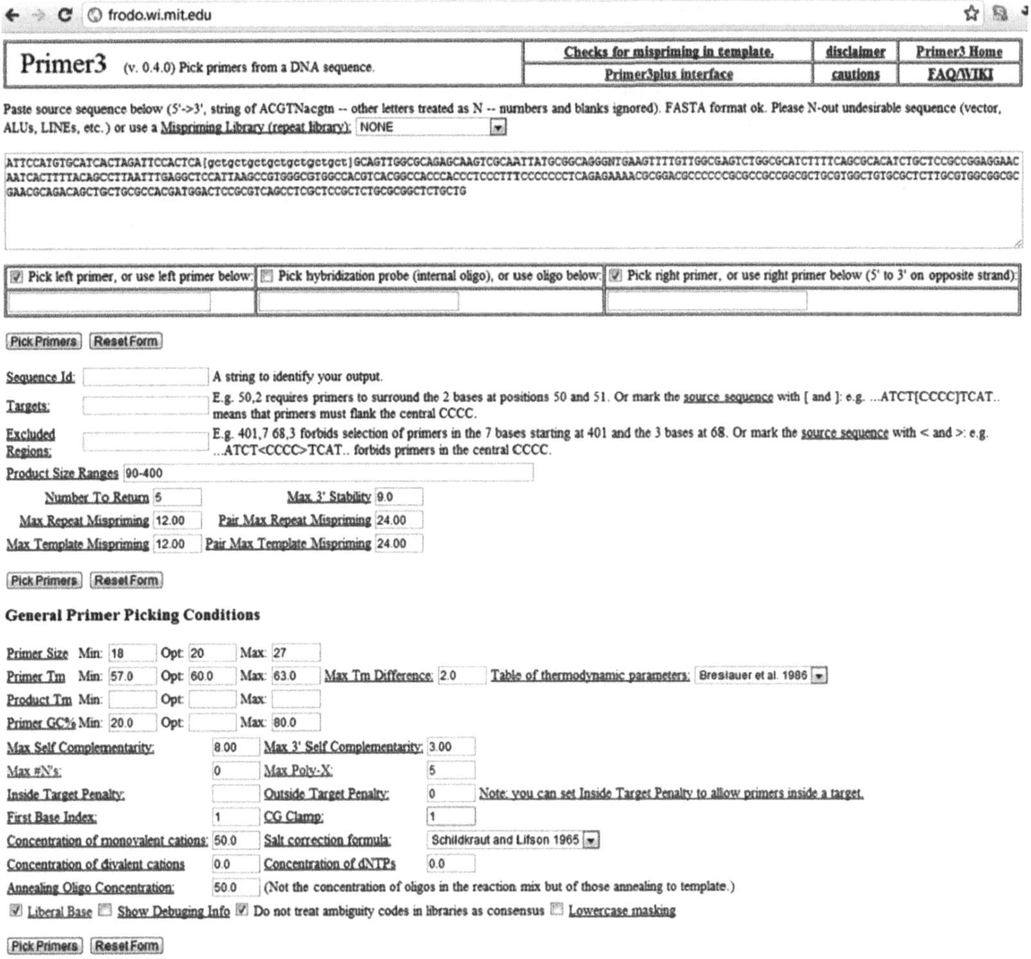

Fig. 5 Screenshot of Primer3. The target sequence, corresponding to the microsatellite, is indicated with *brackets* [ ]. The recommended settings for the design of primers are also shown

First, create a local database with the sequences that you used to design microsatellite primer pairs. This is simply a FASTA file with a list of the source sequences that you are using in Primer3 to find primers. Add the extension *.nt* to the file name (e.g., *usedseq_database.nt*). Format the database using formatdb:

```
$ formatdb -i usedseq_database.nt -p F
```

This will create index files that the standalone BLAST needs to perform the searches and produce results.

Second, create the query list. This is a FASTA file with the extension *.nt* with the list of new sequences that you intend to use for developing new primers (e.g., *newseq_query.nt*).

Finally, use BLASTALL to compare your query (e.g., *newseq_query.nt*) against your database (e.g., *usedseq_database.nt*).

```
$ blastall -p blastn -d usedseq_database.nt -i usedseq_database.nt
    -o results.out -m 8
```

By using the option –m 8, you indicated that you want output formatted as a table in the results file (e.g., *results.out*). The headers for the columns in this file are:(1) Query. The query sequence identifier(2) Subject. The matching subject sequence identifier(3) % identity (4) Alignment length(5) Number of mismatches(6) Gap openings (7) Position (in bp) in the query sequence where the alignment starts (8) Position (in bp) in the query sequence where the alignment ends (9) Position (in bp) in the matching subject sequence where the alignment starts (10) Position (in bp) in the matching subject sequence where the alignment ends (11) *E*-value (12) Bit score.

Inspect the *results.out* file with MORE or GNUMERIC to see if any sequences show similarity to others. Check the length of the alignment to distinguish truly duplicated sequences; these will be pairs of sequences that are similar along most of their length as opposed to sequences that share only a section of similar bases, such as the microsatellite motive itself. Do not use duplicated sequences to design new primers.

4 Notes

1. The default parameters for LUCY are --lucy_bracket=10.0,0.02 --lucy_window=50.0,0.08 10.0,0.3 --lucy_error=0.025,0.02.

   Note that several windows with different parameters can be specified for the *–lucy_window* option.

2. You can usually access the manual or the *help* for any particular program by typing its name followed by --help or -h or invoking the script MAN followed by the program name, as in the following examples:

   ```
   $ clean_reads --help
   $ head –h
   $ man msatfinder
   ```

3. After each step, we encourage to inspect the newly created output files by using the scripts MORE, HEAD, or TAIL followed by the file name:

   ```
   $ head pooled.fastq
   ```

4. FASTQC guesses the encoding method used in the FASTQ file and indicates it in the title of the graph; for 454, it should be Illumina>v1.3 (Phred+33).

5. Although FASTQC shows a quick evaluation of whether the results look normal (*green*), slightly abnormal (*orange*), or very unusual (*red*), the interpretation of these evaluations must be taken in the context of what you expect from your library.

6. In all of the scripts of the FASTX_TOOLKIT where the input file is in FASTQ (e.g., FASTQ_TO_FASTA), it is very important to include the –Q 33 option indicating that the files are coded using Phred+33. This important piece of information is not documented in the FASTX_TOOLKIT website or manual.
7. The script PIPE2.PL of QDD will look for a file with the ending *_pipe2.fas* in the directory "pipe1_xxx" with the highest number.

Acknowledgments

We thank all the members of the ToBo and Karl labs and the Hawai'i Institute of Marine Biology EPSCoR core genetics facility and staff for feedback, discussion, and assistance with this protocol. This project was funded by a Fullbright Fellowship award to I.F.S. and National Science Foundation grants (Bio OCE-0623699, OCE-0929031) to R.J.T. and B.W.B. This is contribution #1521 from the Hawai'i Institute of Marine Biology and 8755 from the School of Ocean and Earth Sciences and Technology (SOEST).

References

1. Selkoe KA, Toonen RJ (2006) Microsatellites for ecologists: a practical guide to using and evaluating microsatellite markers. Ecol Lett 9:615–629
2. Andrés JA, Bogdanowicz SM (2011) Isolating microsatellite loci: looking back, looking ahead. In: Methods in molecular biology, vol 772. Part 3, pp 211–232, doi:10.1007/978-1-61779-228-1_12.
3. Estoup A, Turgeon J (1996) Microsatellite markers: Isolation with non-radioactive probes and amplification. Version of 12/1996 Laboratoire de Génétique des Poissons, INRA 78352 Jouy-en-Josas France.
4. Glenn TC (1996) The microsatellite manual version 6, July 27, 1996 Laboratory of Molecular Systematics—MRC 534. MSC Smithsonian Institution, Washington, DC 20560
5. Toonen RJ (1997) Microsatellites for ecologists: non-radioactive isolation and amplification protocols for microsatellite markers, Unpublished manuscript, available from the author or via anonymous FTP from http://biogeek.ucdavis.edu/Msats/ or http://www2.hawaii.edu/~toonen/files/MsatsV1.pdf
6. Glenn TC, Schable NA (2005) Isolating microsatellite DNA loci. In: Zimmer EA, Roalson E (eds) Molecular evolution: producing the biochemical data, part B. Academic Press, San Diego, USA, pp 202–222
7. Zane L, Bargelloni L, Patarnello T (2002) Strategies for microsatellite isolation: a review. Mol Ecol 11:1–16
8. Neff BD, Gross MR (2001) Microsatellite evolution in vertebrates: inference from AC dinucleotide repeats. Evolution 55:1717–1733
9. Abbott CL, Ebert D, Tabata A et al (2010) Twelve microsatellite markers in the invasive tunicate, *Didemnum vexillum*, isolated from low genome coverage 454 pyrosequencing reads. Conserv Genet Resour 3:79–81
10. Castoe TA, Poole AW, Gu W et al (2010) Rapid identification of thousands of copperhead snake (*Agkistrodon contortrix*) microsatellite loci from modest amounts of 454 shotgun genome sequence. Mol Ecol Resour 10:341–347. doi:10.1111/j.1755-0998.2009.02750.x
11. Lepais O, Bacles DFE (2011) Comparison of random and SSR-enriched shotgun pyrosequencing for microsatellite discovery and single multiplex PCR optimization in *Acacia harpophylla* F. Muell Ex Benth Mol Ecol Resour 11:711–724. doi:10.1111/j.1755-0998.2011.03002.x
12. Malausa T, Gilles A, Meglecz E et al (2011) High-throughput microsatellite isolation through 454 GS-FLX Titanium pyrosequencing of enriched DNA libraries. Mol Ecol Resour 11:638–644. doi:10.1111/j.1755-0998.2011.02992.x

13. Perry JC, Rowe L (2011) Rapid microsatellite development for water striders by next-generation sequencing. J Hered 102(1):125–129. doi:10.1093/jhered/esq099
14. Whitney JL, Karl SA (2012) Development of 38 microsatellite loci from the Arceye hawkfish, *Paracirrhites arcatus*, using next-generation sequencing and cross-amplification in other Cirrhitid species. Cons Genet Resour. doi:10.1007/s12686-011-9589-y
15. Roche Technical Bulletin No. 2010-010 August 2010 Multiplex Identifier (MID) Adaptors for Rapid Library Preparations. http://ftp.genome.ou.edu/pub/454/TCB-10010_MIDAdaptorsforRapidLibraryPreparations.pdf
16. SFF_EXTRACT and CLEAN_READS (http://bioinf.comav.upv.es/)
17. FASTX_TOOLKIT (http://hannonlab.cshl.edu/fastx_toolkit/)
18. FASTQC (http://www.bioinformatics.bbsrc.ac.uk/projects/fastqc/)
19. Megelcz E, Costedoat C, Dubut V et al (2010) QDD: a user-friendly program to select microsatellite markers and design primers from large sequencing projects. Bioinformatics 26(3):403–404. doi:10.1093/bioinformatics/btp670, http://gsite.univ-provence.fr/gsite/Local/egee/dir/meglecz/QDD.html
20. Kohany O, Gentles AJ, Hankus L et al (2006) Annotation, submission and screening of repetitive elements in Repbase: RepbaseSubmitter and Censor. BMC Bioinformatics 25(7):474, http://www.girinst.org/repbase/index.html
21. Thurston MI, Field D (2005) Msatfinder: detection and characterization of microsatellites. Distributed by the authors at http://www.genomics.ceh.ac.uk/msatfinder/. CEH Oxford, Mansfield Road, Oxford OX1 3SR.
22. Rozen S, Skaletsky H (2000) Primer3 on the WWW for general users and for biologist programmers. Methods Mol Biol 132:365–386, http://frodo.wi.mit.edu/
23. Haddock S, Dunn C (2010) Practical computing for biologists, 1st edn. Sinauer Associates, Inc., Sunderland, MA

Chapter 8

Identification of DNA-Microsatellite Markers for the Characterization of Somatic Embryos in *Quercus suber*

Arancha Gómez-Garay, Ángeles Bueno, and Beatriz Pintos

Abstract

Nuclear DNA-microsatellite markers led the possibility to characterize individually both *Quercus suber* trees and somatic embryos. The genotype inferred by SSR markers opens the possibility to obtain a fingerprint for clonal lines identification. Furthermore, allow to infer the origin of somatic embryos from haploid cells (microspores) or from diploid tissues. Using few SSR markers from other *Quercus* species and an automatic system based in fluorescence, it is possible to obtain a high discrimination power between genotypes. This method is sufficient to assign tissues to an individual tree with high statistical certainty. Nevertheless, it is necessary to take care to select the adequate DNA extraction method to avoid PCR inhibitors present in diverse *Q. suber* tissues.

Key words Fingerprint, Dyes, Haploid, Doubled haploid, Germplasm, Clonal lines

1 Introduction

Powerful insights have been gained into the knowledge of forest tree genetics from the analysis of microsatellite markers. There are two types of information supplied by microsatellites according to their nuclear or organelle genomes origin. Nuclear microsatellites are codominant markers and therefore are more informative for genotyping individuals, and for linkage mapping, by the other side, organelle microsatellites are more suitable for phylogenetic studies.

Codominant molecular markers differ between homozygotic and heterozygotic individuals. Due to their codominant inheritance, simple sequence repeats (SSRs) have become the preferred tool for investigations of critical importance for germplasm managers, such as the establishment of unique genetic identities or fingerprints.

Other attribute of microsatellites markers is their use, generally restricted the species for which that are designed, due to the high degree of homology necessary between primers and sample DNA. Sometimes there are amplifications available for one species derived

Stella K. Kantartzi (ed.), *Microsatellites: Methods and Protocols*, Methods in Molecular Biology, vol. 1006,
DOI 10.1007/978-1-62703-389-3_8, © Springer Science+Business Media, LLC 2013

from closely related species during evolution for which those primers were designed (1–6). But, often, these markers may not be highly conserved across species in some genera (7, 8).

Most simple sequence repeats, $(GA)_n$ microsatellites, localized by Steinkellner et al. (9, 10) and Kampfer et al. (11) in *Q. petraea* and *Q. robur* can be PCR amplified using the same primers in other oaks (*Q. pubescens*, *Q. cerris*, *Q. palustris*, *Q. rubra*, *Q. suber*, *Q. ilex*), and even some SSRs have been found in other species of the *Fagaceae* family (*Fagus sylvatica* and *Castanea sativa*). Those markers have been used for tree identification, genotypic characterization, heterozygosity evaluation, and determination of the ploidy level in anther induced embryos by stress treatments in *Q. suber*. Furthermore, in the case of highly heterozygotic parents, a test of parental exclusion may be used for identification of the father tree of the gametic embryos and genotype identification based on the analysis of its haploid progeny (4).

The question of the cellular origin of embryos, gametic or parental (sporophytic) type, rises on microspore embryogenesis post stress treatment. The embryogenic process originated from anther culture may be induced from different origins, e.g., haploid cells such as microspores or pollen grains or somatic cells from anther tissues. The type of origin, haploid, doubled haploid, or diploid, of anther-derived embryos has been studied in forest tree species or even in other plant species.

The isolation of high-quality DNA is a key step because contaminants such as proteins, polyphenols, and polysaccharides may interfere with Taq DNA polymerase. In this sense, embryo and leave tissues from *Q. suber* show large polyphenols. Furthermore, in culture medium, it produced the oxidation of polyphenols and formation of quinines. Thus, diverse methods must be used for different materials. The protocols for DNA extraction by Doyle and Doyle (12) and Ziegenhagen et al. (13) have been used as point of beginning to obtain optimum results.

Microsatellite polymorphisms have provided a new approach to the genetic analysis of oaks and a tree identification system due to the high discrimination power obtained for genotypic differentiation (i.e., Craft et al. (14) used only four SSRs in *Quercus* with forensic applications). The high rate of polymorphisms observed also permitted the identification of the parent tree by parental exclusion. The principle of parental exclusion could be applied in embryo cultures, and few *loci* were sufficient for parental identification. The homozygotic genome for all *loci* tested, in haploid and doubled-haploid embryos, was revealed by the presence of a unique allele per *locus*. This result confirms the applicability of microsatellite markers as indicators of the ploidy level in embryo regeneration from anther cultures. This way, microsatellite markers have proved again to be an interesting tool for in vitro culture management.

2 Materials

All solution must be prepared using ultrapure water (Milli-Q water) and analytical grade reagents. Reagents must be stored at room temperature (unless indicated otherwise). Disposal regulations must be followed when disposing waste materials.

2.1 Solutions for DNA Extraction from Embryos

1. 1 M Tris–HCl, pH 8 and 1 M Tris–HCl, pH 7.5: Tris (hydroxymethyl) aminomethane (FW 121.4 g/mol). Weigh 60.57 g Tris in 0.5 l water. Mix and adjust adequate pH with HCl (see Note 1). Store at 4 °C.
2. 3.8 g/l sodium bisulfite (sodium hydrogen sulfite): Weigh 3.8 g sodium bisulfite and make up to 1 l water.
3. Extraction buffer 1, pH 8: 0.35 M sorbitol (FW 182.17 g/mol), 0.10 M Tris–HCl (pH 8), 5 mM EDTA (ethylenediaminetetraacetic acid, FW 372.24 g/mol). Weigh 15.94 g sorbitol and 0.47 g EDTA. Transfer to 25 ml 1 M Tris–HCl (pH 8) from previous step and make up to 250 ml with water. Mix and adjust pH with HCl or NaOH.
4. Lysis buffer, pH 7.5: 200 mM Tris (pH 7.5), 50 mM EDTA (ethylenediaminetetraacetic acid, FW 372.24 g/mol), 2 M NaCl (FW 58.44 g/mol), 20 g/l CTAB (cetyltrimethylammonium bromide, FW 364.45). Weigh 4.7 g EDTA, 29.22 g NaCl, and 5 g CTAB. Transfer to 50 ml 1 M Tris–HCl (pH 7.5) from item 1 and make up to 250 ml with water. Mix and adjust pH with HCl or NaOH.
5. 5 % Sarkosyl: 50 g/l *N*-laurylsarcosine (FW 293.38). Weigh 25 g *N*-laurylsarcosine, make up to 0.5 l with water, and mix.
6. TE buffer, pH 8: 10 mM Tris–HCl (pH 8) and 1 mM EDTA (ethylenediaminetetraacetic acid, FW 372.24 g/mol). Weigh 0.37 g EDTA. Transfer to 10 ml Tris–HCl (pH 8) from item 1 and make up to 1 l with water. Mix and adjust pH with HCl or NaOH. EDTA will not be soluble until pH reaches 8.0.
7. 70 % Ethanol.

2.2 Solutions for DNA Extraction from Leaves

1. 0.3 M sodium acetate, pH 5: Weigh 24.6 g sodium acetate (FW 82.03) and fill up to 1 l with water. Mix 28.82 ml of 1 M acetic acid (60.05 g/l) and 273.3 ml of 0.3 M sodium acetate and fill up to 1 L with water.
2. 0.5 M EDTA, pH 8: Weigh 18.6 g EDTA (ethylenediaminetetraacetic acid, FW 372.24 g/mol) and dissolve in 100 ml water. Adjust pH to 8.0 using NaOH. EDTA will not be soluble until pH reaches 8.0.
3. Extraction buffer 2, pH 5.5: 100 mM sodium acetate (pH 5), 50 mM EDTA (pH 8), 500 mM NaCl, 2 % PVP (polyvinylpyrrolidone). Adjust pH to 5.5 and add 1.4 % SDS. Mix 330 ml

Table 1
Characteristics of the microsatellite loci amplified in *Quercus suber*

Locus	Repeat motif	Fluorescent dye	Primer sequences (5′–3′)	Annealing temp (°C)	Allele size range
QpZAG46	$(GA)_{13}$	HEX	cccctattgaagtcctagccg tctcccatgtaagtagctctg	48	188–192
QpZAG110	$(GA)_{15}$	TET	ggaggcttccttcaacctact gatctcttgtgtgctgtattt	48	224–236
QpZAG36	$(GA)_{23}$	TET	gatcaaaatttggaatattaagagag actgtggtggtgagtctaacatgtag	50	211–225
QrZAG7	$(TC)_{17}$	FAM	caacttggtgttcggatcaa gtgcatttcttttatagcattcag	52	116–124
QrZAG20	$(TC)_{18}$	HEX	ccattaaaagaagcagtattttgt gcaacactcagcctatatctagaa	52	162–170
QrZAG119	$(GA)_{24}$	FAM	gatcagtgatagtgcctctc gatcaacaagcccaaggcac	46	274–276
QrZAG75	$(GA)_{57}$	TET	accgcctatctcaaccagag gtccgagaatcatcattaaagg	54	158–192

of 0.3 M sodium acetate (pH 5) from item 1 and 100 ml of 0.5 M EDTA (pH 8) from item 2. Add 29.22 gr. NaCl (FW 58.44 g/mol) and 20 g PVP. Make up to 1 l with water, adjust pH to 5.5, and add 14 g SDS (sodium dodecyl sulfate).

4. 5 M Potassium acetate: Weigh 49.07 g potassium acetate (FW 98.14) and add water to 100 ml.
5. TE buffer, pH 8: 10 mM Tris–HCl (pH 8) and 1 mM EDTA (ethylenediaminetetraacetic acid, FW 372.24 g/mol). Weigh 0.37 g EDTA. Transfer to 10 ml Tris–HCl (pH 8) from item 1 and make up to 1 l with water. Mix and adjust pH with HCl or NaOH. EDTA will not be soluble until pH reaches 8.0.
6. Isopropyl alcohol (isopropanol, propan-2-ol, 2-propanol, rubbing alcohol). See Note 2.
7. Chloroform. See Note 2.
8. Phenol (carbolic acid, phenic acid). See Note 2.

2.3 Buffer for DNA Quantification

1. TE buffer, pH 8: as referred in Subheading 2.2, item 5.

2.4 Chemicals for SSR Amplifications

1. Taq DNA polymerase with the buffer supplied for Taq DNA polymerase amplification (including Tris–HCl pH 9, KCl, $MgCl_2$).
2. dNTPs (dATP, dCTP, dGTP, and dTTP).
3. Fluoro-labeled oligonucleotide primers. The forward primer of each pair (see Table 1) was labeled with fluorescent dye

(i.e., FAM, HEX, and TET) to allow detection of the polymerase chain reaction (PCR) products by an automatic DNA sequencer.

2.5 Equipment

1. Microcentrifuge.
2. Shaker.
3. Water bath.
4. Fume hood.
5. Micropipettes to handle from 1 μl to 1 ml.
6. Spectrophotometer.
7. PCR thermocycler.
8. Automatic DNA sequencer.
9. Software for peaks analysis.

2.6 Plastic Ware (See Note 3)

1. 0.2 ml (PCR) and 2 ml (microcentrifuge) tubes.
2. Filter tips.

3 Methods

Carry out all procedures at room temperature unless otherwise specified.

3.1 DNA Extraction from Embryos

1. Weight out 70 mg embryo tissue in a 2 ml microcentrifuge tube. Add 100 μl sodium bisulfite (3.8 g/l).
2. Add 300 μl extraction buffer 1 and grind the embryo with a sterilized tip.
3. Add 300 μl lysis buffer. Mix gently.
4. Add 120 μl sarkosyl (5 %) and mix vigorously.
5. Incubate for 15 min at 65 °C in a water bath.
6. Outside the bath, add 600 μl chloroform. Cap the tube and mix vigorously by vortex to obtain an emulsion.
7. Centrifuge for 10 min at 13,523 × *g* and 4 °C. Transfer upper phase to a clean microcentrifuge tube, add 400 μl volume of −20° isopropyl alcohol, and gently invert the tube several times to mix.
8. To precipitate the DNA, place the tube at −20 °C for 30 min.
9. To pellet the DNA, centrifuge the tubes at 13,523 × *g* for 5 min at 4 °C.
10. Leave the pellet air-drying. Add a drop of 70 % ethanol to wash the pellet. Air-dry the pellet again.
11. Redissolve the pellet in 50 μl of TE pH 8 in the water bath for 15 min at 65 °C.
12. Store the DNA at −20 °C.

3.2 DNA Extraction from Leaves

1. Weight out 50 mg young leave tissue in a 2 ml microcentrifuge tube. Add 1,000 μl extraction buffer 2 and grind the leave tissue with a sterilized tip.
2. Cap the tube and Incubate for 20 min at 65 °C in a water bath.
3. Centrifuge for 10 min at 9,391 ×*g* and 4 °C. Discard upper phase and add approximately 1/3 volume of sodium acetate and gently invert the tube several times to mix. Place at 4 °C for 30 min.
4. Centrifuge for 10 min at 9,391 ×*g* and 4 °C. Discard upper phase and add approximately 0.6 volume −20° isopropyl alcohol and gently invert the tube several times to mix. Place at −20 °C for 30 min.
5. Centrifuge for 10 min at 9,391 ×*g* and 4 °C. Discard upper phase and leave the pellet air-drying.
6. Redissolve the pellet in 200 μl of TE pH 8 for 30 min at t_m.
7. Add 200 μl phenol and mix vigorously.
8. Centrifuge for 10 min at 9,391 ×*g* and 4 °C. Discard upper phase, add 100 μl phenol and 100 μl chloroform, and gently invert the tube several times to mix.
9. Centrifuge for 10 min at 9,391 ×*g* and 4 °C. Discard upper phase, add 200 μl chloroform, and gently invert the tube several times to mix.
10. Centrifuge for 5 min at 9,391 ×*g* and 4 °C. Discard upper phase and add 300 μl sodium acetate and 2.5 volume of 96 % ethanol. Place at −20 °C overnight.
11. To pellet the DNA, centrifuge the tubes at 9,391 ×*g* for 10 min at 4 °C.
12. Leave the pellet air-drying. Redissolve the pellet in 20 μl of TE pH 8 in the water bath for 15 min at 65 °C.
13. Store the DNA at −20 °C.

3.3 DNA Quantification

The concentration of the extracted DNA was determined spectrophotometrically (see Note 4):

1. 2 μl of extracted DNA must be diluted 1:10 in water.
2. Absorption is measured for both blank (TE 0.1 diluted in water) and diluted DNA solution at 260 nm.
3. DNA concentration is calculated based on the assumption that an OD of 1 corresponds to 50 μg/ml DNA.

3.4 SSR Amplifications by PCR

1. Preparing the reaction mixes (see Note 5 and Table 2).
2. Running the PCR: PCR must be conducted in a PCR thermocycler following the manufacturer's instructions and with cycling conditions listed in Table 3.

Table 2
Amplification reaction mixture in 25 μl final volume/concentration per reaction for *Q. suber* SSR amplification

Step	Reagent	Final concentration
1	Taq DNA polymerase	1.75 U
2	Tris–HCl (pH 9.0)	10 mM
3	KCl	50 mM
4	$MgCl_2$	1.5 mM
5	dNTPs	200 μM
6	Primer forward (1 μM)	200 nM
7	Primer reverse (1 μM)	200 nM
8	Sterile Milli-Q water	Up to 23 μl
9	DNA template (10 ng/μl)	2 μl

Table 3
Thermocycling profiles for amplification of *Quercus* SSRs

Locus	Amplification profiles
QpZAG46	95 °C, 6′/[92 °C, 1′/48 °C, 30″/72 °C, 1′]×30/72 °C, 8′/4 °C, ∞
QpZAG110	95 °C, 6′/[92 °C, 1′/48 °C, 30″/72 °C, 1′]×30/72 °C, 8′/4 °C, ∞
QpZAG36	95 °C, 6′/[92 °C, 1′/50 °C, 30″/72 °C, 1′]×30/72 °C, 8′/4 °C, ∞
QrZAG7	95 °C, 6′/[92 °C, 1′/52 °C, 30″/72 °C, 1′]×30/72 °C, 8′/4 °C, ∞
QrZAG20	95 °C, 6′/[92 °C, 1′/52 °C, 30″/72 °C, 1′]×30/72 °C, 8′/4 °C, ∞
QrZAG119	95 °C, 6′/[92 °C, 1′/46 °C, 30″/72 °C, 1′]×30/72 °C, 8′/4 °C, ∞
QrZAG75	95 °C, 6′/[92 °C, 1′/54 °C, 30″/72 °C, 1′]×30/72 °C, 8′/4 °C, ∞

3.5 Analyzing the Data

1. Three microsatellite markers are analyzed together by the automatic sequencer; each of them must have a different dye in order to discriminate the results.
2. Peak sizes are quantified by comparison with internal size standards using software provided by the automatic sequencer manufacturer (see Note 6 and Table 1).
3. The genotype of each sample (leaves from parent trees and haploid, diploid, and doubled-haploid embryos) is defined as the combination of the analyzed fragments (see Note 7).

4 Notes

1. Dissolve the Tris into water, 1/3–1/2 of the desired final volume. Mix in HCl until the pH meter gives the desired pH for the Tris buffer solution. Dilute the buffer with water to reach the desired final volume of solution.
2. Phenol, chloroform, isoamyl alcohol, and isopropanol are hazardous chemicals. Follow safety guidelines, under fume hood.
3. All plastic ware has to be sterile and free of DNAs and nucleic acids.
4. Each DNA extract must be measured twice, and the two values must be averaged.
5. If necessary, thaw all reagents (a 37 °C water bath is recommended). Store all reagents on ice once thawed. Be sure to thoroughly mix each reagent before use. Reaction mixes must be prepared consisting of all components of the PCR, except DNA template, in sufficient quantities for all reactions to be performed.
6. The software provides estimates of fragment sizes reliable to two decimal places. The distribution of fragment sizes for each microsatellite *locus* was not continuous but displayed discontinuities or breaks which were used to define sets of peaks, i.e., alleles or variants. For example, fragment sizes 187.80–188.54 bp might be designated as allele "188," in the case that relatively large breaks separate them from neighboring peaks. We applied two criteria in defining peaks: (a) the range of fragment sizes within a peak should not exceed 1 bp, and (b) the gaps between peaks should be substantially greater than the gaps between fragment sizes within peaks.
7. For embryos derived from anther culture: If the parent tree is heterozygous for one SSR, the diploid embryos derived from this tree can show two alleles for this SSR, meaning the embryo is diploid. If the parent tree is heterozygous for one SSR, the diploid embryos derived from this tree can show only one allele for this SSR, meaning the embryo is haploid or doubled haploid.

References

1. Primmer CR, Moller AP, Ellegren H (1996) Polymorphisms revealed by simple sequence repeats. Trends Plant Sci 1:215–222
2. Sun HS, Kirkpatrick BW (1996) Exploiting dinucleotide microsatellites conserved among mammalian species. Mamm Genome 7: 128–132
3. Fields RL, Scribner KT (1997) Isolation and characterization of novel waterfowl microsatellite loci: cross-species comparisons and research application. Mol Ecol 6:199–202
4. Gómez A, Pintos B, Aguiriano E et al (2001) SSR markers for *Quercus suber* tree identification and embryo analysis. J Hered 92(3):292–295

5. Gómez A, Manzanera JA, Alía R et al (2004) Microsatellite diversity in forest trees. Recent Res Devel Genet Breeding Res Signpost Trivandrum Kerala India 1:425–448
6. González-Martínez SC, Mariette S, Ribeiro MM et al (2004) Genetic resources in maritime pine (*Pinus pinaster* Aiton): patterns of differentiation and correlation between molecular and quantitative measures of genetic variation. For Ecol Manage 197: 103–115
7. Echt CS, May-Marquardt P, Hseih M et al (1996) Characterization of microsatellite markers in eastern white pine. Genome 39:1102–1108
8. Karhu A, Dieterich J-H, Savolainen O (1999) Rapid extension of microsatellites in pines. Mol Biol Evol 17:259–265
9. Steinkellner H, Fluch S, Turetschek E et al (1997) Identification and characterization of (GA/CT)n—microsatellite loci from *Quercus petraea*. Plant Mol Biol 33:1093–1096
10. Steinkellner H, Lexer C, Turetschek E et al (1997) Conservation of (GA)n microsatellite loci between *Quercus* species. Mol Ecol 6:1189–1194
11. Kampfer S, Lexer C, Glössl J et al (1998) Characterization of (GA)n microsatellite loci from *Quercus robur*. Hereditas 129:183–186
12. Doyle JJ, Doyle JL (1990) Isolation of plant DNA from fresh tissue. Focus 12:13–15
13. Ziegenhagen B, Guillemaut P, Scholz F (1993) A procedure for mini-preparations of genomic DNA from needles of Silver Fir (*Abies alba* Mill.). Plant Mol Biol Rep 11(2):117–121
14. Craft KJ, Owens JD, Ashley MV (2007) Application of plant DNA markers in forensic botany: genetic comparison of *Quercus* evidence leaves to crime scene trees using microsatellites. Forensic Sci Int 165:64–70

Part II

Amplification and Visualization

Chapter 9

Simple Sequence Repeats Amplification

Kundapura V. Ravishankar and Padmakar Bommisetty

Abstract

The technique of SSR amplification is a prerequisite to generate the molecular profiles of various alleles of an individual or genotype. Amplification is the multifold duplication and accumulation of a targeted region which is achieved by polymerase chain reaction. It needs ingredients such as buffer, $MgCl_2$, dNTPs, primers, and DNA polymerase enzyme. The utilization of these essential PCR components in optimal concentrations determines the success of amplification. Thus SSRs, as primers, play an important role in enhancing the amplification and thereby generating the genotype profile. With the advent of technology, fluorophore-labeled primers along with automated capillary electrophoresis system have enhanced the efficiency of detection.

Key words Simple sequence repeats, Polymerase chain reaction, Amplification, Genotyping, Fluorophores

1 Introduction

Simple sequence repeats (SSR) or microsatellites are tandem repeats of nucleotide motifs ranging from 1 to 6 and are evenly distributed throughout the genome. They exhibit allelic variation at a locus due to variation in the number of repeat motifs. The flanking regions of these motifs are conserved within the species and sometimes across the species within the genus. Using this property, primers are designed and standardized. These primers can amplify individuals in the species. SSR markers are PCR-based molecular markers; they have many desirable attributes such as hypervariability, multiallelic nature, codominant inheritance, reproducibility, relative abundance, extensive genome coverage (including organellar genomes), chromosome-specific locations, amenability to automation, and high-throughput genotyping (1). The allelic variation, existing due to replication slippage and/or unequal crossing over during meiosis, of an individual organism can be converted to a specific molecular genotype profile through a method called genotyping. The genotyping is achieved by amplifying specific loci of an individual with the help of SSR primers.

Stella K. Kantartzi (ed.), *Microsatellites: Methods and Protocols*, Methods in Molecular Biology, vol. 1006, DOI 10.1007/978-1-62703-389-3_9, © Springer Science+Business Media, LLC 2013

Amplification of SSR is a PCR-based technique involving the exponential increment of DNA of the target region. Repeat motif SSR amplification is amenable to high-throughput genotyping and has proven to be a useful tool for paternity analysis, construction of high-density genome maps, mapping of useful genes, marker-assisted selection, and for establishing genetic and evolutionary relationships (2).

Amplification is achieved with the help of either labeled or unlabeled SSR primers. The annealing temperature of PCR plays a crucial role for accurate priming of the SSR markers to the template. In most studies, SSR amplification was successfully achieved within the temperature range of 45–60 °C (3–8). Here we describe three different methodologies being implemented in SSR amplification.

2 Materials

The basic materials required for SSR amplification are:

1. Thermocycler.
2. PCR ingredients:
   (a) Taq DNA polymerase.
   (b) Taq DNA polymerase buffer (10×).
   (c) $MgCl_2$ (100 mM).
   (d) Deoxyribose nucleotide triphosphates (dNTPs; see Note 4).
   (e) Primers; forward and reverse (SSRs; labeled or unlabeled or M13 tailed; see Note 3).
3. Nuclease-free molecular biology grade water.
4. Template DNA.
5. PCR coolers (see Note 1).
6. PCR tubes or plates (sterilized).
7. PCR tube storage racks (96-well).
8. Ice-making machine.
9. Ice bucket.
10. Micropippetes.
11. Micro-tips (sterile).

The template DNA, primers, and dNTPs are diluted in nuclease-free water according to the working concentrations from the respective stocks.

3 Methods

The PCR setup for SSR amplification has to be planned depending upon the requirement, i.e., the number of PCR and the volume of the reaction. It has to be done carefully at cold temperature in order to maintain the integrity of the PCR ingredients (see Note 1 and Note 8).

3.1 Amplification Using Unlabeled SSR Primers

1. The initial step is the preparation of a master mix containing all the PCR ingredients except one component which is variable. In general, the component that is excluded is either template DNA or primer. The thumb rule for excluding the PCR component from the master mix is its quantity/number, i.e., the smaller number components are included in the master mix by excluding the component which is variable. For example, if there are 32 template DNA samples to be screened with 3 SSR primers, then SSR primers are included into the master mix and template DNA is excluded. The master mix is made to minimize pipetting, thereby error. Here three master mixes are prepared separately for three SSR primers. For example, if 20 μl is the reaction volume, then total volume for 32 samples is going to be $20 \times 32 = 640$ μl. Master mix is prepared by addition of the following components (Table 1; see Note 8).
2. First the master mix is prepared by adding the PCR ingredients in the following order (see Note 2):
   (a) Nuclease-free water.
   (b) Taq buffer (provided by supplier of Taq polymerase).
   (c) $MgCl_2$ (added if we need to increase its concentration above what is there in buffer; see Note 5).
   (d) dNTPs (see Note 4).

Table 1
Master-mix preparation

PCR components	Volume (in μl) for single reaction	Volume (in μl) for 32 reactions included in master mix	
Complete buffer (10×) (includes $MgCl_2$)	2	64	Master mix
dNTPs (1 mM)	2	64	
Forward primer (5 μM)	2	64	
Reverse primer (5 μM)	2	64	
Nuclease-free water	9.8	314.7	
Taq polymerase (3 U/μl)	0.166 (0.5 U)	5.3	
Template DNA (20 ng/μl)	2		Dispensed individually

(e) Primers (see Note 6)/template DNA (see Note 7).

(f) Taq DNA polymerase (see Note 6).

3. The required amount of master mix is dispensed to the PCR tubes; here, in this example, 18 μl is transferred to PCR tubes.
4. Then the required amount of variable PCR component (template DNA or primers) is dispensed to each PCR tube containing 18 μl master mix (see Note 8).
5. The tubes are spun for proper mixing of the ingredients.
6. PCR tubes are placed in the thermal cycler and run using appropriate program.
7. Amplified products can be detected on either agarose or PAGE gels or using automated microchip electrophoresis or automated capillary electrophoresis system.

3.2 Amplification Using Labeled SSR Primers

1. The methodology remains the same as mentioned above, the only exception being the use of labeled primers, either forward or reverse.
2. In this case, either of the primers is initially modified at the 5′ end with the fluorophores FAM, PET, NED, TET, HEX, etc., depending upon the system used for detection of amplified products.
3. The amplified product with labeled primers are used for high-throughput genotyping employing automated capillary electrophoresis system. They have precision of detection with 1 bp difference.

3.3 Amplification Using M13-Tailed SSR Primers and M13-Labeled Probes

1. In order to reduce the cost incurred in labeling of each primer with fluorophores, M13-tailed PCR has been developed (9).
2. This is an economic method for fluorescent labeling of PCR products.
3. This methodology remains the same as mentioned in above section with the exception of using labeled M13 probe, i.e., M13 sequence labeled with different fluorophores FAM, PET, NED, TET, HEX, etc., at its 5′ end in addition to unlabeled forward and reverse primers.
4. Forward primer is modified with the addition of M13 sequence (21-mer) at its 5′ end.
5. Fluorescent dye labeling of PCR product is done in a single reaction.
6. PCR is performed with three primers: a sequence-specific forward primer with M13 sequence (21-mer) at its 5′ end, a sequence-specific reverse primer, and the universal fluorescent-labeled M13 sequence (21-mer) primer.

7. The ratio of primer (sequence-specific forward primer) to probe (fluorescent-labeled M13 primer) plays a key role in achieving successful amplification and incorporation of label (fluorophore). In general, 1:5, 1:2, and 1:1 of forward-specific primer: universal fluorescent-labeled primer ratios are used.
8. Cost-effectiveness of this method over the second method described earlier is its advantage.

3.4 Multiplex PCR

Multiplex PCR involves the amplification of more than one target region by using more than one set of primers in a single reaction. The main objective of multiplexing is to combine all markers into smallest number reaction. The throughput of routine SSR analysis is very low as it yields genotype information at only one locus per PCR. However multiplex PCR can enhance genotyping by reducing work, time, and cost (10). Multiplex PCR is a sensitive technique where careful standardization of all steps is required especially DNA concentration should be standardized (11) and is discussed in detail in Chapter 11.

4 Notes

1. As all the PCR ingredients play an important role in successful amplification, therefore, each ingredient's integrity is essential. Hence, PCR setup has to be done under cold conditions without lapse of much time.
2. The concentrations of all the PCR ingredients have to be in optimal level for successful amplification (for details, see Chapter 10).
3. In general, the primer concentration is in the range of 0.1–0.5 μM (12).
4. The dNTPs concentrations are in the range of 0.1–0.5 mM (13).
5. The concentration of $MgCl_2$ is in the range of 1.0–2.5 mM (13).
6. The concentration of Taq DNA polymerase is in the range of 0.5–1.0 unit per 25 μl reaction volume (12).
7. The concentration of template DNA is in the range of 50–100 ng per 25 μl reaction volume (13).
8. Aerosol contamination is a major factor that has to be taken into consideration which results in false-positives. Simple measures such as minimizing pipetting steps by preparing a master mix, using filter tips, closing lids on all tubes and expelling reagents carefully, changing gloves regularly, and having separate working place for DNA isolation and PCR setup will help in preventing the aerosol contamination (14).

References

1. Rajwant KK, Manoj KR, Sanjay K, Rohtas S, Dhawan AK (2011) Microsatellite markers: an overview of the recent progress in plants. Euphytica 177:309–334
2. Parida SK, Kalia SK, Sunita K, Dalal V, Hemaprabha G, Selvi A, Pandit A, Singh A, Gaikwad K, Sharma TR, Srivastava PS, Singh NK, Mohapatra T (2009) Informative genomic microsatellite markers for efficient genotyping applications in sugarcane. Theor Appl Genet 118:327–338
3. Ravishankar KV, Mani BH, Anand L, Dinesh MR (2011) Development of new microsatellite markers from Mango (*Mangifera indica*) and cross-species amplification. Am J Bot 98:e96–e99. doi:10.3732/ajb.1000263
4. Narina SS, d'Orgeix CA, Sayre BL (2011) Optimization of PCR conditions to amplify microsatellite loci in the bunchgrass lizard (*Sceloporus slevini*) genomic DNA. BMC Res Notes 4:26. doi:10.1186/1756-0500-4-26
5. Mishra MK, Patrizia T, De Barbara N, Elisa A, René D, Lorenzo DT, Rajkumar R, Paola R, Alberto P, Giorgio G (2011) Genome organization in coffee as revealed by EST PCRRFLP, SNPs and SSR analysis. J Crop Sci Biotech 14:25–37
6. Wang H, Huan P, Xia L, Baozhong L (2011) Mining of EST-SSR markers in clam *Meretrix meretrix* larvae from 454 shotgun transcriptome. Genes Genet Syst 86:197–205
7. Blair MW, Hurtado N, Chavarro CM, Monica CM, Martha CG, Fabio P, Jeff T, Wing R (2011) Gene-based SSR markers for common bean (*Phaseolus vulgaris L.*) derived from root and leaf tissue ESTs: an integration of the BMc series. BMC Plant Biol 11:50
8. Risterucci AM, Duval MF, Rohde W, Billotte N (2005) Isolation and characterization of microsatellite loci from *Psidium guajava L.* Mol Ecol Notes 5:745–748
9. Schuelke M (2000) An economic method for fluorescent labeling of PCR fragments. Nat Biotechnol 18:233–234
10. Guichoux E, Lagache L, Wagner S, Chaumeil P, Leger P, Lepais O, Lepoittevin C, Malausa T, Revardel E, Salin F, Petit RJ (2011) Current trends in microsatellite genotyping. Mol Ecol Res 11:591–611
11. Livingstone D et al (2009) Improvement of highthroughput genotype analysis after implementation of a dual-curve Sybr Green I-based quantification and normalization procedure. Hort Sci 44:1228–1232
12. Su H, Li Z-G, Song S-H (2009) Optimizing System of SSR-PCR on Soyabean by Orthogonal design and SSR primer selection. Acta Agriculturae Boreali-Sinica 24:99–102
13. Li M, Lu X-L, Luo C-De, Zhang F, Wu Z-X, Zhong J-Y (2009) Optimizing System of SSR-PCR in *Pinus radiata* and *Pinus tabulaeformis*. J Mol Genet 1:44–49
14. Mifflin TE (2003) Setting up a PCR laboratory (Chapter 1). In: Dieffenbach CW, Dveksler GS (eds) PCR Primer, 2nd edn. Cold Spring Harbor Laboratory Press, Cold Spring Harbor, NY

Chapter 10

Microsatellite Amplification in Plants: Optimization Procedure of Major PCR Components

Sana Ghaffari and Nejib Hasnaoui

Abstract

Microsatellites (SSRs) are the most informative and popular class of molecular markers used for diverse purposes, particularly in plants: genetic diversity study, marker assisted selection, breeding, mapping, phylogenetics and phylogeography, systematics, etc. They have become a routine technique practically in each laboratory for studying molecular plant genetics. Despite their wide utilization, however, setup and optimization of various conditions involved in PCR amplification is a prerequisite for reliable inference of results. In this chapter, we describe optimization of SSR-PCR conditions and give ranges of concentrations for different parameters. The protocol provided here is inspired from bench work on the use of microsatellite to study diversity of *Vitis vinifera* germplasm.

Key words DNA, Microsatellites, PCR conditions, Optimization

1 Introduction

Molecular markers based on the polymerase chain reaction (PCR) are widely used in plant breeding and genetic research. Actually, microsatellite markers or Simple Sequence Repeats (SSRs) are regarded to be the best suited methods for these applications and have been widely used for multiple purposes, and are always of great usefulness, especially in plant genetic field (1–6).

SSRs have numerous advantages such as a high discriminatory power, a high information content arising from their multiallelic nature, codominant transmission, a robust and reproducible assay, in addition to their relative abundance and uniform coverage of genome. Also, the PCR-SSR accomplishment needs small amount of DNA template. Besides, their detection is quite easy, actually via the automated systems and capillary electrophoresis (7, 8).

The two authors contributed equally to this work.

Stella K. Kantartzi (ed.), *Microsatellites: Methods and Protocols*, Methods in Molecular Biology, vol. 1006, DOI 10.1007/978-1-62703-389-3_10, © Springer Science+Business Media, LLC 2013

An important limitation, however, regarding their use, is the need of a prior optimization of PCR conditions (9). Practically, each laboratory has to set up its own protocol depending on its bench equipments and reagent supplier. In practice, SSR-PCR can fail for various reasons, in part due to involvement of many factors that can alter PCR amplification such as different types/brands of thermocyclers, reaction components (template DNA, dNTPs, DNA polymerase, Mg^{2+} concentration, etc.), or even minor differences in thickness of walls of PCR tubes (10, 11). Theoretically all reagents included in the PCR-mix influence the outcome of SSR loci amplification. In this context, amount and quality of template DNA strongly influences the PCR result. In order to ensure that there are sufficient products in the PCR system, the template volume cannot be too much; otherwise nonspecific products might be obtained (12).

$MgCl_2$ concentration can have enormous influence on PCR success. Increasing the Mg^{2+} ions enhances Taq activity up to an optimum, above which the former may act as a depressant of the same (13).

As substrate of the PCR reaction, dNTPs' content affects directly the output of PCR amplification. An excess of dNTPs will compete with the polymerase linking to Mg^{2+} and hence inhibit PCR reaction (12).

The dosage of TaqDNA polymerase affects the amplification efficiency and excess dosage will produce a high mismatch rate, while low dosage will affect the good combination of the enzyme and the primer (12, 14).

Although primer concentration does not matter as other parameters for the PCR itself, their optimization is important in economic point of view, to avoid unnecessary wastage. Variable concentrations of forward and reverse primers of different markers are reported in literature (10, 15, 16).

In practice, optimization of conditions is generally achieved by changing one factor at a time, despite the fact that this may lead to suboptimal results since interactions between conditions are difficult to detect with this approach (17). An extensive optimization is typically required for multiplex SSR-PCR, where more than one SSR locus is amplified per reaction, to satisfactorily amplify all the targeted amplicons (18, 19). Although PCR is nowadays a routinely used and quite easy technique, for reproducible and readable patterns of SSR-PCR products, optimization and setup of optimal conditions is often needed (9–11, 20, 21).

The first step of SSR-PCR simply entails mixing template DNA, a PCR buffer, magnesium chloride ($MgCl_2$), forward and reverse primers of SSR locus, deoxyribonucleoside triphosphates (dNTPs), and Taq or other thermostable DNA polymerases. Once assembled, the mixture is cycled many times (usually 30) in temperature conditions that allow denaturation, primer annealing, and synthesis of DNAs. This led to an exponentially amplified microsatellite sequence flanked by primers with an expected size. The SSR-PCR products are then migrated through an appropriate gel, to control the amplification specificity and allele sizing.

In this chapter we describe a detailed method for microsatellite (SSR) amplification, including quick optimization of conditions. Details and given concentration range of different reagents are based on previous work successfully optimized for PCR amplification of *Vitis vinifera* SSRs.

2 Materials

1. Template DNA solution: ~25 ng/μl of genomic DNA in sterile water (see Note 1).
2. PCR buffer: 10× $MgCl_2$-free PCR buffer. Store at −20 °C.
3. Magnesium chloride solution: Provided as 25 mM $MgCl_2$ in sterile water. $MgCl_2$ solution will remain stable at −20 °C in a constant-temperature freezer.
4. Forward microsatellite primer: 10 μM in sterile water. Store at −20 °C (see Notes 2 and 3).
5. Reverse microsatellite primer: 10 μM in sterile water. Store at −20 °C (see Notes 2 and 3).
6. dNTP Mix (Deoxynucleotide Mix) containing dATP, dCTP, dGTP, dTTP at a final concentration of 10 mM of each dNTP (see Note 4); Ultrapure quality greater than 99 % triphosphate purity by HPLC, free of DNase, RNase, Protease, and no nicking activity. Store the dNTP Mix at −20 °C or −70 °C, in a constant temperature freezer. Avoid multiple freeze–thaw cycles. Aliquoting is recommended.
7. Taq DNA polymerase: 5 U/μl; in native or recombinant form. Store at −20 °C. Avoid exposure to frequent temperature changes.
8. Sterile water: Ultrapure quality, nuclease-free (see Note 5).
9. Mineral oil (optional). Store indefinitely at room temperature.

3 Methods

Carry out all procedures on ice.

Thaw all frozen solutions, including PCR buffer, template DNA, sterile water, and the primer mix, and mix well before use (see Note 6).

3.1 Optimize Reaction Components

3.1.1 Optimize Amount of $MgCl_2$

1. Prepare reaction master mix I according to the recipes given in Table 1 (see Note 7).
2. Mix the reaction mix gently but thoroughly, for example by pipetting up and down few times. Keep on ice.

Table 1
Master mixes for optimizing reaction components

Components	Master mix[a] (μl)			
	I	II	III	IV
10× $MgCl_2$-free PCR buffer	20	20	20	20
10 mM 4dNTPs mix	3	3	3	3
25 mM $MgCl_2$	Va[b]	Optimal	Optimal	Optimal
5 U Taq DNA polymerase	1.5	Va[c]	Optimal	Optimal
10 μM Forward microsatellite primer	15	15	Va[d]	Optimal
10 μM Reverse microsatellite primer	15	15	Va[d]	Optimal
25 ng/μl Template DNA	20	20	20	Va[e]
MilliQ water	*qs* up to 20 μl/reaction			

[a]Final volume enough for $n+1$ reactions, $n=4$
[b]Variable amount for $MgCl_2$
[c]Variable amount for Taq DNA polymerase
[d]Variable amount for microsatellite primers
[e]Variable amount for template DNA

3. Dispense 14.9 μl master mix I into each of four 0.5-ml thin-walled PCR tubes labeled I-A, I-B, I-C, and I-D. Add 3.2 μl of 25 mM $MgCl_2$ into tube labeled I-A (4 mM final concentration). Similarly, aliquot 2 μl, 1.2 and 0.8 μl of 25 mM $MgCl_2$ to tubes labeled respectively I-B, I-C, and I-D (2.5, 1.5, and 1 mM final concentrations respectively). Keep on ice.
4. Add sterile water up to 20 μl in each tube. Keep on ice.
5. Spin tubes if necessary.
6. Overlay the reaction mixture with 15–20 μl mineral oil (see Note 8).
7. Program the thermal cycler according to the following profile: an initial step of 4 min at 94 °C followed by 35 cycles of 20 s at 94 °C, 1 min at 56 °C, 2 min at 72 °C with a final extension time of 5 min at 72 °C.
8. Place the PCR tubes in the thermal cycler and start the cycling program.
9. Recover the tubes from the PCR machine; if electrophoresis is to be carried out later, PCR products could be stored at 4 °C. Alternatively, a final step of 4 °C can be added to the PCR program for holding samples overnight (at the end of step 7.) (see Note 9).
10. PCR running check (see Subheading 3.1.5).

3.1.2 Optimize Amount of Taq DNA Polymerase

1. Prepare reaction master mix II using the optimal $MgCl_2$ concentration determined in step 10 and according to the recipes given in Table 1.
2. Aliquot appropriate volumes of master mix II into four PCR tubes labeled II-A, II-B, II-C, and II-D. Add 0.4 μl of 5 U/μl Taq DNA polymerase into the tube labeled II-A (2 U final concentration). Similarly, aliquot 0.3, 0.2, and 0.1 μl of 5 U/μl Taq DNA polymerase in tubes labeled, respectively, II-B, II-C, and II-D (1.5, 1 and 0.5 U final concentrations, respectively) (see Note 10). Keep on ice.
3. Consider steps 4–6 in the previous Subheading 3.1.1.
4. Begin amplification of all four reactions, using the same cycling parameters as before.
5. Electrophoresis check (see Subheading 3.1.5).

3.1.3 Optimize Amount of Microsatellite Primer

1. Prepare reaction master mix III using the optimal $MgCl_2$ and Taq DNA polymerase concentrations as determined firstly in previous subheadings (cf. see Table 1).
2. Aliquot master mix III into appropriately labeled tubes III-A, III-B, III-C, and III-D. Add, respectively, 3, 2, 1.2, and 0.6 μl of each forward and reverse 10 μM Microsatellite Primer into PCR tubes (1.5, 1, 0.6, and 0.3 μM final concentrations, respectively) and add sterile water up to 20 μl. Keep on ice.
3. Amplify samples.
4. PCR running check (see Subheading 3.1.5).

3.1.4 Optimize Amount of Template DNA

1. Prepare reaction master mix IV using the optimized $MgCl_2$, Taq DNA polymerase, and Microsatellite Primer concentrations determined in Subheadings 3.1.1–3.1.3. Consider recipes given in Table 1.
2. Add, respectively, 4, 3, 2, and 1 μl of template DNA solution (25 ng/μl) to the four aliquots prepared from master mix IV to obtain a final amount of 100, 75, 50, and 25 ng of DNA/reaction. Add sterile water up to 20 μl.
3. Amplify samples.
4. PCR running check (see Subheading 3.1.5).

3.1.5 Checking of PCR Amplification

1. Electrophorese 10 μl of each PCR product through agarose, nondenaturing polyacrylamide, or sieving agarose gel stained with ethidium bromide, and check the amplification running, the band quality as well as the SSR-PCR product size (see Notes 11–13).
2. Generally, retained conditions are those resulted in the greatest amount of PCR product. Our work was carried out on grapevine; the optimal conditions for the different parameters tested

were as follows: (1) 2.5 mM of $MgCl_2$, (2) 1 U/20 μl reaction, (3) 0.6 μM of microsatellite primer/reaction, and (4) 100 ng of template DNA.

4 Notes

1. Both the quality and quantity of nucleic acid starting to be amplified affect SSR-PCR, in particular the sensitivity and efficiency of amplification. Template DNA is normally stored at 4 °C. It can also be frozen, but regular freezing and defrosting damages DNA through "shearing." Once DNA has been frozen, it should only be defrosted to take working stocks for PCR, which should be stored at 4 °C.
2. Primers should be purchased from an established oligonucleotide manufacturer. They are supplied as lyophilized powder, with sheet data containing the requested volume of TE (10 mM Tris, bring to pH 8.0 with HCl; 1 mM EDTA) in which they should be dissolved to get stock solutions of 50 or 100 μM; concentration could be checked by spectrophotometry. Stock solutions are divided into several aliquots and conserved at −20 °C. Primer working solutions of 10 μM are obtained by simple dilution. It is worth to note that primer integrity is a crucial factor for successful SSR-PCR. Problems encountered in SSR-PCR are frequently due to the use of incorrect primer concentrations, low-quality primers, or degraded primers (old primer solution, thawing–freezing cycles).
3. When capillary electrophoresis is used for allele separation and sizing, primers are labeled with fluorescent dyes. When labeled, primers should always be kept in dark to prevent bleaching of the fluorescent dye.
4. The four nucleotides, dATP, dCTP, dGTP, and dTTP, are mixed in equal amounts to form a stock solution in sterile distilled water, which is stored in small aliquots at −20 °C and defrosted when required.
5. Sterile distilled water makes up the "volume" of a SSR-PCR, in which the other components can properly function. Sterile distilled water can be purchased or, normally, is produced in the laboratory by distilling and then autoclaving tap water.
6. It is important to mix all the solutions (PCR buffer, sterile water, $MgCl_2$, dNTP Mix, primer mix, and the template DNA) completely before use.
7. Prepare a volume of reaction mix 10 % greater than that required for the total number of reactions to be performed to take in account pipetting inaccuracies and/or losses. Add Taq to reaction mix lastly.

8. Alternatives to mineral oil include silicone oil and paraffin beads. Nowadays, most of the used thermocyclers are designed to obviate the need for an oil overlay.
9. After amplification, samples can be stored overnight at 2–8 °C or at −20 °C for long-term storage.
10. Taq DNA polymerase, although known to be a heat-stable enzyme, is generally stored at −20 °C. The concentration of *Taq* polymerase used in reactions needs to be optimized for the species in question. As *Taq* is generally the most expensive component of a reaction, it is important to ensure that not too much of enzyme is being used. Use of low concentrations leads to inconsistency, and generation of weak band becomes probable, and sometimes non-amplification can occur.
11. An alternative to ethidium bromide, SYBR Gold Nucleic Acid Gel Stain (Molecular Probes), is 25–100 times more sensitive than ethidium bromide, is more convenient to use, and permits optimization of 10–100-fold lower starting template copy number.
12. SSRs are normally scored by separation on polyacrylamide gels and bands are commonly detected through fluorescent or silver staining. Often, small aliquots will first be run on agarose gels that are stained with ethidium bromide to check if reactions have been unsuccessful. This saves the time and expense of running polyacrylamide gels. SSRs are not normally scored on agarose gels because of insufficient resolution.
13. More usual actually, SSR-PCR products are scored using high-resolution sequencing instruments (capillary electrophoresis), such as the ABI PRISM® 3100, Applied Biosystems® 3130 or 3130XL, or Applied Biosystems 3730 or 3730XL Genetic Analyzer, which via the supplied programs (software) allow much more accurate sizing of SSR alleles.

References

1. Dreisigacker S, Zhang P, Warburton ML et al (2004) SSR and pedigree analyses of genetic diversity among CIMMYT wheat lines targeted to different megaenvironments. Crop Sci 44:381–388
2. McCouch SR, Chen X, Panaud O et al (2004) Microsatellite marker development, mapping and applications in rice genetics and breeding. Plant Mol Biol 35:89–99
3. Ruiz C, Breto MP, Asíns MJ (2004) A quick methodology to identify sexual seedlings in citrus breeding programs using SSR markers. Euphytica 112:89–94
4. Song QJ, Marek LF, Shoemaker RC et al (2004) A new integrated genetic linkage map of the soybean. Theor Appl Genet 109: 122–128
5. N'Diaye A, Van de Weg WE, Kodde LP et al (2008) Construction of an integrated consensus map of the apple genome based on four mapping populations. Tree Genet Genomes 4:727–743
6. Hasnaoui N, Buonamici A, Sebastiani F, Mars M, Zhang D, Vendramin GG (2012) Molecular genetic diversity of Punica granatum L. (pomegranate) as revealed by microsatellite DNA markers (SSR). Gene 493:105–112
7. Rafalsky JA, Tingey SV (1993) Genetic diagnostics in plant breeding: RAPDs, microsatellites and machines. Trends Genet 9:275–279

8. Powell W, Gordon MC, Provan J (1996) Polymorphism revealed by simple sequence repeats. (Reviews). Trends Plant Sci 1: 215–222
9. Ogliari JB, Boscariol RL, Camargo LEA (2000) Optimization of PCR amplification of maize microsatellite loci. Genet Mol Biol 23: 395–398
10. Doğrar N, Aakkaya MS (2001) Optimization of PCR amplification of wheat simple sequence repeat DNA markers. Turk J Biol 25:153–158
11. Mogali SC, Basavaraj M, Krishna Naik L, Nadaf HL (2011) Optimization of PCR amplification of wheat simple sequence repeat DNA markers. Karnataka J Agric Sci 24:239–240
12. Li M, Lü XL, Luo CD, Zhang F, Wu ZX, Zhong JY (2009) Optimizing system of SSR-PCR in Pinus radiate and Pinus tabulaeformis. J Mol Genet 1:44–49
13. Kramer MF, Coen DM (2004) Enzymatic amplification of DNA by PCR: standard procedures and optimization. In: Ausubel FM, Brent R, Kingston RE, Moore DD, Seidman JG, Smith JA, Struhl K (ed) Current protocols in molecular biology, vol 2. Wiley, New York, pp 1–15
14. Saiki RK (1992) The design and optimization of the PCR. In: Erlich HA (ed) PCR technology: principles and applications for DNA amplification. Oxford University Press, New York, pp 7–8
15. Rahman MH, Jaquish B, Khasa PD (2000) Optimization of PCR protocol in microsatellite analysis with silver and SYBR stains. Plant Mol Biol Reporter 18:339–348
16. Ramsay L, Macaulay M, degli Ivanissevich S et al (2000) A simple sequence repeat-based linkage map of barley. Genetics 156: 1997–2005
17. Niens M, Spijker GT, Diepstra A, te Meerman GJ (2005) A factorial experiment for optimizing the PCR conditions in routine genotyping. Biotechnol Appl Biochem 42:157–162
18. Masi O, Spagnoletti-Zeuli PL, Donini P (2003) Development and analysis of multiplex microsatellite marker sets in common bean (Phaseolus vulgaris L.). Mol Breed 11: 303–313
19. Zhang LS, Becquet V, Li SH, Zhang D (2003) Optimization of multiplex PCR and multiplex gel electrophoresis in sunflower SSR analysis using infrared fluorescence and tailed primers. Acta Bot Sin 45:1312–1318
20. Bencina M (2002) Optimization of multiple PCR using a combination of full factorial design and three dimensional simplex optimization methods. Biotechnol Lett 24:489–495
21. Ahmed I, Islam M, Mannan A, Naeem R, Mirza B (2009) Optimization of conditions for assessment of genetic diversity in barley (Hordeum vulgare L.) using microsatellite markers. Barley Genet Newslett 39:5–12

Chapter 11

Development of a Multiplex PCR Assay for Characterization of Embryonic Stem Cells

Rajarshi Pal, Murali Krishna Mamidi, Anjan Kumar Das, Mahendra Rao, and Ramesh Bhonde

Abstract

Several molecular methods like real-time PCR (Q-PCR), expression sequence tag (EST) scan, microarray and microRNA analysis, and massively parallel signature sequencing (MPSS) have proved to be increasingly sensitive and efficient for monitoring human embryonic stem cell (hESC) differentiation. However, most of these high-throughput tests have a limited use due to high cost, extended turnaround time, and the involvement of highly specialized technical expertise. Hence, there is a need of rapid, cost-effective, robust, yet sensitive method for routine screening of hESCs. A critical requirement in hESC cultures is to maintain a uniform undifferentiated state and to determine their differentiation capacity by showing the expression of germ-layer-specific gene markers. To determine the modulation of gene expression in hESCs during propagation, expansion, and differentiation via embryoid body (EB) formation, we developed a simple, rapid, inexpensive, and definitive multimarker, semiquantitative multiplex RT-PCR (mxPCR) platform technology. Among the 15 gene primers tested, 4 were pluripotent markers comprising of set 1; and 3 lineage-specific markers from each ecto-, meso-, and endoderm layers were combined as sets 2, 3, and 4, respectively. In summary, this study was performed to characterize hESCs on a molecular level and to determine the quality and degree of variability among hESC and their early progenies (EB). This single-reaction mxPCR assay was flexible and, by selecting appropriate reporter genes, can be designed for characterization of different hESC lines during routine maintenance and directed differentiation.

Key words Human embryonic stem cells, Embryoid body, Multiplex PCR, Molecular characterization, Pluripotency, Differentiation

1 Introduction

Human embryonic stem cells (hESCs) are derived from the inner cell mass (ICM) of blastocysts and possess the capacity of extensive undifferentiated proliferation in vitro (1, 2). In addition to their spontaneous differentiation ability, they can be guided to embrace specific signaling pathways leading to the formation of specialized cell types (3). Besides their importance in basic research, it is well established that hESC derivatives hold enormous promise

Stella K. Kantartzi (ed.), *Microsatellites: Methods and Protocols*, Methods in Molecular Biology, vol. 1006, DOI 10.1007/978-1-62703-389-3_11,

in regenerative medicine (4) and drug screening (5). Conventionally, undifferentiated hESCs have been grown on feeder layers derived from mouse embryonic fibroblasts (MEF), human embryonic or foreskin fibroblasts (HEF/HFF) (1, 2, 6), and feeder-free culture systems (7). Nevertheless, growth and expansion of hESC depends on the time and method (enzymatic, manual, laser assisted) of passaging the cells, media composition, and other culture conditions and even varies across different cell lines (3). Despite standardized methods, spontaneous differentiation in hESC cultures is very common, rendering their maintenance and expansion technically challenging. Differentiation in hESC can be identified based on changes in the morphology and distribution of cells/colonies accompanied by downregulation of pluripotent stem cell-specific markers accompanied by upregulation of markers associated with the differentiated phenotypes (8). Differentiated cells are often diffused within or toward the edge of the colonies, and hence it is difficult to detect them only by inspection of morphological features or by staining techniques.

1.1 Need for Alternate Methods Toward Simple and Cost-Effective Characterization of hESC

An in-depth characterization scheme of hESC, including RT-PCR, immunochemistry, karyotype, human leukocyte antigen (HLA) and short tandem repeat (STR) analyses, telomerase assay, teratoma formation in severe combined immunodeficient (SCID) mice, and focused cDNA microarray, miRNA, and mitochondrial DNA analysis (9, 10) has been proposed and validated earlier. Several tedious, lengthy, and expensive assays can be performed to demonstrate uniformity among hESC lines (11–13). One of the most common and effective methods of characterizing pluripotent stem cells is through reverse transcriptase (RT) PCR employing novel stage-specific genes that distinguish between undifferentiated hESC and their differentiated counterpart, the embryoid bodies (EB) (14, 15). Furthermore, the number and nature of genes that can be considered as reliable molecular markers for undifferentiated hESC is relatively concise and also well documented (16). Differentiated progenies, on the other hand, can be identified by a number of lineage- and tissue-specific gene markers.

Chamberlain et al. (17) had demonstrated that PCR could simultaneously amplify multiple loci in the human dystrophin gene. Since then, mxPCR has been widely employed as a routine technique for pathogen identification, gender screening, linkage analysis, forensic studies, template quantitation, and genetic disease diagnosis. Likewise we illustrate an approach, based on single-reaction multiplex RT-PCR, for semiquantitative evaluation of hESC during ex vivo expansion. By comparing the relative mRNA levels of a set of carefully selected 15 markers in the hESC and EB samples, we could successfully discriminate between undifferentiated hESCs and their differentiated derivatives. Hence, the combination of

RT-PCR and related hESC-based technologies may provide a useful tool for setting a standard for hESC characterization in a cost-effective manner.

1.2 How Can Multiplex PCR Become a Suitable Screening Tool?

Unequivocal readouts from gene expression analysis of hESC lines at different stages of their development, irrespective of the origin and culture conditions, is a critical piece of information required at regular passages. In the present study, the multiplex PCR was developed primarily as a potential screening tool for molecular characterization of hESCs. It may be the endpoint of analysis or preliminary to further analyses such as sequencing, hybridization, or real-time PCR depending on the objective and importance of the study. This is great consequence owing to the recent shift in paradigm from hESC to induced pluripotent stem cells (iPSC).

This mxPCR assay may emerge as a promising tool in determining spontaneous differentiation during routine maintenance of hESC and iPSC. This method permits a clear distinction between undifferentiated and differentiated cells displaying differential gene expression. It may also facilitate assessment of a contaminating population of undifferentiated cells in hESC-derived differentiated phenotypes during preclinical or clinical studies. Hence, this assay can be employed as a reliable quality test for monitoring the purity and authenticity of specialized cells in regenerative medicine and drug-screening applications. It is quick, accurate, and sensitive, and unlike other advanced molecular methods, it is affordable, especially in countries with limited economic resources and highly skilled expertise.

1.3 Specific Advantages and Limitations of This Method

1.3.1 Indication of Template Quality

The quality of the template may be determined more effectively in multiplex than in uniplex PCR. Degraded templates give weaker signals for long bands than for short (18). A loss in amplification efficiency due to PCR inhibitors in the template samples can be indicated by reduced amplification of an abundant control sequence in addition to the amplification of rarer target sequences in an otherwise standardized reaction.

1.3.2 Indication of Template Quantity

The majority of multiplex quantitation assays compare the signal intensity of a reference sequence to the signal from another sequence in the same reaction, either directly or by extrapolating the result to standard curves. The exponential amplification and internal standards of multiplex PCR can be used to assess the amount of a particular template in a sample. To quantitate templates accurately by multiplex PCR, the amount of reference template, the number of reaction cycles, and the minimum inhibition of the theoretical doubling of product for each cycle must be calculated (19).

1.3.3 Internal Controls

Potential problems in uniplex PCR include false-negatives due to reaction failure or false-positives due to contamination. False-negatives are often revealed in multiplex amplification because each amplicon provides an internal control for the other amplified fragments. For example, multiple exons may be amplified in assays that survey for gene deletion. Unless the entire region scanned by the multiplex PCR is deleted, amplification of some fragment (s) indicates that the reaction has not failed. Complete PCR failure can be distinguished from an informative no-amplification result by adding a control amplicon external to the target sequence to the reaction (20, 21). In addition to monitoring PCR failure and artifacts, internal control amplicons can be designed to verify the presence of target template.

1.3.4 Use as a Molecular Weight Ladder

Multiplex PCR products can be of used as a molecular weight ladder (standard DNA marker) because the lengths of the amplified fragments are known. Compared with conventional methods like Touchdown PCR combined with hot-start PCR for producing DNA marker, multiplex PCR method could reduce costs and improve production in laboratory and industry scale.

1.3.5 Efficiency

The expense of reagents and preparation time is less in multiplex PCR than in systems where several tubes of uniplex PCR reactions are used. A multiplex reaction is ideal for conserving costly enzymes (polymerase) and templates in short supply. For maximum efficiency of preparation time, the reactions can be prepared in bulk, randomly tested for quality, and stored frozen without enzyme or template until use.

1.3.6 Limitations

An aspect of PCR that may be exacerbated in multiplex is competition for resources and resulting artifacts. Differences in the yields of unequally amplified fragments are enhanced with each cycle (22). Sets of amplicons of varying lengths but similar sequence may show preferential amplification of the shortest, particularly if they share a common primer sequence. This may be due to limited processivity or suppressed amplification of the outer, longer amplicon by the inner, shorter one when primers anneal on the same strand (23). This effect can be evaded by initiating PCR with the long amplicon primers and by adding the shorter primer after certain cycles (24) or by using a low concentration of the short amplicon primer (23).

Further suppressed amplification of one amplicon by another has been noted in a multiplex in which sequence and primers were not shared (25), but co-amplification was resolved by initiating the limited amplicon several cycles before the other. Primer-template mismatches have been noted to be at a disadvantage relative to perfect matches in multiplex, presumably due to competition for binding to the polymerase (26). Multiple sets of primers increase the possibility of primer complementarity at the 3′ ends, leading to "primer dimers." These artifacts deplete the reaction of dNTPs and primers and outcompete the multiplex amplicons for polymerase (24, 27). This effect can be reduced by titrating primer concentrations and cycling conditions.

2 Materials

2.1 Equipments, Consumables, and Cell Lines

1. Hood for cell culture with vertical laminar flow and equipped with UV light for decontamination (PC-2 certified).
2. CO_2 incubator with temperature display.
3. Inverted microscope with phase-contrast equipment.
4. Stereomicroscope.
5. Tabletop centrifuge.
6. Water bath with temperature control.
7. Magnetic orbital stirrer.
8. Sterile Teflon-coated magnetic beads.
9. Hemocytometer and replacement cover slips.
10. ND-1000 spectrophotometer (NanoDrop Technologies).
11. PCR thermal cycler.
12. Way microtube racks.
13. Centrifuge tubes 15 ml.
14. Centrifuge tubes 50 ml.
15. Pipettes 25 ml.
16. Pipettes 10 ml.
17. Pipettes 5 ml.
18. Pipettes 2 ml.
19. Aspiration pipettes.
20. Tissue culture dishes 35 mm.
21. Tissue culture dishes 60 mm.
22. Tissue culture dishes 100 mm.
23. Tissue culture flasks T25.
24. Tissue culture flasks T75.
25. 6-welled tissue culture plates.
26. 250 ml vacuum filter units.
27. 500 ml vacuum filter units.
28. 1.8 ml cryovials/cryotubes.
29. Cell scraper—large, PE blade—sterile.
30. 1.5 ml clear microtubes.
31. 0.1–10 μl tips—extra long.

32. 200 μl tips.
33. 1,000 μl tips.
34. 100–1,000 μl variable pipette.
35. Sterile 1,000 μl pipette tips.
36. Syringes 50 ml.
37. Syringes 30 ml.
38. Syringes 20 ml.
39. Syringes 10 ml.
40. Pasteur pipettes.
41. Syringe-driven filter units.
42. Cell strainer with 100 μM nylon meshes.
43. Membrane filter—0.22 μm.
44. hESCs lines HUES-7 and HUES-9 were obtained as a generous gift from Harvard University Stem Cell Institute, Harvard University (Prof Douglas Melton).
45. Human teratocarcinoma cell line, NTERA-2 from ATCC.
46. In-house derived mouse embryonic fibroblasts (MEFs).

2.2 Culturing Human ES Cells

1. The hES culture medium used in this study consists of 80 % DMEM/F-12, 20 % ES-tested fetal bovine serum (HyClone), 1 % nonessential amino acid solution, 1 mM glutamine, 0.1 % β-mercaptoethanol, and 2 ng/ml human basic fibroblast growth factor (bFGF) (Sigma).
2. EB media: Same as hESC media but without (−) bFGF.
3. MEF media containing 10 % fetal bovine serum, 89 % Dulbecco's modified Eagle's medium (DMEM)-high glucose, 1 mM L-glutamine, 1 % nonessential amino acids, and 0.1 mM β-mercaptoethanol.
4. Trypsin 0.25 % (1×) with EDTA·4Na.
5. TrypLE™ Express (Life Technologies).
6. Dulbecco's Phosphate-Buffered Saline—with Ca, Mg.
7. Dulbecco's Phosphate-Buffered Saline—without Ca, Mg.
8. Gelatin type-A porcine.
9. Mitomycin C (Sigma).
10. Dimethyl sulfoxide (DMSO).
11. 0.4 % (W/V) trypan blue in 1× PBS.

2.3 RNA Isolation and cDNA Synthesis

1. TRIzol reagent.
2. Chloroform, isopropanol, and ethanol.
3. Diethylpyrocarbonate-treated water (DEPC water).
4. Oligo dT.

5. dNTP mix.
6. RNaseOUT.
7. SuperScript II Reverse Transcriptase (Life Technologies).

2.4 PCR and Primers/Oligos

1. 0.2 ml PCR tubes.
2. AB gene 2× master mix (Thermo Scientific).
3. Respective forward and reverse primers (Table 1; Fig. 1).
4. Agarose.
5. Ethidium bromide (EtBr).
6. 100 bp ladder.
7. 5× DNA-loading buffer orange.

3 Methods

3.1 Culture and Propagation of Human ES Cell Lines

1. The cryopreserved MEF feeder cells (P1) were thawed and grown till confluence.
2. MEF cells were inactivated with 10 μg/ml mitomycin C for two and half hours at 37 °C and 5 % CO_2 incubator as per standard protocol.
3. The inactivated cells were then plated on 0.2 % gelatin-coated 35 mm tissue culture dishes in MEF media. These feeder plates were used for growing hESC culture from third day of plating.
4. HUES-7 and HUES-9 cell lines were cultured routinely on mitomycin C-inactivated MEF feeder layers.
5. Manual passaging was preferred over the enzymatic method to guarantee the best quality of hESCs for downstream characterization.
6. Manual passaging was performed by mechanical dissociation of undifferentiated hESC colonies into small clumps of about 100–200 cells using the sharp edge of a flame-pulled Pasteur pipette under the stereomicroscope.
7. The undifferentiated hESC colonies that were identified by morphological features, including large compacted cells with a higher nucleus-to-cytoplasm ratio, and shiny borders were selectively picked.
8. During every passage, utmost caution was adopted to ensure exclusion of spontaneously differentiated portions of the hESC colonies demarcated by their loosened distribution of relatively darkened cells lacking shiny borders and prominent nucleoli.
9. Media were replenished every day, and passaging was done on the fourth or fifth day in culture.

Table 1
Represents the list of gene primers divided into four sets along with the forward and reverse sequences, annealing conditions, and region of amplification by mxPCR technique

Gene symbol	Primer sequences	T_m (°C)	Primer conc. (nm)	Product size	NCBI accession ID
Set 1: pluripotent/self-renewal					
GAPDH	5′-TGAAGGTCGGAGTCAACGGATT-3′ 5′-CATGTGGGCCATGAGGTCCACCAC-3′	57	2	983	NM_002046.3
Oct-4	5′-CGACCATCTGCCGCTTTGAG-3′ 5′-CCCCCTGTCCCCCATTCCTA-3′	57`	3	572	NM_203289.3
TDGF-1	5′-GCCCGCTTCTCTTACAGTGTGATT-3′ 5′-AGTACGTGCAGACGGTGGTAGTTCT-3′	57	3	497	NM_003212.1
Sox-2	5′-CCCCCGGCGGCAATAGCA-3′ 5′-TCGGCGCCGGGGAGATACAT-3′	57	8	447	NM_003106.2
NANOG	5′-TCCTCCATGGATCTGCTTATTCA-3′ 5′-CAGGTCTTCACCTGTTTGTAGCTGAG-3′	57	3	259	NM_024865.2
Set 2: ectoderm					
GAPDH	5′-TGAAGGTCGGAGTCAACGGATT-3′ 5′-CATGTGGGCCATGAGGTCCACCAC-3′	58	2	983	NM_002046.3
Neurofilament	5′-ACGCTGAGGAATGGTTCAAG-3′ 5′-GCCTCAATGGTTTCC-3′	58	5	555	NM_006158.3
Sox-2	5′-CCCCCGGCGGCAATAGCA-3′ 5′-TCGGCGCCGGGGAGATACAT-3′	58	8	447	NM_003106.2
Nestin	5′-CAGCGTTGGAACAGAGGTTGG-3′ 5′-TGGCACAGGTGTCTCAAGGGTAG-3′	58	7	388	NM_006617.1
Beta-III tubulin	5′-AACAGCACGGCCATCCAGG-3′ 5′-CTTGGGGCCCTGGGCCTCCGA-3′	58	3	242	NM_006086.2

Set 3: mesoderm					
GAPDH	5′-TGAAGGTCGGAGTCAACGGATT-3′ 5′-CATGTGGGCCATGAGGTCCACCAC-3′	59	2	983	NM_002046.3
Cardiac actin	5′-TCTATGAGGGCTACGCTTTG-3′ 5′-CCTGACTGGAAGGTAGATGG-3′	59	3	668	NM_005159.4
MEF-2	5′-GATGCGGACGATTCCGTAGG-3′ 5′-TGGTGCCTGCACCAGACGTG-3′	59	4	327	NM_002397.3
GATA-2	5′-AGCCGGCACCTGTTGTGCAA-3′ 5′-TGACTTCTCCTGCATGCACT-3′	59	4	243	NM_032638.3
hTERT	5′-AGCTATGCCCGGACCTCCAT-3′ 5′-GCCTGCAGCAGGAGGATCTT-3′	59	3	184	NM_198253.2
Set 4: endoderm					
GAPDH	5′-TGAAGGTCGGAGTCAACGGATT-3′ 5′-CATGTGGGCCATGAGGTCCACCAC-3′	58	2	983	NM_002046.3
AFP	5′-AGAACCTGTCACAAGCTGTG-3′ 5′-GACAGCAAGCTGAGGATGTC-3′	58	10	675	NM_001134.1
BMP-4	5′-GTCCTGCTAGGAGGCGCGAG-3′ 5′-GTTCTCCAGATGTTCTTCG-3′	58	3	338	NM_130851.2
HNF-3 beta	5′-GACAAGTGAGAGtAGCAAGTG-3′ 5′-ACAGTAGTGGAAACCGGAG-3′	58	5	234	NM_153675.1

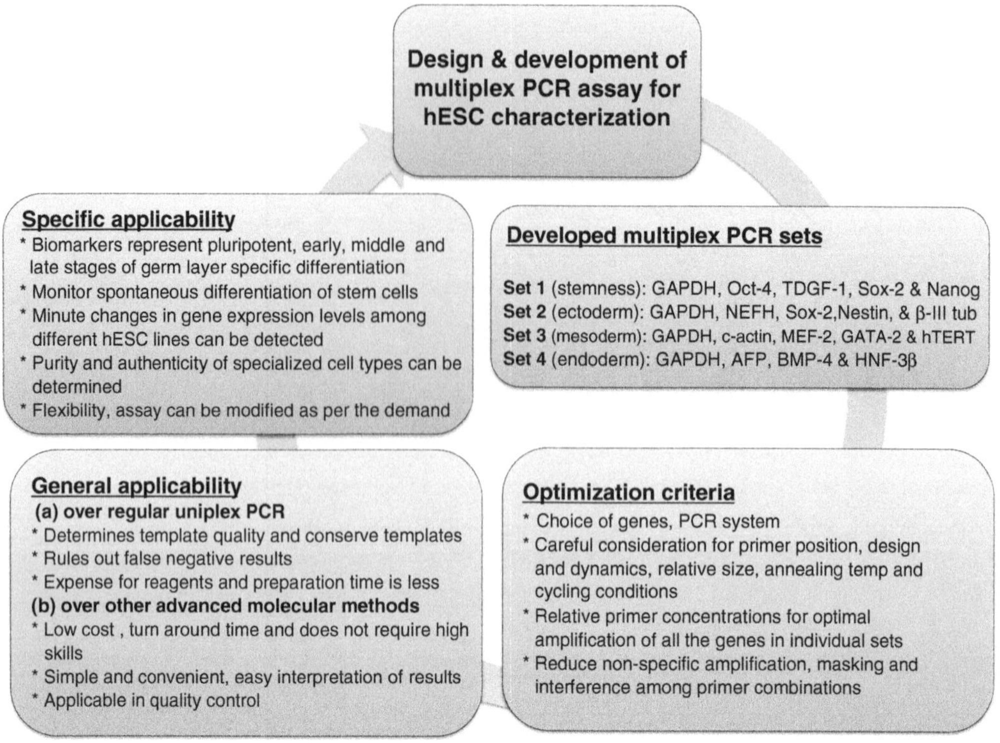

Design & development of multiplex PCR assay for hESC characterization

Developed multiplex PCR sets

Set 1 (stemness): GAPDH, Oct-4, TDGF-1, Sox-2 & Nanog
Set 2 (ectoderm): GAPDH, NEFH, Sox-2,Nestin, & β-III tub
Set 3 (mesoderm): GAPDH, c-actin, MEF-2, GATA-2 & hTERT
Set 4 (endoderm): GAPDH, AFP, BMP-4 & HNF-3β

Optimization criteria
* Choice of genes, PCR system
* Careful consideration for primer position, design and dynamics, relative size, annealing temp and cycling conditions
* Relative primer concentrations for optimal amplification of all the genes in individual sets
* Reduce non-specific amplification, masking and interference among primer combinations

General applicability
(a) over regular uniplex PCR
* Determines template quality and conserve templates
* Rules out false negative results
* Expense for reagents and preparation time is less
(b) over other advanced molecular methods
* Low cost , turn around time and does not require high skills
* Simple and convenient, easy interpretation of results
* Applicable in quality control

Specific applicability
* Biomarkers represent pluripotent, early, middle and late stages of germ layer specific differentiation
* Monitor spontaneous differentiation of stem cells
* Minute changes in gene expression levels among different hESC lines can be detected
* Purity and authenticity of specialized cell types can be determined
* Flexibility, assay can be modified as per the demand

Fig. 1 Schematic representation elucidating the important aspects toward the development of the mxPCR assay for characterization of hESC. Focus has been given on the general applicability of mxPCR along with its applicability in routine screening of hESC. The optimization criteria and combination of gene markers in the four sets are also mentioned

10. The cultures were maintained at 37 °C and 5 % CO_2 in air.
11. The same MEF media was used for culturing the NTERA-2 cell line.

3.2 Differentiation Induction by EB Formation

1. The hES colonies were manually cut into small clumps of approximately 50–100 cells/clump.
2. These aggregates were plated onto bacteriological plates (nonadherent) in EB formation media.
3. Undifferentiated hESCs spontaneously form EBs in suspension starting from day 2 at 37 °C and 5 % CO_2 incubator, indicating the onset of differentiation leading to the formation of three germ layers.
4. Media was replaced every alternate day until the EBs had grown in size and maturity for up to 10–14 days.

3.3 Total RNA Extraction and cDNA Synthesis

1. Test samples included hESC lines HUES-7 and HUES-9, NTERA-2, and MEF.
2. Cells were harvested and pellets were collected; total RNA was isolated by the TRIzol method following the manufacturer's protocol.

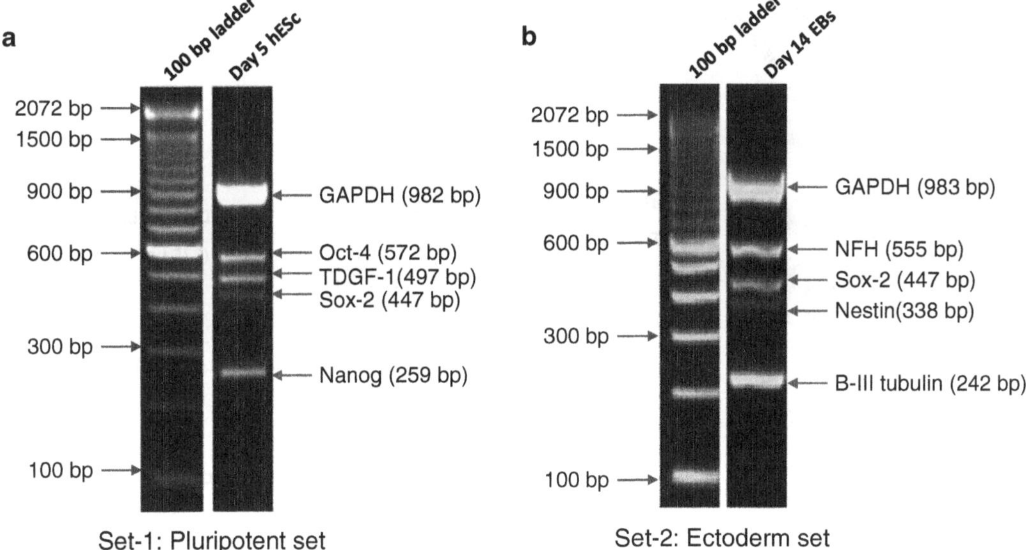

Fig. 2 Standardization of pluripotent and ectoderm sets, wherein each set constitutes a group of prevalidated gene primers representing different stages of hESC differentiation. The name of the gene markers mentioned below follows the order in which they appear in the gel picture from *top* to *bottom*. Images (**a**) represent 2 % agarose gel pictures of set 1 (Oct-4, NANOG, TDGF-1, and Sox-2). Images (**b**) represent 2 % gel pictures of set 2 (nestin, β-III tubulin, NFH, and Sox-2). For each mxPCR set, GAPDH was used as an internal control. 100 bp ladder was used as molecular marker

3. After RNA estimation, 1 μg of RNA treated with RNaseOUT ribonuclease inhibitor was used for cDNA synthesis.
4. Broadly, 40–50 medium-sized hESC colonies yielded approximately 3–5 μg of total RNA.
5. Reverse transcription using SuperScript II Reverse Transcriptase and Oligo dT to prime the reaction was carried out in 20 μl of reaction mix.

3.4 Standardizing Independent Sets of Multiplex PCR and in Combinations

1. Several permutation and combinations of primers and their concentrations were evaluated for simultaneous amplification of target sequences (see Note 4.1).
2. After repeated attempts, we have successfully developed the mxPCR technique for 4 separate sets of markers (Table 1; Fig. 1) using minimal sample volume (28).
3. Among them, set 1 constitutes of pluripotent ESC markers Oct-4, NANOG, Sox-2, and Rex1, and the rest of the 3 sets comprise lineage-specific markers associated with the formation of 3 germ layers (Table 1; Figs. 2 and 3). The housekeeping gene GAPDH was used as an internal control to show constitutive expression in all samples (see Note 4.1).
4. GAPDH, when compared to the target gene primers, has different amplification kinetics, and these differences were circumvented by optimizing the primer concentrations (see Note 4).

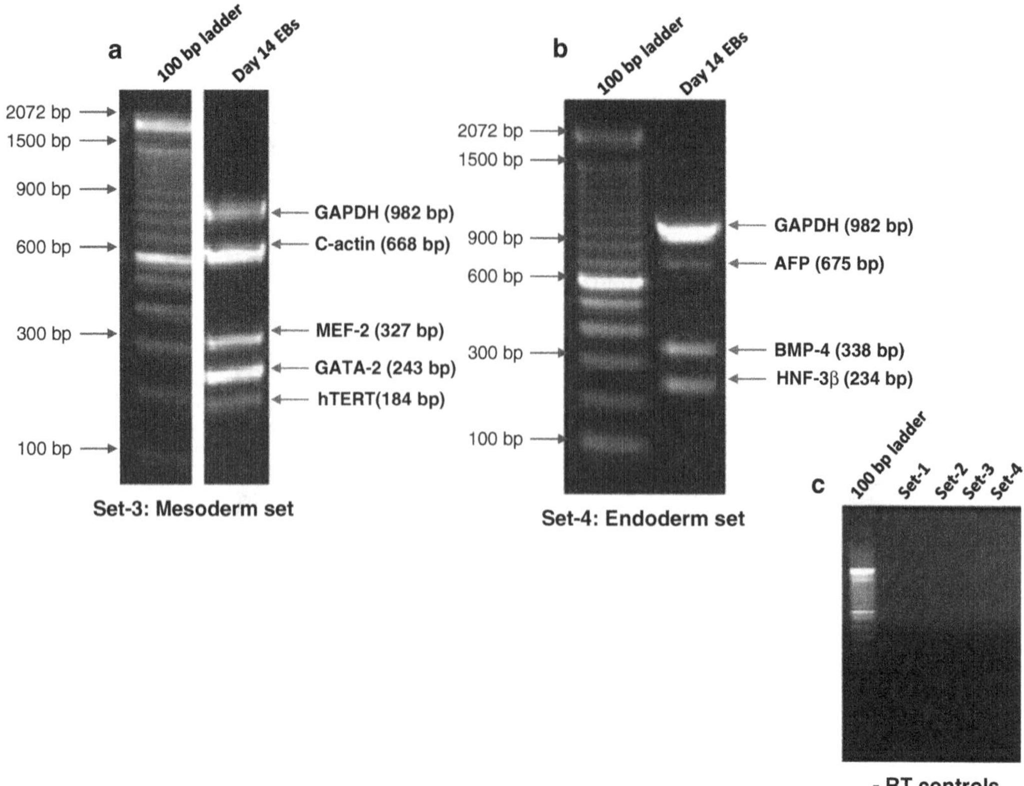

Fig. 3 Standardization of mesoderm and endoderm sets. Images (**a**) represent 2 % agarose gel pictures of set 3 consisting of three mesoderm markers and one pluripotent marker such as MEF-2, c-actin, GATA-2, and hTERT. Whereas figure (**b**) represents the 2 % agarose gel picture of set 4 comprising of three endoderm gene markers, namely, AFP, BMP-4, and HNF-3β. GAPDH was used as an internal control. (**c**) 2 % agarose gel showing RT controls for all four sets of mxPCR sets highlighting the specificity of the gene primers

5. To determine the sensitivity of this assay, we first set up the mxPCR with HUES-7 and then performed the same tests using HUES-9.
6. High concordance in the gene expression profile and pattern was observed, which indicates simultaneous amplification of more than one pluripotent and differentiation gene marker in a single-reaction tube without any interference between primers.
7. As a second criterion, we performed mxPCR (sets 1–4) using the cDNA from MEF and demonstrated that none of the markers was expressed except GAPDH.
8. Therefore, the primers selected for this study are specific for humans except GAPDH (which showed about 60–70 % homology with mouse).

3.5 Validation of Multiplex PCR

1. Initially we standardized the assay with HUES-7 and subsequently reproduced similar results with HUES-9.

2. This mxPCR assay was then validated using NTERA-2, an independent cell line which shows similar gene expression pattern.
3. It is well known that HUES-9 and NTERA-2 can readily form neuroectodermal lineage (3), whereas HUES-7 tends to form mesoderm and endoderm lineage more easily.
4. Likewise, majority of the gene markers associated with endoderm formation in set 4 were downregulated in day 14 HUES-9 and NTERA-2 compared to HUES-7.
5. Therefore we anticipated that this mxPCR system would be able to detect even such small changes in mRNA transcript levels which may facilitate understanding differential gene regulation among cell lines (see Notes 4.2, 4.3) including other molecular biology as well as microbiological applications (see Note 4.4).

4 Notes

4.1 Precautions to Be Taken

4.1.1 Positioning of Primers

Complete sequence information about the primer sites are important to eliminate nonspecific amplification that might occur at sites with similar sequences (18) and at mismatched primer-template sites (31). Primers for exon-amplifying multiplexes are ideally placed in intronic sequences adjacent to the exons. This provides some margin for adjustment of fragment length or amplification quality as well as possible information about alterations affecting splice sites. Fragment sizes should be selected carefully so that they may be separated easily from each other, when multiplex PCR product is to be resolved electrophoretically. At the same time, the range of band sizes should not be so wide that all fragments cannot be resolved well on the same gel. However, with the use of fluorescent-labeled primers, product ranges may overlap and yet be distinguished by color (20, 30, 33, 34). Fluorescent-labeled multiplex primers aid diagnostics by representing product amounts more accurately than EtBr stain and also reduce reaction time and nonspecific amplification with the less number of PCR cycles in order to obtain sufficient amplification signal (18).

4.1.2 Developing Primers and Reaction Conditions

Equimolar primer concentrations may not yield uniform amplification signals for all fragments. In such cases, the concentration of some primer pairs can be reduced in relation to others. This is particularly important in samples where one target is more abundant than others (37). When all primer pairs are not compatible, it may be necessary to subgroup them in smaller multiplexes. Primer sequences should be designed so that their predicted hybridization kinetics is similar to those of other primers in the multiplex reaction. Forty to sixty percent of G/C content and a length of

23–28 nucleotides are suggested as general guidelines for specific annealing at moderate temperatures (35). Primer annealing temperatures and concentrations may be calculated to some extent, but conditions will almost certainly have to be refined empirically in multiplex. Conditions for each set of primers should be developed individually and modified if necessary as primer sets are added (Table 1; Fig. 1). Primer pairs that work separately but not when combined may be improved by a prior ethanol precipitation in 0.3 M sodium acetate (36). The possibility of nonspecific priming and other artifacts is increased with each additional primer. Thus, primer pairs that produce "clean" signal alone but produce artifact bands in multiplex may benefit from "hot-start" PCR (31), addition of organics, annealing at the highest possible temperature, reselection of the primer sequence (18).

4.1.3 Titration of Reaction Components

It may be necessary to adjust concentrations of various reaction components to achieve a robust multiplex PCR. Magnesium and dNTP requirements generally increase with the number of amplicons in the multiplex, but the concentrations must be optimized because each primer pair may have different requirements (18). Likewise, polymerase requirements generally increase with the size of the multiplex. Buffer systems may affect amplification dramatically. DMSO was found to be a beneficial ingredient (22, 38) or an inhibitor (31) in different multiplex systems. Other additives that minimize nonspecific binding in multiplex PCR are Tween 20 and Triton X-100 (6), β-mercaptoethanol (1, 15, 41), and tetramethylammonium chloride (22).

4.1.4 Selection of Multiplex Loci

The regions selected for multiplex amplification may be determined by the nature of the analysis. The type of analysis to be done may lend itself to a PCR technique that was originally applied to uniplex amplification but that can be adapted for multiplex purposes.

4.1.5 Adapting Thermocycling Conditions

Thermocycling parameters are also determined largely by the sequence of the primer sets. Generally, extension times should be increased with the number of loci amplified in the reaction (18). However, nonspecific amplification may occur because of long extension and annealing times.

4.1.6 Competition and Interference

Another crucial aspect of PCR that may be aggravated in multiplex reactions is competition for resources and resulting artifacts. Differences in the yields of unequally amplified fragments are enhanced with each cycle (27). Sets of amplicons of varying lengths but similar sequence may show preferential amplification of the shortest, particularly if they share a common primer. This can be avoided by initiating PCR with the long amplicon primers and by adding shorter amplicon primers after some cycles (42) or by using a low concentration of the short amplicon primer (32). Suppressed amplification of one amplicon by another has been noted in a multiplex in which sequence and primers were not shared (7), but co-amplification was resolved by initiating the limited amplicon several cycles before the other. Primer-template mismatches have been noted to be at a disadvantage relative to perfect matches in multiplex, presumably due to competition for binding to the polymerase (31). Also multiple sets of primers increase the possibility of primer complementarity at the 3′ ends, leading to "primer dimers." These artifacts deplete the reaction of dNTPs and primers and compete for polymerase in multiplex reaction (38, 42). Such effects can be reduced by titrating primer concentrations and cycling conditions.

4.1.7 Post-PCR Analysis

Many of the techniques used for product analysis of uniplex PCR can be applied directly to multiplex PCR. However in some systems of multiplex products, more extensive analysis is required than gel electrophoresis. Additionally, the complexity of some multiplex reactions make verification of specific PCR. A second multiplex reaction can be generated by using the product of the first as a template when high specificity is required (31). Alternatively, the second reaction may be based on the results of the first. The product of a multiplex PCR may be sequenced to reveal new mutations or small alterations where major deletions are not present. This may be done directly from the multiplex reaction product (43, 47), or the product may require further preparation prior to sequencing. The introduction of biotinylated and universal-tailed primers in nested PCR following multiplex allows solid-phase sequencing of exons and flanking intronic sequence for small alterations (46).Other multiplex reactions have been subcloned prior to sequencing in the development phase (32, 48).

4.2 Detection Sensitivity Compared to Other Similar Methods

Detection sensitivity of mxPCR assay is much higher when compared to semiquantitative RT-PCR and may hence emerge as a promising tool in determining spontaneous differentiation during routine maintenance of hESC and iPSC. This method permits a clear distinction between undifferentiated and differentiated cells through diverse gene regulation. It may also facilitate assessment of a contaminating population of undifferentiated cells in hESC-derived differentiated phenotypes during preclinical or clinical studies. Hence, this assay may be employed as a reliable test for monitoring the purity and authenticity of specialized cells in regenerative medicine and drug-screening applications. It is quick, accurate, and sensitive, and unlike other advanced molecular methods, it is affordable, especially in countries with limited economic resources but highly skilled expertise.

4.3 Problems Encountered: Primer Interface, Cross-Reactivity, and Cycling Conditions

1. We encountered several hurdles during the inclusion of various gene markers highly expressed in hESC such as Oct-4, nestin, β-III tubulin, GATA-2, AFP, and HNF-3β.
2. In the pluripotent and endoderm sets (sets 1 and 4), GAPDH levels were slightly lower in the undifferentiated hESCs as compared to EBs. We reasoned that because Oct-4 is the most prolific marker for pluripotent ESCs, despite using the same amount of RNA for both samples, Oct-4 emerged stronger to mask GAPDH expression at least to some extent. Moreover, we could not correct this even by cutting down the primer concentration of Oct-4 considerably.
3. It was also interesting to note that the specific primer combinations in set 3, such as c-actin, GATA-4, and HAND1 or c-actin, MEF-2, and HAND1, are not suitable, although the reason remains unclear.
4. In all the differentiation sets (sets 2–4), we successfully combined one pluripotent ESC marker, which led to multiple issues of primer interference and cross-reactivity.
5. Furthermore, we noticed that during cDNA synthesis, just before the addition of reverse transcriptase (RT) enzyme, incubating the RNA mix for 2–5 min at 42 °C improves the quality of cDNA. We speculated this may be due to the time provided for RNA stabilization, which subsequently enhanced the substrate-enzyme reaction kinetics.
6. Finally, we witnessed that the best amplification of our mxPCR products, for the majority of the primer sets, was between 30 and 35 cycles, which was in concurrence with the theory that the probability of nonspecific products is aggravated with an increase in the number of amplification cycles.

4.4 Other Molecular Biology Applications of Multiplex PCR

4.4.1 Gene Deletion and Mutation Detection

Traditionally multiplex PCR technique is employed for the detection of X-linked human diseases such as Lesch-Nyhan syndrome (29), Fabry disease (30), Duchenne/Becker muscular dystrophy (DMD/BMD) (31), and others. Mutations and small deletions in genes are detected by multiplex assays either directly by PCR or by subsequent analysis of PCR products. Several mutation types may be examined simultaneously, as a multiplex reaction can detect a point mutation, a 4-base deletion, and complete deletion of the α-globin genes (32). Other mutation-amplifying multiplexes rely on post-PCR manipulation of the reaction product for diagnosis. SSCP detects human p53 tumor suppressor gene mutations associated with breast cancer (33).

Genotyping by multiplex PCR employs similar techniques. ABO blood group alleles are distinguished by allele-specific primers (34) or by enzymatic digestion of amplified product (35). HLA-DR4 variants, associated with autoimmune diseases, are typed by multiplex ARMS (36).

Multiplex PCR of sequence tagged sites has aided the physical mapping of breakpoints and loci on chromosome 16 using somatic cell hybrids (37) and of the X chromosome in deletion patients (38).

4.4.2 Polymorphic Repetitive DNA

Repetitive DNA polymorphisms are multiplexed for mapping, disease linkage, gender determination, and DNA typing/identification. STRs of 1–6 bp are convenient for multiplexing because they are numerous, highly polymorphic (39), and may be co-amplified without overlapping size ranges (40). Multiplexes of relatively close repeats are employed for disease linkage, but chromosomally unlinked repeats are used for the identification of individuals (41). Multiplex PCR is an ideal technique for DNA typing because the probability of identical alleles in two individuals decreases with the number of polymorphic loci examined. Reactions have been developed with potential applications in paternity testing, forensic identification, and population genetics (40–42).

4.4.3 Microbe Detection and Characterization

PCR analysis of bacteria is advantageous, as the culturing of some pathogens has been a lengthy process or impossible. Bacterial multiplexes indicate a particular pathogen among others or distinguish species or strains of the same genus. An amplicon of sequence conserved among several groups is often included in the reaction to indicate the presence of phylogenetically or epidemiologically similar, or environmentally associated, bacteria and to signal a functioning PCR.

Multiplex assays with this format distinguish species of Legionella (25), Escherichia coli, Shigella (43), major groups of Chlamydia (44), Mycobacterium (45), and Salmonella (46) from other genus members or associated bacteria. An assay for Mycobacterium leprae co-amplifies human and pathogen DNA (47). Viral DNA is amplified by multiplex PCR to screen tissue samples or to examine associations of infection with disease. A fragment from the host genomic DNA is generally co-amplified in these assays (26, 27, 48). Human papilloma virus (HPV) associations with carcinomas or lesions (49, 50) and adenovirus12 with celiac disease (51) have been examined. Multiplex assays detect or screen for HPV (27), human immunodeficiency virus type 1 (HIV-1) and human T-cell leukemic viruses (48), human T-lymphotropic virus types I and II (52), hepatitis B virus (23), parvovirus B19 (53), and hog cholera viruses (54). HIV-1 infection can be detected by nested multiplexes of conserved regions (26).

Acknowledgment

The authors are grateful to Stempeutics Research Malaysia for facilitating our research on stem cells and regenerative medicine.

References

1. Thomson J, Itskovitz-Eldor J, Shapiro SS, Waknitz MA, Swiergiel J, Marshall VS, Marshall VS, Jones JM (1998) Embryonic stem cell lines derived from human blastocysts. Science 282:1145–1147
2. Reubinoff BE, Pera MF, Fong CY, Trounson A, Bongso A (2000) Embryonic stem cell lines from human blastocysts: somatic differentiation in vitro. Nat Biotech 18:399–404
3. Pal R, Totey S, Krishna M, Bhat VS, Totey SM (2009) Distinct propensity of human embryonic stem cell during early stage of lineage specification controls their terminal differentiation into mature cell types. Exp Biol Med 234:1230–1243
4. Pera MF, Reubinoff B, Trounson A (2000) Human embryonic stem cells. J Cell Sci 113:5–10
5. Davila JC, Cezar GG, Thiede M, Strom S, Miki T, Trosko J (2004) Use and application of stem cells in toxicology. Toxicol Sci 79:214–223
6. Richards M, Tan S, Fong CY, Biswas A, Chan WK, Bongso A (2003) Comparative evaluation of various human feeders for prolonged undifferentiated growth of human embryonic stem cells. Stem Cells 21:546–556
7. Xu C, Inokuma MS, Denham J, Golds K, Kundu P, Gold JD, Carpenter MK (2001) Feeder-free growth of undifferentiated human embryonic stem cells. Nat Biotech 19:971–974
8. Draper JS, Fox V (2003) Human embryonic stem cells: multilineage differentiation and mechanisms of self-renewal. Arch Med Res 34:558–564
9. Mandal A, Tipnis S, Pal R, Ravindran G, Bose B, Patki A, Rao MS, Khanna A (2006) Characterization and in vitro differentiation potential of a new human embryonic stem cell line, ReliCell®hES1. Differentiation 74:1–10
10. Pal R, Mandal A, Rao HS, Rao MS, Khanna A (2007) A panel of tests to standardize the characterization of human embryonic stem cells. Regen Med 2:179–192
11. Carpenter MK, Rosler E, Rao MS (2003) Characterization and differentiation of human embryonic stem cells. Cloning Stem Cells 5:79–88
12. Bhattacharya B, Miura T, Brandenberger R, Mejido J, Luo Y, Yang AX, Joshi BH, Ginis I, Thies RS, Amit M, Lyons I, Condie BG, Itskovitz-Eldor J, Rao MS, Puri RK (2004) Gene expression in human embryonic stem cell lines: unique molecular signature. Blood 103:2956–2964
13. Josephson R, Sykes G, Liu Y, Ording C, Xu W, Zeng X, Shin S, Loring J, Maitra A, Rao MS, Auerbach JM (2006) A molecular scheme for improved characterization of human embryonic stem cell lines. BMC Biol 4:28
14. Maitra A, Arking DE, Shivapurkar N, Ikeda M, Stastny V, Kassauei K, Sui G, Cutler DJ, Liu Y, Brimble SN, Noaksson K, Hyllner J, Schulz TC, Zeng X, Freed WJ, Crook J, Abraham S, Colman A, Sartipy P, Matsui S, Carpenter M, Gazdar AF, Rao M, Chakravarti A (2005) Genomic alterations in cultured human embryonic stem cells. Nat Genet 37:1099–1103
15. Cai J, Chen J, Liu Y, Miura T, Luo Y, Loring JF, Freed WJ, Rao MS, Zeng X (2006) Assessing self-renewal and differentiation in hESC lines. Stem Cells 3:516–530
16. Bhattacharya B, Cai J, Luo Y, Miura T, Mejido J, Brimble SN, Zeng X, Schulz TC, Rao MS, Puri RK (2005) Comparison of the gene expression profile of undifferentiated human embryonic stem cell lines and differentiating embryoid bodies. BMC Dev Biol 5:22
17. Chamberlain JS, Gibbs RA, Ranier JE, Nguyen PN, Caskey CT (1988) Deletion screening of the Duchenne muscular dystrophy locus via multiplex DNA amplification. Nucleic Acids Res 16:11141–11156
18. Chamberlain JS et al (1992) Diagnostic of Duchenne and Becker muscular dystrophies by polymerase chain reaction: a multicenter study. J Am Med Assoc 267:2609–2615
19. Ferre F (1992) Quantitative or semi-quantitative PCR: reality vs. myth. PCR Methods Appl 2:1–9
20. Ballabio A, Ranier JE, Chamberlain JS, Zollo M, Caskey CT (1990) Screening for steroid sulfatase (STS) gene deletions by multiplex DNA amplification. Hum Genet 84:571–573
21. Levinson G, Fields RA, Harton GL, Palmer FT, Maddelena A, Fugger EF, Schulman JD (1992) Reliable gender screening for human preimplantation embryos, using multiple DNA target-sequences. Hum Reprod 7:1304–1313
22. Fettle RM, Schwartz MJ, Robertson NH, Vaudin S, Super M, Malone G, Little S (1992) Development, multiplexing, and application of ARMS tests for common mutations in the CFTR gene. Am J Hum Genet 51:251–262
23. Repp R, Rhiel S, Heermann KH, Schaefer S, Keller C, Ndumbe P, Lambert F, Gerlich WH (1993) Genotyping by multiplex polymerase chain reaction for detection of endemic hepatitis B virus transmission. J Clin Microbiol 31:1095–1102

24. Bourque SN, Vatero JR, Mercier J, Lavoie MC, Lavesque RC (1993) Multiple polymerase chain reaction for detection and differentiation of the microbial insecticide *Bacillus thuringiensis*. Appl Environ Microbiol 59:523–527
25. Bej AK, Mahbubani MH, Miller R, DiCesare JL, Haft L, Atlas RM (1990) Multiplex PCR amplification and immobilized capture probes for detection of bacterial pathogens and indicators in water. Mol Cell Probes 4:353–365
26. Zazzi M, Romano L, Brasini A, Valensin PE (1993) Simultaneous amplification of multiple HIV-1 DNA sequences from clinical specimens by using nested-primer polymerase chain reaction. AIDS Res Hum Retroviruses 9: 315–320
27. Vandenvelde C, Verstraete M, Van Beers D (1990) Fast multiplex polymerase chain reaction on boiled clinical samples for rapid viral diagnosis. J Virol Methods 30:215–227
28. Mamidi MK, Pal R, Bhonde R, Zakaria Z, Totey S (2010) Application of multiplex PCR for characterization of human embryonic stem cells (hESCs) and its differentiated progenies. J Biomol Screen 15:630–643
29. Gibbs RA, Nguyen PN, Edwards A, Civitello AB, Caskey CT (1990) Multiple DNA deletion detection and exon sequencing of the hypoxanthine phosphoribosyltransferase gene in Lesch-Nyhan families. Genomics 7: 235–244
30. Kornreich R, Desnick RJ (1993) Fabry disease: detection of gene rearrangements in the human alpha-galactosidase A gene by multiplex PCR amplification. Hum Mutat 2:108–111
31. Chamberlain JS, Gibbs RA, Ranier JE, Nguyen PN, Caskey CT (1989) Multiple PCR for the diagnosis of Duchenne muscular dystrophy. In: Gelfand DH, Innis MA, Sninsky JJ, White TJ (eds) PCR protocols, a guide to methods and applications. Academic Press, San Diego, CA, pp 272–281
32. Chehab FF, Kan YW (1989) Detection of specific DNA sequences by fluorescence amplification: a color complementation assay. Proc Natl Acad Sci 86:9178–9182
33. Runnebaum IB, Nagarajan M, Bowman M, Soto D, Sukumar S (1991) Mutations in p53 as potential markers for human breast cancer. Proc Natl Acad Sci U S A 88:10657–10661
34. Uggozoli L, Wallace B (1992) Application of an allele-specific polymerase chain reaction to the direct determination of ABO blood group genotypes. Genomics 12:670–674
35. O'Keefe DS, Dobrovic A (1993) A rapid and reliable method for genotyping the ABO blood group. Hum Mutat 2:67–70
36. Jawaheer D, Oilier WE, Thomson W (1993) Multiple ARMS-RFLP: a simple and rapid method of HLA-DR4 subtyping. Eur Immunogen 20:175–187
37. Richards IR, Holman K, Lane S, Sutherland GR, Callen DF (1991) Human chromosome 16 physical map: mapping of somatic cell hybrids using multiplex PCR deletion analysis of sequence tagged sites. Genomics 10: 1047–1052
38. Worley KC, Towbin JA, Zhu XM, Barker DF, Ballabio A, Chamberlain J, Biesecker LG, Blethen SL, Brosnan P, Fox JE, Rizzo WB, Romeo G, Sakuragawa N, Seltzer WK, Yamaguchi S, McCabe ERB (1992) Identification of new markers in Xp21 between DXS28 (C7) and DMD. Genomics 13: 957–961
39. Beckmann S, Weber JL (1991) Survey of human and rat microsatellites. Genomics 12:627–631
40. Edwards A, Hammond HA, Jin L, Caskey CT, Chakroborty R (1992) Genetic variation at five trimeric and tetrameric tandem repeat loci in four human population groups. Genomics 12:241–253
41. Edwards A, Civitello A, Hammond HA, Caskey CT (1991) DNA typing and genetic mapping with trimeric and tetrameric tandem repeats. Am J Hum Genet 49:746–756
42. Klimpton CP, Gill P, Walton A, Urquhart A, Millican ES, Adams M (1993) Automated DNA profiling employing multiplex amplification of short tandem repeat loci. PCR Methods Appl 3:13–21
43. Bej AK, McCarty SC, Atlas RM (1991) Detection of coliform bacteria and *Escherichia coli* by multiplex polymerase chain reaction: comparison with defined substrate and plating methods for water quality monitoring. Appl Environ Microbiol 57:1473–1479
44. Kaltenboek B, Kansoulas KG, Storz J (1992) Two-step polymerase chain reactions and restriction endonuclease analyses detect and differentiate *ompA* DNA of the *Chlamydia* spp. J Clin Microbiol 30:1098–1104
45. Wilton S, Cousins D (1992) Detection and identification of multiple mycobacterial pathogens by DNA amplification in a single tube. PCR Methods Appl 1:269–273
46. Way JS, Josephson KL, Pillai SD, Abbaszadegan M, Gerba CP, Pepper IL (1993) Specific detection of *Salmonella* spp. by multiplex polymerase chain reaction. Appl Environ Microbiol 59:1473–1479
47. Vander Vliet GM, Hermans CJ, Klatser PR (1993) Simple colorimetric microtiter plate

hybridization assay for detection of amplified *Mycobacterium leprae* DNA. J Clin Microbiol 31:665–670

48. Sunzeri FJ, Lee T-H, Brownlee RG, Busch MP (1991) Rapid simultaneous detection of multiple retroviral DNA sequences using the polymerase chain reaction and capillary DNA chromatography. Blood 77:879–886
49. Soler C, Allibe P, Chardonnet Y, Cros P, Matrand B, Thivolet J (1991) Detection of human papilloma virus types 6, 11, 16, and 18 in mucosal and cutaneous lesions by the multiplex polymerase chain reaction. Virol Methods 3S:143–157
50. Toh Y, Kuwano H, Tanaka S, Baba K, Matsuda H, Sugimachi K, Mori R (1992) Detection of human papillomavirus DNA in esophageal carcinoma in Japan by polymerase chain reaction. Cancer 70:2234–2238
51. Vesy CJ, Greenson JK, Papp AC, Snyder PJ, Qualman SJ, Prior TW (1993) Evaluation of celiac disease biopsies for adenovirus 12 DNA using a multiplex polymerase chain reaction. Mod Pathol 6:61–64
52. Wattel E, Mariotti M, Agis F, Gordien E, Prou O, Courouce AM, Rouger P, Wain-Hobson S, Chen ISY, Lefrere JJ (1992) Human T lymphotropic virus (HTLV) type I and II DNA amplification in HTLV-I/II-seropositive blood donors of the French West Indies. J Infect Dis 165:369–372
53. Sevall JS (1990) Detection of parvovirus B19 by dot-blot and polymerase chain reaction. Mol Cell Probes 4:237–246
54. Wirz B, Traschin JD, Muller HK, Mitchell DB (1993) Detection of hog cholera virus and differentiation from other pestiviruses by polymerase chain reaction. J Clin Microbiol 31:1148–1154

Chapter 12

Agarose Gel Electrophoresis and Polyacrylamide Gel Electrophoresis for Visualization of Simple Sequence Repeats

James Anderson, Drew Wright, and Khalid Meksem

Abstract

In the modern age of genetic research there is a constant search for ways to improve the efficiency of plant selection. The most recent technology that can result in a highly efficient means of selection and still be done at a low cost is through plant selection directed by simple sequence repeats (SSRs or microsatellites). The molecular markers are used to select for certain desirable plant traits without relying on ambiguous phenotypic data. The best way to detect these is the use of gel electrophoresis. Gel electrophoresis is a common technique in laboratory settings which is used to separate deoxyribonucleic acid (DNA) and ribonucleic acid (RNA) by size. Loading DNA and RNA onto gels allows for visualization of the size of fragments through the separation of DNA and RNA fragments. This is achieved through the use of the charge in the particles. As the fragments separate, they form into distinct bands at set sizes. We describe the ability to visualize SSRs on slab gels of agarose and polyacrylamide gel electrophoresis.

Key words Deoxyribonucleic acid, Ribonucleic acid, Agarose, Polyacrylamide, Simple sequence repeats, Gel electrophoresis

1 Introduction

Gel electrophoresis is a common technique that is used to visualize proteins and DNA. Gel electrophoresis has allowed for the visualization of DNA and RNA with the use of markers (1). This process is achieved by sorting of a sample by size and charge (2). A gel slab is prepared with either a specific concentration of agarose or a polyacrylamide. The gel forms a matrix through which the sample travels through (3). The matrix is a cross-linked polymer which contains and separates the sample (4). The higher the concentration the more tightly the matrixes' mesh is. The different concentrations will determine the length in kilobase pairs (KB) which are visualized through the slab. The lower percentage of agarose (0.7 %) will visualize a higher KB (5–10 KB), while a higher percent agarose (2 %)

Stella K. Kantartzi (ed.), *Microsatellites: Methods and Protocols*, Methods in Molecular Biology, vol. 1006,
DOI 10.1007/978-1-62703-389-3_12, © Springer Science+Business Media, LLC 2013

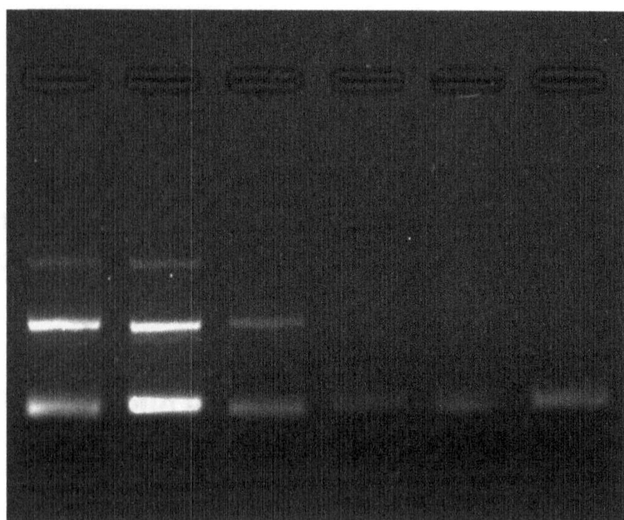

Fig. 1 Gel electrophoresis

will visualize a lower KB (0.2–1 KB) (5). The use of polyacrylamide is done for visualization of smaller KB pairs rather than using agarose at a higher percentage (6).

Ethidium bromide is added to the gel in order to increase the visibility of DNA in the gel. It was known that gel electrophoresis can be used to visualize virus DNA (7). Originally, ultracentrifuges were used to detect DNA quality (8). This method involved machinery that broke down often due to the constant high speed motion involved with the centrifuges. It was during one of these breakdowns that Piet Borst and Cees Aaij got the idea to use gel electrophoresis to try and visualize mitochondrial DNA. While gel electrophoresis would allow for easy visualization of hot DNA, one would not be able to visualize cold DNA (8). In order to visualize it, ethidium bromide was added to it, which allowed for consistent visualization (8).

One of the concerns about the use of ethidium bromide as a dye for gel electrophoresis is its toxicity. The chemical is a known carcinogen as well as a mutagenic chemical. Because of this, alternate methods for staining are being undertaken. The use of different dyes has been looked at. Dyes such as GelRed and GelGreen can be used instead of ethidium bromide for the staining process (9).

The visualization occurs due to both the buffer in the electrophoresis chamber and the power supply that is connected to it (10, 11). The buffers that are used are either Tris/Acetate/EDTA (TAE) or Tris/Borate/EDTA (TBE) (10). An electrical charge is conducted through the buffer and allows for the samples electrical charge to be attracted to the electrical current and moves towards the positive terminal based on their weight (1). The sample will then coalesce into bands at different lengths along the gel based on their size (see Fig. 1) (12). These bands can be identified by the use of ladder markers in order to determine the KB size of the band (8). The KB size of the ladder markers is known, which allows for the identification of KB sizes in the sample.

The main problem with the TAE and TBE buffers that are used in gel electrophoresis is the production of heat with the Tris-base in the gels (13). A method to alleviate this problem is the use of a sodium boric acid to act as a buffer. This buffer is cheaper and offers a larger voltage range and better resolution for high voltage than TBE and TAE buffers (13). There are issues with runaway positive feedback that may occur from this method, but that may be controlled and minimized.

This same process can be utilized to identify known traits through the use of molecular markers. Molecular markers are short bits of DNA that are associated with a specific section of a genome (14). These sections will allow for the selection of one or many traits in the interested organism (15). By loading the marker into the slab and running the gel, the presence of a section of the genome can be determined. This may be used to either identify a trait or determine genetic heritage (14, 15). SSR markers are used to identify the presence of a specific genomic section (16). SSR and other markers are loaded into a slab along with samples and the ensuing bands are analyzed to determine if the region of interest is present or not.

Polyacrylamide gel electrophoresis uses the same principle of agarose gel electrophoresis. It has better resolution than gel electrophoresis. The polyacrylamide is a less compact media that is generally used to visualize larger sequences such as proteins and bacteria due to its ability to allow larger sized sequences to travel through the matrix (17). The gel forms a series of channels in the gel which will allow materials to pass through it. When placed into an electrified buffer solution (such as a TBE buffer), a loaded sample will travel from one end to the other, with the smaller parts traveling at a faster rate.

A concern about polyacrylamide gel electrophoresis is the formation of unpolymerized acrylamide in the gel. Acrylamide is a neurotoxin, so it should be handled with care. It has also been shown to interact with proteins during electrophoresis (18). Because of this, special care should be taken whenever polyacrylamide is used.

An alternative to using polyacrylamide gel electrophoresis is the use of TreviGel 500. This method allows for the same superior visualization of proteins while allowing the ease of use that is used in agarose gel electrophoresis (19).

The most common method of staining polyacrylamide gels for visibility is the use of silver staining (20). This technique allows for the visualization of bands through imaging machines (see Fig. 4). One of the down sides of this technique is that it uses formaldehyde in the visualization process. An alternative to this method is one that uses glucose in alkaline borate buffer (21). This allows for visualization within the polyacrylamide gels without the use of harmful chemicals which may cause damage.

2 Materials

2.1 Agarose Gel Electrophoresis

Reagents

10× DNA loading buffer.

1× TBE.

Agarose stock solution for DNA gels.

Deionized water

Equipment

Gloves.

Safety glasses.

Laboratory wipes.

Ethanol, soap, and glass cleaner.

Siliconizing agent.

Pipette.

Beaker.

Microwave.

Combs (see Fig 2).

Imaging equipment (see Fig 3).

Gel plates (see Fig 5).

Electrophoresis apparatus (see Fig 6).

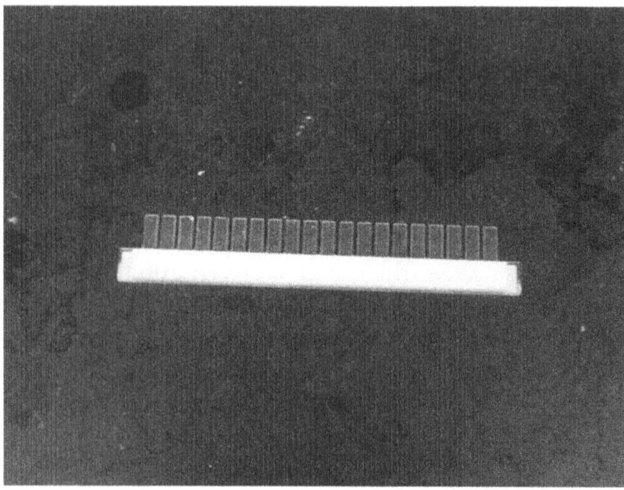

Fig. 2 Comb

Fig. 3 Ethidium bromide imaging system

2.2 Polyacrylamide Gel Electrophoresis

Reagents

Acrylamide/bisacrylamide stock solution for DNA gels.

Urea.

1× TBE electrophoresis buffer.

Deionized water.

Ammonium persulfate.

Tetramethylethylenediamine (TEMED) electrophoresis grade.

Equipment

Gloves.

Safety glasses.

Laboratory wipes.

Pipette.

Ethanol, soap, and glass cleaner.

Beaker.

Water bath.

Clips and spacers.

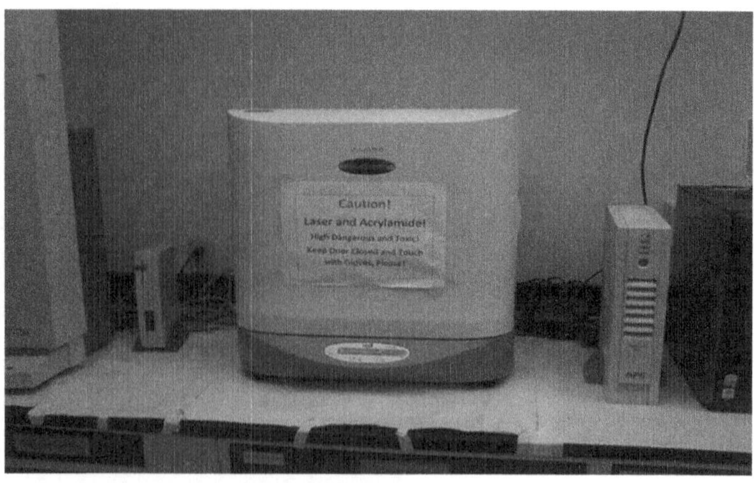

Fig. 4 Acrylamide imagining system

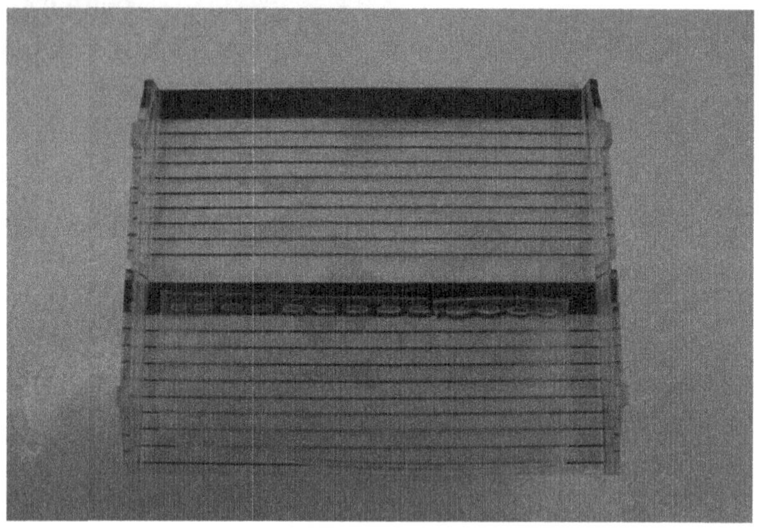

Fig. 5 Agarose gel plate

Squirt bottle.

Combs (see Fig 2).

Plastic wrap.

Siliconizing agent.

Stir bar.

Syringe.

Whatman paper.

Imaging equipment (see Fig 4).

Gel plates (see Fig 5).

Electrophoresis apparatus (see Fig 6).

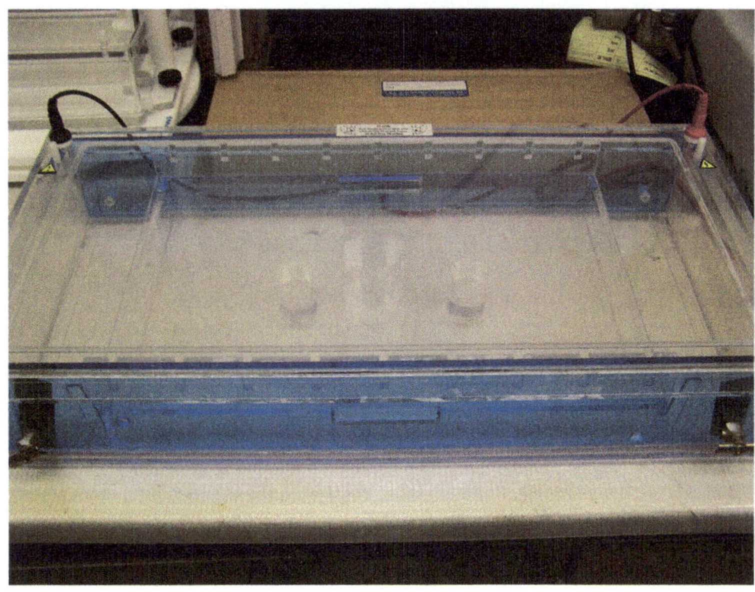

Fig. 6 Electrophoresis apparatus

Table 1
Agarose solution

% Agarose	Agarose (g)	1× TBE (ml)	Visualization size (bp)
0.5	0.5	100	1,000–30,000
0.7	0.7	100	800–12,000
1	1.0	100	500–10,000
1.2	1.2	100	400–7,000
1.5	1.5	100	200–3,000
2	2.0	100	50–2,000

3 Methods

3.1 Agarose Gel

1. Clean gel plates with soap and water cleaning away any particulate or dried acrylamide.
2. Treat the shorter of the plates with a siliconizing agent or a common commercial product such as Rainex. Apply one of these agents with a cloth or chem wipe as a thin even layer.
3. Rinse plates with water (making sure that agent on smaller plate makes water bead) and allow to air-dry before use.
4. Use Table 1 to create the desired gel solution based on the visualization desired.
5. Stir gel solution.

6. Heat in microwave till agarose is completely dissolved. Stop the microwave every 30 s and stir solution. Use caution when stirring. Wear protective gloves, goggles, and lab coat when stirring as solution may boil over and cause serious burns.
7. Cool the agarose to 60 °C.
8. 50 μg of ethidium bromide may be added to molten agarose to help visualization.
9. Prepare plates for gel solution.
   (a) Place spacer on inside edge of long plate.
   (b) Put second plate with agent side down on top of long plate.
   (c) Secure with binder clips.
   (d) Securely fit comb.
   (e) Tape edge to make sure that plates are secure.
10. Fill the assembled plates quickly before gel polymerizes.
    (a) Tip plate to allow solution to run down plate and fill until it gets to edge closest to you then lay horizontally.
    (b) Make sure that solution fills the whole area.
11. Make sure that there are no bubbles and clamp down comb securely between two plates.
12. Wait 30 min and make sure that the separations between the wells formed by the comb are formed correctly.
13. Wait another 30 min for complete polymerization.
14. Once the gel has polymerized remove all clips and items used to secure plates.
15. Once the plates are separated rinse the top of the gel.
16. Remove the comb and rinse comb area.
17. Attach the gel plates to the gel apparatus.
18. Add 1× TBE buffer.
19. Flush wells with 1× TBE.
20. Allow to pre-run for 60 min at constant Watts (80 W).
21. Turn power supply off.
22. Flush wells again with 1× TBE.
23. Remove lid and add SSR marker and samples.
24. Run gel based on size of samples.
25. Remove gel from apparatus and spacers.
26. Separate plates carefully using spatula so that gel is retained on smaller plate.
27. Visualize in a gel imaging system.

Table 2
Polyacrylamide gel solution

Gel %	Acrylamide (g)	Bisacrylamide (g)	Urea (g)	TBE 10× buffer (ml)	Deionized water (ml)	10 % APS (μl)	TEMED (μl)
6	5.7	0.3	48	10	40	500	50
8	7.6	0.4	48	10	40	500	50
10	9.5	0.5	48	10	40	500	50
12	11.4	0.6	48	10	40	500	50
16	15.2	0.8	48	10	40	500	50

3.2 Polyacrylamide Gel

1. Clean gel plates with soap and water cleaning away any particulate or dried acrylamide.
2. Treat the shorter of the plates with a siliconizing agent or a common commercial product such as Rainex. Apply one of these agents with a cloth or chem wipe as a thin even layer.
3. Rinse plates with water (making sure that agent on smaller plate makes water bead) and allow to air-dry before use.
4. Use Table 2 to create the desired gel solution.
5. Stir gel solution.
6. Heat briefly in a 60 °C water bath (do not heat for long because gel solution needs to stay at room temperature for the polymerization process).
7. Filter gel solution through three layers of Whatman No.1 paper.
8. Prepare plates for gel solution.
   (a) Place spacer on inside edge of long plate.
   (b) Put second plate with agent side down on top of long plate.
   (c) Secure with binder clips.
   (d) Securely fit comb.
   (e) Tape edge to make sure that plates are secure.
9. Add ammonium persulfate and TEMED to gel solution before filling a 60 ml syringe with it.
10. Fill the assembled plates quickly before gel polymerizes.
    (a) Tip plate to allow solution to run down plate and fill until it gets to edge closest to you then lay horizontally.
    (b) Make sure that solution fills the whole area.
11. Make sure that there are no bubbles and clamp down comb securely between two plates.

12. Wait 30 min and make sure that the separations between the wells formed by the comb are formed correctly.
13. Wait another 30 min for complete polymerization.
14. Once the gel has polymerized remove all clips and items used to secure plates.
15. Once the plates are separated rinse the top of the gel.
16. Remove the comb and rinse comb area.
17. Attach the gel plates to the gel apparatus.
18. Add TBE buffer.
19. Flush wells with TBE.
20. Allow to pre-run for 60 min at constant Watts (80 W).
21. Turn power supply off.
22. Flush wells again with TBE.
23. Remove lid and add SSR marker and samples.
24. Run gel based on size of samples.
25. Remove gel from apparatus and spacers.
26. Separate plates carefully using spatula so that gel is retained on smaller plate.
27. Now is the final step to subject the gel to whatever visualizing technique was selected such as silver staining or CYBR Green.

References

1. Stuber CW, Lincoln SE, Wolff DW, Helentjaris T, Lander ES (1992) Identification of genetic factors contributing to heterosis in a hybrid from two elite maize inbred lines using molecular markers. Genetics 132:823–839
2. Lehran H, Diamond D, Wozney JM, Boedtker H (1977) RNA molecular weight determinations by gel electrophoresis under denaturing conditions, a critical reexamination. Biochemistry 16:4743–4751
3. Eckerskorn C, Lottspeich F (1989) Internal amino acid sequence analysis of proteins separated by gel electrophoresis after tryptic digestioninpolyacrylamidematrix.Chromatographia 28:92–94
4. Viovy JL, Duke T (2005) DNA electrophoresis in polymer solutions: ogston sieving, reptation and constraint release. Electrophoresis 14: 322–329
5. Johnson PH, Grossman LI (1977) Electrophoresis of DNA in agarose gels. Optimizing separations of conformational isomers of double- and single-stranded DNAs. Biochemistry 16:4217–4225
6. Rio DC, Ares M Jr, Hannon GJ, Nilsen TW (2010) Polyacrylamide gel electrophoresis of RNA. Cold Spring Harb Protoc. doi:10.1101/pub.prot5444
7. Thorne HV (1966) Electrophoretic separation of polyoma virus DNA from host cell DNA. Virology 29:234–239
8. Borst P (2005) Ethidium DNA agarose gel electrophoresis: how it started. IUBMB Life 57(11):745–747
9. Schmidt F, Schmidt J, Riechers A, Haase S, Bosserhoff A, Heilmann J, Konig B (2010) DNA staining in agarose gels with ZN^{2}+-cyclen-pyrene. Nucleosides Nucleotides Nucleic Acids 29(10):748–759
10. Stellwagen NC, Gelfi C, Righetti PG (1998) The free solution mobility of DNA. Biopolymers 42:687–703
11. Aebersold PB, Winans GA, Teel DJ, Milner GB, Utter FM (1987) Manual for starch gel electrophoresis: a method for the detection of genetic variation. NOAA technical report NMFS 61
12. Tenover FC, Arbeit RD, Goering RV, Mickelson PA, Murray BE, Persing DH,

Swaminathan B (1995) Interpreting chromosomal DNA restriction patterns produced by pulsed-field gel electrophoresis: criteria for bacterial strain typing. J Clin Microbiol 33: 2233–2239

13. Brody J, Kern S (2004) Sodium boric acid: a Tris-free, cooler conductive medium for DNA electrophoresis. Biotechniques 36:214–216
14. Avise JC (1994) Molecular markers, natural history and evolution. Chapman & Hall, New York
15. Mohan M, Nair S, Bhagwat A, Krishna TG, Yano M, Bhatia CR, Sasaki T (1997) Genome mapping, molecular markers and marker-assisted selection in crop plants. Mol Breed 3:87–103
16. Han Y, Teng W, Yu K, Poysa V, Anderson T, Qiu L, Lightfoot DA, Li W (2008) Mapping QTL tolerance to phytophthora root rot in soybean using microsatellite and RAPD/SCAR derived markers. Euphytica 162:231–239
17. Johnson W, Silhavy T, Boos W (1975) Two-dimensional polyacrylamide gel electrophoresis of envelope proteins of *Escherichia coli*. Appl Microbiol 1975:405–413
18. Bonaventura C, Bonaventura J, Stevens R, Millington D (1994) Acrylamide in polyacrylamide gels can modify proteins during electrophoresis. Anal Biochem 222:44–48
19. Vanek P, Fabian S, Fisher C, Chirikjian J, Collier G (1995) Alternative to polyacrylamide gels improves the electrophoretic mobility shift assay. Biotechniques 18(4):704–706
20. Cong W, He H, Zhu Z, Ye C, Ysng X, Choi J, Jin L, Li X (2010) Improved conditions for silver–ammonia staining of DNA in polyacrylamide gel. Electrophoresis 31:1662–1665
21. He H, Cong W, Jiang C, Pu J, You W, Gao H, Zhu Z, Jin L, Li X (2010) A user-friendly alternative to formaldehyde-based DNA silver-staining method on polyacrylamide gels. Electrophoresis 31:2416–2421

Part III

Automated Capillary Sequencers

Chapter 13

Microsatellite Fragment Analysis Using the ABI PRISM® 377 DNA Sequencer

Mark A. Renshaw, Melissa Giresi, and J. Orville Adams

Abstract

The ABI PRISM® 377 DNA Sequencer is used for a variety of microsatellite-based research. The platform provides researchers with a cost-effective means for high-throughput genotyping, which can be further optimized by multiplexing microsatellite loci or by using a tail-labeling approach to screen large sets of markers. The goals of this chapter are to present a protocol for performing microsatellite-based analyses on the ABI 377 and to provide researchers with information on how to troubleshoot common issues associated with running the ABI 377 sequencers.

Key words Microsatellites, ABI 377 PRISM® DNA Sequencer, Polyacrylamide gels, Genotyping

1 Introduction

Introduced in 1995, the ABI 377 PRISM® DNA Sequencer was a noteworthy development in the progression towards whole genome sequencing (1, 2). While the slab-gel-based system has become somewhat antiquated, it is still used in an assortment of applications including DNA sequencing, amplified fragment length polymorphisms (AFLPs), and microsatellite (SSR) analysis (3–10). For microsatellite fragment analyses, the ABI 377 platform continues to provide researchers with a cost-effective way to obtain high-throughput data. A quick literature search yields a variety of current microsatellite-based applications, including parentage analysis (7), genetic linkage mapping (9), and population genetics studies (8, 10).

To generate samples for analysis on the ABI 377, microsatellite fragments are PCR amplified with locus-specific primers and DNA templates. One of the primers from each locus pair (either forward or reverse) is labeled on the 5′ end with a fluorescent dye, i.e., 6-FAM, HEX, or NED. Amplified products from multiple loci can be analyzed in a single lane on the gel if there are nonoverlapping allele size ranges for markers that utilize the same fluorescent label

Stella K. Kantartzi (ed.), *Microsatellites: Methods and Protocols*, Methods in Molecular Biology, vol. 1006, DOI 10.1007/978-1-62703-389-3_13, © Springer Science+Business Media, LLC 2013

(i.e., fluoresce at the same wavelengths) or utilize different fluorescent labels for markers with overlapping allele size ranges. These multiplex panels can be achieved either by combining products from separate PCR amplifications prior to loading the products onto a gel, by amplifying multiple microsatellites simultaneously in a single PCR mix (11, 12), or by employing some combination of these two methods.

For projects that screen a large number of markers, a tail-labeling approach (13, 14) can be employed to reduce the prohibitive cost associated with fluorescently labeled primers. These PCR amplifications incorporate three oligonucleotides: a fluorescently labeled tail primer, a forward primer (with tail sequence on the 5′ end), and a reverse primer. A reduced quantity of the forward primer is added to the PCR mix relative to the tail and reverse primers, for example, a 1:10 ratio (13) or a 1:15 ratio (14). During the first several PCR cycles, the microsatellite products result from the annealing of the forward primer. Subsequent PCR cycles replace the forward primer with the fluorescently labeled tail primer to generate products that fluoresce on the gel. The higher quantity of tail primer relative to the forward primer ensures the amplification of an adequate number of fluorescently labeled amplicons.

The ABI 377 utilizes a "flow-through" gel electrophoresis principle to analyze these fragments. Using traditional gel electrophoresis, voltage is applied across a gel for a fixed amount of time, moving the negatively charged amplicons through the gel. The amplicons are separated by size because smaller fragments move more quickly through the gel matrix than larger fragments. With the ABI 377, a polyacrylamide gel is placed vertically on the system and samples are introduced at the top of the gel. Voltage is applied across the gel, and amplicons migrate through the gel towards the bottom. As the amplicons are about to "fall off" the gel into the bottom buffer reservoir, an argon laser excites the fluorophores attached to the migrating DNA strands, and a CCD camera detects the amplicons. Since gel electrophoresis is ubiquitous in today's molecular biology laboratory, this system is very easy-to-use for most biologists. The cost of the reagents required to run this system is relatively inexpensive and the system is capable of running up to 96 samples simultaneously, which makes this system a cost-effective option for running a large number of samples. This chapter will focus on the use of ABI 377 Prism® DNA Sequencer for microsatellite fragment analyses.

2 Materials

Prepare all solutions using deionized water. Prepare and store all reagents at room temperature (unless otherwise specified). Follow all Federal, State, and institutional regulations when disposing of waste materials. Wear gloves when handling polyacrylamide gel components.

2.1 Polyacrylamide Gel Casting

1. Urea.
2. Long Ranger™ 50 % (VWR).
3. 10× TBE Buffer: Combine 108 g of Tris base, 55 g of boric acid, and 9.3 g of ethylenediaminetetraacetic acid (disodium salt dihydrate) in 800 ml of water. Stir until all components have gone into solution; solution should be clear and transparent. Add water to bring total volume to 1 l.
4. 10 % Ammonium Persulfate (APS): Dissolve a single 150 mg APS tablet in 1.5 ml of water. This solution should be made fresh daily.
5. TEMED.
6. 1× TBE Buffer: Combine 200 ml of 10× TBE buffer and 1,800 ml of water and mix.

2.2 Electrophoresis of PCR-Amplified Fragments

1. Size Standard Mix: Add 30 μl of GEL LOADING DYE and 40 μl of GS-400HD ROX SIZE STD (GeneScan® 400HD (ROX) Size Standard, Applied Biosystems) to 200 μl of formamide and mix. Store at −20 °C.
2. 96 Lane Loading Tray (The Gel Company).
3. 96 Lane Porous Membrane Comb (The Gel Company).
4. Wedge Plate Separators (VWR).

3 Methods

Carry out all procedures at room temperature. Wear gloves for all steps that involve handling polyacrylamide gel components. Protocols are outlined for a single pair of 36 cm plates and assume that matrix standards have been run and the appropriate matrix file created (see Note 1).

3.1 Polyacrylamide Gel Casting

1. Place one notched front plate and one hipped back plate on separate cassettes (Fig. 1a, b) with the etched serial numbers facing downward (see Note 2). Thoroughly wipe the upward facing sides of the plates with deionized water and Kimwipes to remove all lint, fingerprints, and other debris. Lightly wet two spacers (Fig. 1c) with deionized water (just enough to make them moist) and place the moistened spacers on the outside edges of the back plate (see Note 3). The straight edge of the spacer should be flush with the outside of the plate; the top and bottom of the spacer should be flush with the top and bottom of the plate. Place the notched front plate on top of the hipped back plate with etched side facing upwards, taking

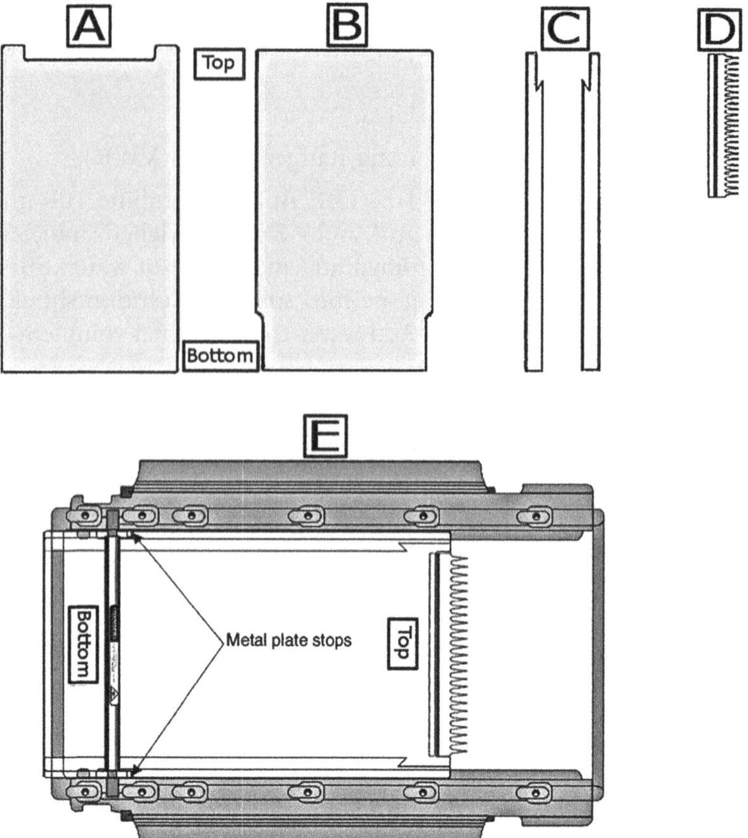

Fig. 1 (**a**) Notched front plate, (**b**) hipped back plate, (**c**) spacers, (**d**) shark's tooth comb, and (**e**) assembled gel in the run position on a cassette

care not to get fingerprints on the non-etched side of the plates. Plates should be flush with one another along all sides. Slide plates into the run position on the cassette (Fig. 1e) and turn the clamps to the locked position. Make sure that the movable arm on the cassette is free to move.

2. In a 50 ml beaker, combine 13 ml of water, 2.5 ml of 10× TBE buffer, 2.5 ml of Long Ranger™ 50 %, and 9 g urea. Mix on a stir plate until urea is dissolved (see Note 4). Filter solution through Whatman® #1 filter paper into a clean 50 ml beaker. Add 125 μl of 10 % APS and 17.5 μl of TEMED and swirl gently for 5–10 s.
3. Draw gel solution into a 20 ml syringe and dispense the solution between the plates from the top. As the solvent front moves towards the bottom of the plates, tap gently on the top plate along the leading edge of the solvent line to prevent bubbles from forming (see Note 5). Stop dispensing the gel solution once the entire area between the plates is full. Insert the straight

edge of a shark's tooth comb (Fig. 1d) between the top of the plates, starting at one corner and laying it down slowly to allow air bubbles to escape. Once the shark's tooth comb is fully inserted in between the plates, clamp the top of the plates (and comb) together with two 2″ binder clips.

4. After the gel has polymerized for 30 min, wrap it in 1× TBE buffer to prevent it from drying out. To do this, saturate paper towels with 1× TBE buffer, remove the binder clips, remove the gel from the cassette, and cover both the top and the bottom of the gel with the saturated paper towels. Wrap the top and bottom of the gel with plastic wrap (i.e., Foodservice Film or Saran Wrap), slide the gel into the run position on the cassette, turn all clamps to the locked position (making sure to fasten the movable cassette arm into the locked position), and refasten the two 2″ binder clips along the top of the gel. Wait an additional 90 min before using the gel, allowing a minimum of 2 h for polymerization.

3.2 Electrophoresis of PCR-Amplified Fragments

1. Unlock and remove the gel from the cassette, remove the binder clips, and discard the plastic wrap and paper towels. Remove the shark's tooth comb from the top of the gel, vacating a large well. Thoroughly rinse the outside of the plates with water, wiping off any dried buffer or gel. Using the teeth on the shark's tooth comb, remove pieces of loose gel lodged between the plates in the well, being careful not to poke into the straight bottom edge of the well. Rinse the well 2–3 times with 1× TBE buffer, holding the gel on its side and inserting a Kimwipe at the edge of the well to remove the buffer. If there is extraneous pieces of gel, use the shark's tooth comb to remove them and rinse the well with buffer again, repeating until all gel fragments have been removed. Place the gel on a clean cassette and dry off the outside of the plates using Kimwipes, paying careful attention to keep the camera scanning region at the bottom of the gel (under the moveable cassette arm) free of foreign material (i.e., dried buffer, pieces of dried gel, and fingerprints). Slide and lock the gel into the run position, clamping down the moveable cassette arm.
2. Open the door on the ABI 377, slide the bottom buffer chamber into position, and plug the buffer chamber electrode into the red receptor. Slide the cassette (with gel) onto the four corner clamps (Fig. 2). Apply pressure on each corner and lock into position by turning the corner clamps, starting in the bottom left corner and moving in a clockwise direction. Push top buffer chamber against the gel plates with the top of the chamber flush with the top of the gel plates. Turn clamps on the cassette to lock upper buffer chamber into position, and plug the buffer chamber electrode into the black receptor.

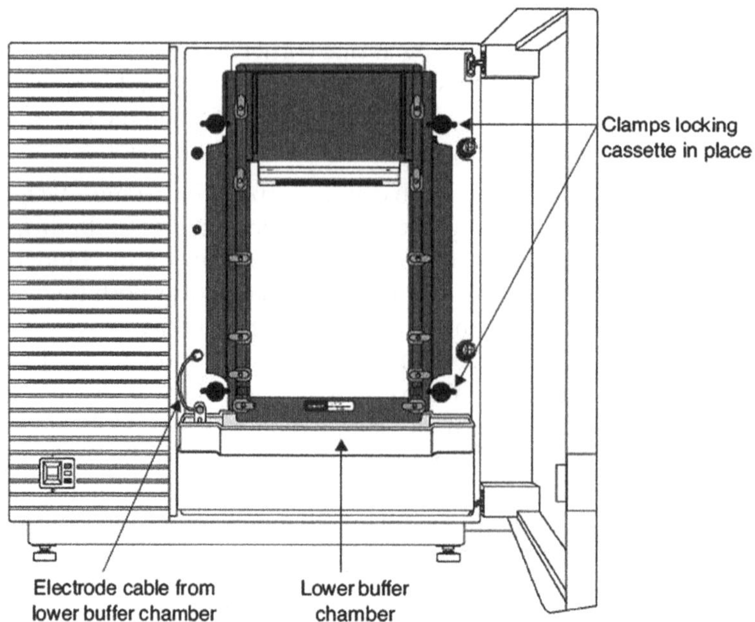

Fig. 2 Inside of the ABI 377 with the cassette mounted on the four corner clamps and inside the lower buffer chamber

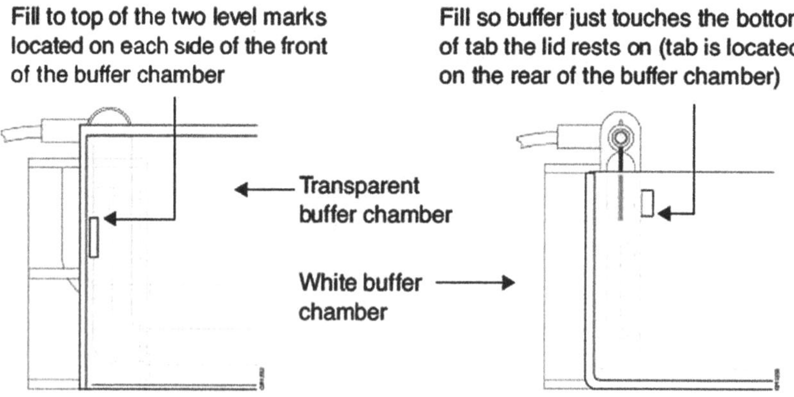

Fig. 3 Fill level for two different types of upper buffer chambers

Add 1× TBE buffer to both the lower buffer chamber (filling the larger space) and upper buffer chamber (Fig. 3). Place lid on the upper buffer chamber and close the door on the ABI 377.

3. Open the data collection software program on the computer (i.e., "ABI Prism™ 377-96 Collection"), select "New" from the File menu (upper left, Fig. 4), and "GeneScan™ Run." This will open the Run window (Fig. 4a) and create a new run folder ("Run Folder—Date Time") in the Runs folder. Once the Run window opens, select "GS PR 36D—2400" from the

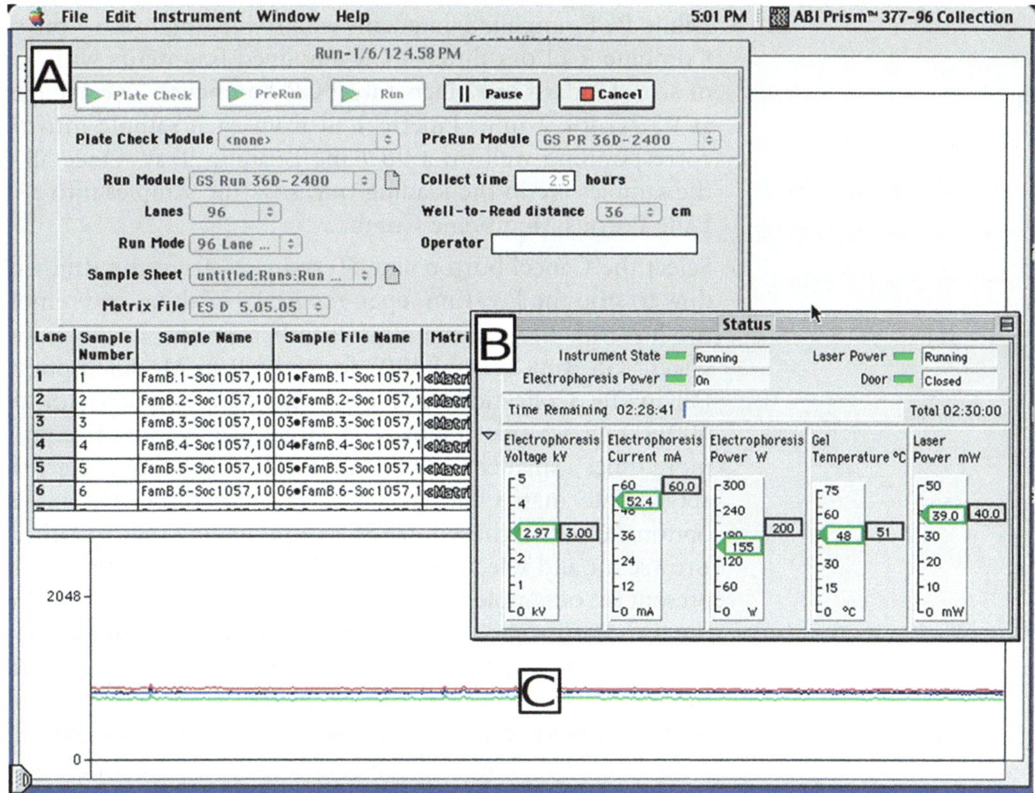

Fig. 4 Data collection software program with (**a**) the Run window, (**b**) the Status window, and (**c**) the Scan window

Pre-Run Module options. Click the Pre-Run button. Select "Status" from the Windows menu and watch the countdown until the electrophoresis turns on. If the countdown is interrupted, select the Cancel button and "Terminate" from the Run window, and then click on the Pre-Run button again. The Status window displays the gel temperature (Fig. 4b); the current temperature is marked by a green box, and the target temperature is marked by a black box. Once the gel temperature reaches 51 °C (20–30 min), the gel is ready for sample loading.

4. To prepare the sample sheet, select "New" and "GeneScan™ Sample" from the File menu. For the first sample under the "Pres" column, click on the fluorescent dye colors that need to be analyzed for the run; be sure to include the size standard dye (i.e., *R*ed (Rox)). Click the heading of the "Pres" column to highlight the entire column. Select "Fill Down" from the Edit menu, applying the settings to the entire sample sheet. Enter the information for each sample in the "Sample Name" column. Save the sample sheet in the newly created run folder (located in the Runs folder).

5. Dilute PCR-amplified fragments with sterile water (see Note 6). Combine 2 μl of diluted PCR-amplified fragments with 1 μl of size standard mix. Denature PCR fragments/size standard at 95 °C for 2 min. Pipette 1 μl from each sample into the corresponding well on a 96 Lane Loading Tray. Once all of the samples are in the loading tray, soak the samples into a 96 Lane Porous Membrane Comb.
6. Select the Cancel button and "Terminate" from the Run window to stop the Pre-Run, open the ABI 377 door, and remove the lid from the upper buffer chamber. In the Run window, select "GS Run 36D—2400" from the Run Module options. Fill in the Collection time box with "2.5" (hours) for the 400HD (Rox) size standard. A larger size standard requires more time; a smaller size standard requires less time. Select the appropriate matrix file (see Note 1) under the Matrix File options. Click on the Sample Sheet options and select "Other." Browse for and select the sample sheet that was created for the present set of samples, importing it into the Run window. Now the Run button should be available. Take the membrane comb with the samples, and carefully slide it into the well between the top of the front and back gel plates. Keep the membrane comb centered to prevent samples from running outside the camera scanning region. Do not slide the comb back and forth! Replace the lid on the upper buffer chamber, close the ABI 377 door, and click the Run button. Save the Gel File (default name) in the run folder that was just created for this run, keeping all files from the run (gel file, sample sheet, log file, and run file) together in the same folder. Immediately after the run begins, select "Status" from the Window menu and monitor the countdown. If the countdown is altered, select the Cancel button and "Terminate." Then, click on the Run button again. Once the countdown is complete, the electrophoresis should begin. After about 30 s, four lines (blue, black, green, and red) should show up in the Scan window (Fig. 4c). If no lines appear or only a single blue line appears, the run must be aborted and started again. Select "Gel Image" from the Window menu to see a real-time image of the run.
7. The run requires 2–2.5 h for the size standard to run through the gel. The run can be terminated before the full 2.5 h by clicking the Cancel button and "Terminate" at any time. While the run is in progress, prepare a second sample set (and corresponding sample sheet) in the same fashion as the first sample set and sheet. Store the sample set at 4 °C. The sample sheet can be saved in the Runs folder.
8. Once the first run is complete, restart the computer (see Note 7), open the ABI 377 collection program, select "New," "GeneScan™ Run" from the File menu. Transfer your second

sample sheet from the Runs folder into the newly created run folder, and import the second sample sheet into the Run window. Denature samples at 95 °C for 2 min, transfer 1 μl of each sample to the appropriate well in the loading tray, and soak samples into a membrane comb. Open the ABI 377 door, remove the lid from the top buffer chamber, and remove the comb from the first run. Carefully insert the comb for the second load into the well between the plates, keeping the comb centered. Close the ABI 377 door and click the Run button. Save the Gel File in the run folder with the corresponding sample sheet. Monitor the Status window to ensure the run starts and the scan shows up as expected. Select "Gel Image" from the Windows menu to see the real-time image of the run.

9. Once the second run is complete, restart the computer, open the ABI 377 door, and unplug the top and bottom buffer chambers. Unlock the four corner clamps, and carefully remove the cassette and upper buffer chamber as a single unit. Carry the unit to a sink where the buffer can be poured out of the upper chamber. Carefully carry the bottom buffer chamber to a sink where the buffer can be poured out. Rinse the buffer chambers and lid with water. Allow the buffer chambers to air-dry. Rinse the combs from both the first and second runs with water, removing any pieces of gel that may be attached. The combs can be air-dried and reused for future runs. Remove the gel plates from the cassette, and rinse the cassette with water. Pry the gel plates apart using a VWR. Press a paper towel onto the gel; pat down and peel the paper towel off of the plates; this should remove most of the gel from the plates. Thoroughly wash the plates with 1 % Alconox, being careful to remove any pieces of gel or dried buffer. Rinse plates with water, being careful to remove all Alconox. Store plates in a rack, allowing them to air-dry for future use.

3.3 Sizing and Scoring Microsatellite Fragments (See Note 8)

1. Open the Gel File in GeneScan®, select "Track Lanes" from the Gel menu (upper left, Fig. 5), and select "Auto-Track Lanes" (see Note 9). Select "Extract Lanes" from the Gel menu, producing the Analysis Control window (Fig. 5b). Choose a size standard for the first sample (see Note 10), click on the size standard column heading, and select "Fill Down" from the Edit menu to apply the size standard to all of the samples. Click on the header for each dye color that needs to be analyzed: *B*lue (6-FAM), *G*reen (HEX), *Y*ellow (NED), and *R*ed (ROX). This should fill the entire column for each color selected. Click the Analyze button, save the project, and close GeneScan®. This creates a folder ("Run Folder—Date Time"), with a file for each sample analyzed, in the same folder as the Gel File. Additional information regarding the GeneScan® software can be found in the User's Manual (15).

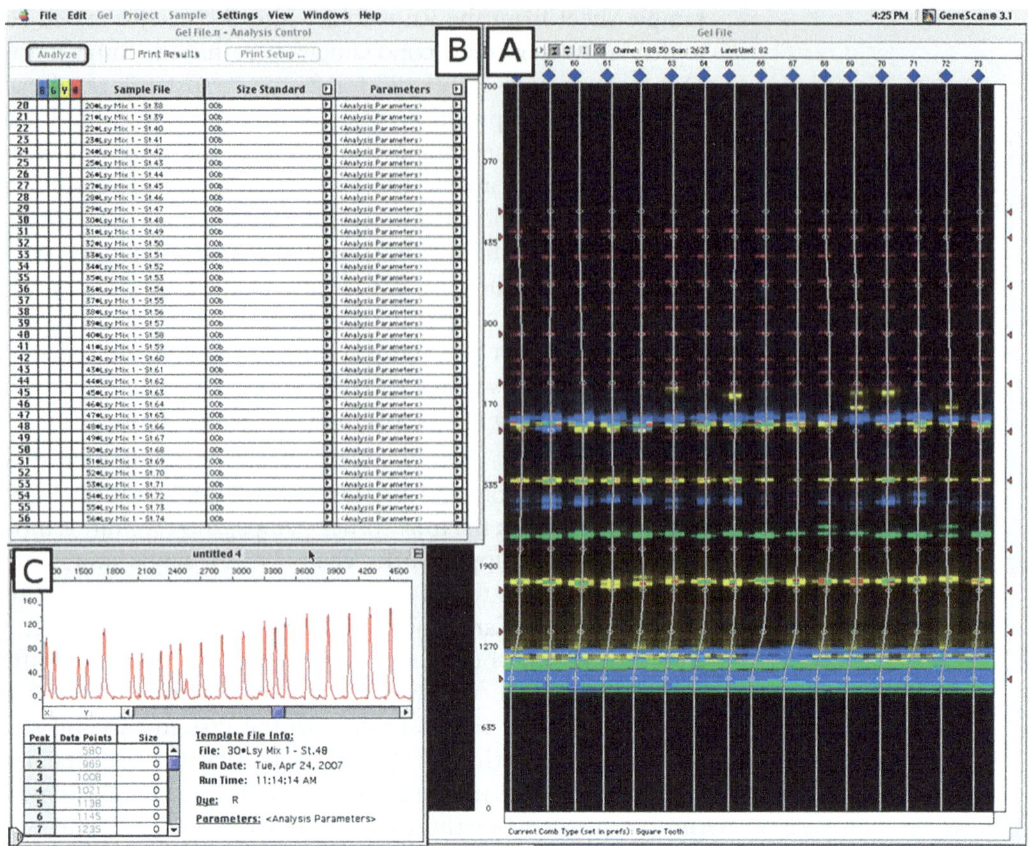

Fig. 5 Example of a GeneScan® file, including (**a**) Gel File with tracker lines in *white*, (**b**) analysis Control window, and (**c**) size standard window

2. Open Genotyper® and select "New" from the File menu; this opens the Main window (Fig. 6a). Select "Import," "From GeneScan File" from the File menu; browse for and open the folder created by GeneScan®, and select "Import All." This imports the GeneScan® output files into the Main window. Select the dye color button (top left of the Main window) that corresponds to the fluorescent label of the marker to be analyzed, and click on the Plot window icon (Fig. 6), opening the Plot window (Fig. 6b) and sample-specific plots. Select "Zoom," "Zoom Out (Full Range)" from the Views menu. Click and drag a box around a size range that includes all the peaks for the microsatellite marker. Select "Zoom" and "Zoom In (Selected Range)" from the Views menu. Click on allele peaks in the Plot window to produce raw size estimates to two decimal places. From the Category menu, select "Add Multiple Categories"; fill in the Add Multiple Categories window (Fig. 6c): "Starting size" is the median of the binning range for the smallest allele, "Category tolerance" is the binning range

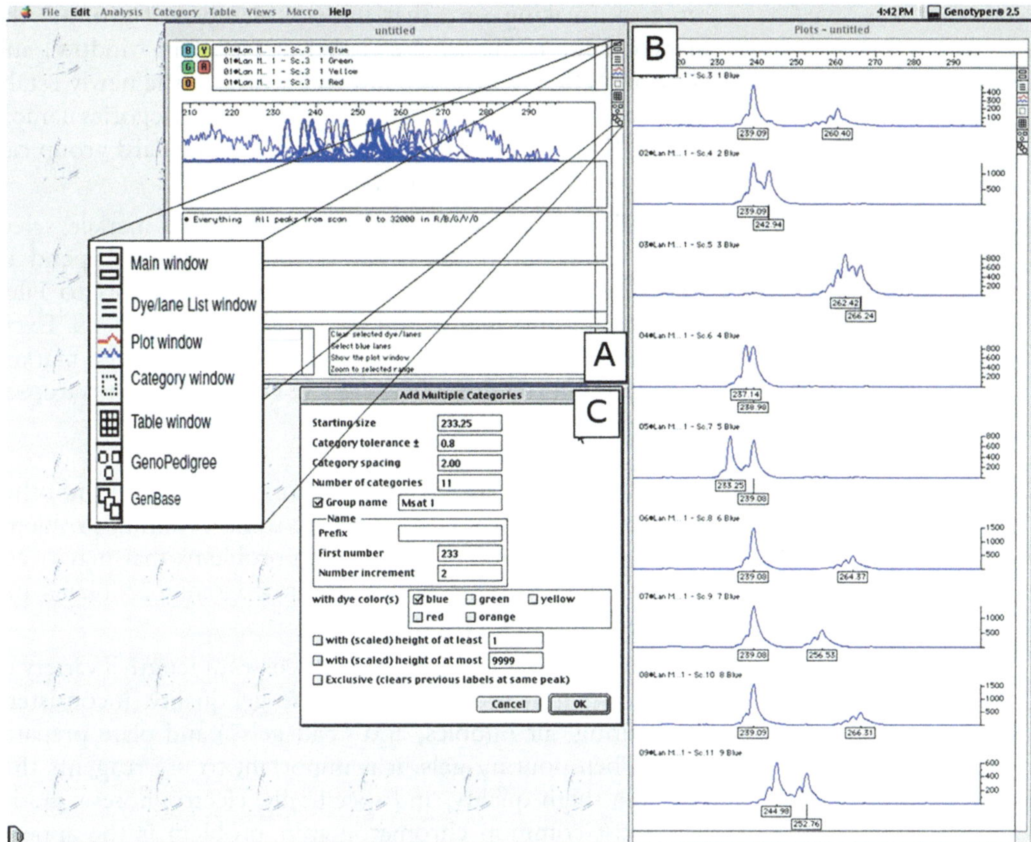

Fig. 6 Genotyper® program with (**a**) the Main window, (**b**) the Plot window, and (**c**) the Add Multiple Categories window. The *icon bar* is enlarged to highlight the Plot window icon

on either size of the median (see Note 11), "Category spacing" is the size increment between alleles, "Number of categories" is the number of alleles for the marker, "Group name" is the marker name, the "Name" box is for naming the binned alleles (this can also be accomplished by clicking on the allele names in the Main window and selecting "Edit Categories" from the Category menu), and "with dye colors" is the appropriate dye color for the marker. Click "OK" to add the marker and alleles to the Main window.

3. Double-click on the marker name in the Main window to select the marker and its alleles. There should be a dot to the left of the marker name and each allele. Select "Label Peaks," "the categories name," and "OK" from the Analysis menu. This changes the raw allele scores, in the Plot window, to the binned scores. Look through all the samples, and click on peaks that need to be added and labels that need to be removed. To add a new category to an existing marker, select "Add Category" from the Category menu. Fill in the information for the new

category (making sure that under "Member of group," the marker name is the same as listed in the Main window) and click on "OK." To change raw size scores to the newly established category, select "Change Labels," "the categories name," and "OK" from the Analysis menu. A size standard group can be set up the same way (see Note 12).

4. Once all of the peaks are labeled properly for a marker, select "Clear Table" from the Analysis menu. Select "Append to Table" from the Table menu. Then, select "Export to File" from the Table menu and save. This produces an Excel compatible spreadsheet of genotypes for the analyzed marker. For more information on using the Genotyper® software, see the User's Manual (16).

3.4 Troubleshooting

1. The ABI 377 is a rugged platform for microsatellite and other genetic analysis; however, there are some common problems that may occur. The most common problems that may occur fall into two broad categories, chromatography problems and hardware (machine) problems.
2. The ABI 377 user's manual (17) discusses at length a variety of chromatography problems relating to gel quality (inconsistent gel pouring, air bubbles, and "bad gels") and plate preparation. When pouring gels, it is important to use reagents that are fresh, high-quality, and specifically, electrophoresis grade. The most common chromatography problem is the appearance of vertical red lines on the gel image, commonly referred to as "red rain" (see Note 13).
3. Although it is a rugged and robust platform, the ABI 377 machine does occasionally fail. In order to troubleshoot hardware problems, it is important to understand that the instrument is comprised of a group of subsystems controlled by a single microprocessor PC board that also receives and processes data from the host computer. The subsystems include the following: (1) the power control subsystem which uses 200–250 V A/C responsible for feeding the appropriate voltage to the various systems; (2) the electrophoresis subsystem which consists of the electrophoresis power supply; (3) the temperature control subsystem which employs a static heater and a chiller to heat and cool the antifreeze solution, heat plates that transfer the hot and cold to the gel, and a pump which circulates the antifreeze solution; (4) the detection subsystem which consists of a 40 mW multiline argon laser used to excite the fluorophores attached to the DNA fragments, emitting color that a spectrograph focuses on discrete pixels of a cooled CCD camera, providing pictures that are combined to form the gel image; and (v) the CPU, a single microprocessor PC board that controls and coordinates all of the subsystems,

receiving and processing data and communicating with the host computer.

4. The single most common problem with the ABI 377 is the downloadable firmware that is stored in battery-backed static memory on the main PC board. As it ages, the firmware becomes corrupt and problems manifest themselves in a number of ways. The most common manifestation of corrupt firmware is the loss of control of one or more of the subsystems; temperature control, electrophoresis, or even detection may be compromised. Fixing this problem requires depressing the "Reset" button, located on the back of the instrument next to the communication cable, for 3 s and then letting go. There should be an audible whining sound when the button is depressed. After the firmware is reset, the lights near the power switch on the front of the instrument will begin to flash yellow. Open the data collection software program to reload the firmware.
5. The second most common problem with the ABI 377 is laser failure. The laser is a gas-filled tube, and like common fluorescent lights, the gas leaks out over time, causing the laser to fail. When this happens, the laser needs to be changed. As a laser ages and its resistance changes, an increase in current is required to achieve the desired power (i.e., 40 mW). Laser current can be monitored to determine when a laser is about to fail. Maintaining power at 40 mW, a new laser will draw approximately 5.8 Amps, while a laser at the end of its life will draw approximately 8.0 Amps. You can see the laser current draw in the Diagnostics program installed on the 377 computer. CAUTION: the laser is cooled with a high-output fan. The fan draws air across the laser and expels the heat out of a vent in the back of the instrument. If the laser cooling fan fails, the laser will heat rapidly and fail within minutes. It is best to periodically check the back of the instrument to make sure that the fan is working.
6. Other than the two most common problems, other subsystems are rugged and fail infrequently. The ABI PRISM® 377 DNA Sequencer was discontinued in 2002, but there are still hundreds if not thousands in operation because of their sturdy reliability to continue running.

4 Notes

1. The matrix file estimates the overlapping fluorescence emission from a single dye in the detection ranges of other fluorescent dyes. For a single fluorescent dye, the matrix file removes the fraction of the detected signal that is due to the fluorescence emission of each of the other dyes. For directions on how to

establish a matrix file and determine issues that may be due to a poor or incorrect matrix file, refer to the ABI 377 manual (17).

2. The same surfaces of the plates should be used repeatedly as the outside and inside (gel side) surfaces. Etched marks on the outside surface are an easy means for identification. This is vital for the front notched plate as the upper buffer chamber leaves a permanent, hydrophobic mark on the outside surface (17).
3. Use enough water to allow the spacers to act as a temporary adhesive between the front and back plates. Too much water will cause the excess to be squeezed out into the gel region when the two plates are put together, creating bubbles when the gel is poured.
4. The solution can be warmed slightly to facilitate the dissolution of the urea but needs to be cooled to room temperature prior to filtration.
5. A rubber stopper works well for tapping on the plates, but making a fist and using the bottom of the hand works as well. The tapping needs to be gentle so as not to harm the plates but the vibrations are helpful in preventing the formation of bubbles in the gel.
6. A dilution between 1/5 and 1/10 is often sufficient, but finding the ideal dilution is a locus-specific process. Some loci may produce stronger amplifications and require more of a dilution, while other loci may produce weaker amplifications and require less of a dilution. Making notes when running novel markers can help fine-tune the dilution process for future loads.
7. Restarting the computer before each run reduces memory fragmentation in the computer as well as closes any additional programs that may be running (17).
8. The default programs (GeneScan® and Genotyper®) for sizing and scoring microsatellite fragments are Mac-based. STRand (18) is a Windows-based software program available for download from the UC Davis' Veterinary Genetics Lab (http://www.vgl.ucdavis.edu/informatics/strand.php). The web site also contains a link to the User's Manual.
9. GeneScan® also has the option to track the lanes manually ("Straight Track") which is helpful when the auto-track feature struggles to correctly identify lanes. If changes need to be made following the auto-tracking, lanes can be moved by clicking/dragging the diamonds at the top of each lane. The diamonds within each lane allow for adjustments to the path of individual lanes.
10. A previously saved size standard can be applied by selecting it from the pull-down list; a new size standard can be established

by selecting "New" from the pull-down list, clicking on each size standard peak (Fig. 5c), typing in the corresponding size for each peak (see manufacturer's instructions for chosen size standard), and then saving the newly established size standard. It is critical to incorporate accurate size standards as the estimates for the microsatellite fragment sizes are based on the size standards. It may be necessary to establish size standards for individual lanes if a general application fails to provide adequate estimates.

11. Size ranges for alleles need to be large enough to allow for some size differences from sample to sample and gel to gel, but the size ranges also need to be small enough to ensure there is no overlap between alleles. For example, dinucleotide repeat size ranges must be less than 2 bp in size. Using a "Category tolerance" of 0.5–0.8 allows for some variation in the sizes while not producing overlap between alleles.
12. A quick check of the size standards can help identify lanes with errors in the estimation of standard peak sizes. A set of categories can be set up for the size standard following the same protocol as outlined for microsatellite loci in the text. Clicking the *R*ed (ROX) button and applying the categories allows for a quick scan through all the samples to ensure that the size standard peaks have been properly identified within each lane. Size standards in problematic lanes can be changed by opening the Analysis Control window in GeneScan®, establishing unique size standards for the problematic lanes, clicking on the colors for those lanes, and then clicking on "Analyze." Save the GeneScan® output files and then reimport all of the output files into Genotyper®.
13. The cause of the red lines is the presence of bubbles between the plates in the scanning region of the gel. There are many proposed causes and solutions for the bubbles. The gel may dry out quickly, a problem that may be mitigated by wrapping the bottom of the gel with a moist paper towel immediately after pouring the gel. Reducing the run temperature from 51 to 48 °C may also mitigate the issue. A buildup dirt, oil, and/or fluorescent contaminants on the plates may also cause red rain. Soaking the plates for 1–2 h in 2 M NaOH or 2 M HCl and then thoroughly rinsing with deionized water can help to remove buildup on the plates, thus removing the red rain. Do not soak the plates for more than 2 h because strong acids and bases can compromise the integrity of the plates. Another way to reduce red rain is to degas the polyacrylamide solution prior to gel pouring; this will reduce or eliminate the introduction of air bubbles into the gel. Other suggestions include utilizing 1.2× TBE in the gel and 1× TBE for running to minimize the effects of ion depletion and using 7.5 μl TEMED and 175 μl

of 10 % APS to attach the gel more tightly to the plates. There doesn't appear to be a singular solution that works in all cases, but following through on one or more of these suggestions may reduce or even remove the red rain.

References

1. Stephenson F (2006) Twenty-Five Years of Advancing Science. Applied Biosystems
2. MacBeath JRE, Harvey SS, Oldroyd NJ (2001) Automated fluorescent DNA sequencing on the ABI PRISM 377. In: Graham CA, Hill AJM (eds) Methods in molecular biology: DNA sequencing protocols, vol 167. Humana, Totowa, NJ, pp 119–152
3. Faria PJ, Lazarus CM, van Oosterhout C, Harris PD, Cable J (2011) First polymorphic microsatellites for the gyrodactylids (Monogenea), an important group of fish pathogens. Conservat Genet Res 3:177–180
4. Mason RAB, Browning TL, Eldridge MDB (2011) Reduced MHC class 2 diversity in island compared to mainland populations of the black-footed rock-wallaby (*Pterogale lateralis lateralis*). Conservat Genet 12:91–103
5. Lee B-Y, Coutanceau J-P, Ozouf-Costaz C, D'Cotta H, Baroiller J-F, Kocher TD (2011) Genetic and physical mapping of sex-linked AFLP markers in Nile tilapia (*Oreochromis niloticus*). Marine Biotechnol 3:557–562
6. Asmussen-Lange CB, Maunder M, Fay MF (2011) Conservation genetics of the critically endangered Round Island bottle palm, *Hyophorbe lagenicaulis* (Arecaceae): can cultivated stocks supplement a residual population? Bot J Linn Soc 167:301–310
7. Byrne RJ, Avise JC (2012) Genetic mating system of the brown smoothhound shark (*Mustelus henlei*), including a literature review of multiple paternity in other elasmobranch species. Mar Biol. doi:10.1007/s00227-011-1851-z
8. Loxterman JL (2011) Fine scale population genetic structure of pumas in the Intermountain West. Conservat Genet 12:1049–1059
9. Portnoy DS, Renshaw MA, Hollenbeck CM, Gold JR (2010) A genetic linkage map of red drum, *Sciaenops ocellatus*. Anim Genet 41:630–641
10. Croteau EK, Heist EJ, Nielsen CK (2010) Fine-scale population structure and sex-biased dispersal in bobcats (*Lynx rufus*) from southern Illinois. Can J Zool 88:536–545
11. Renshaw MA, Saillant E, Gold JR (2006) Microsatellite multiplex panels for genetic studies of three marine fishes: red drum (*Sciaenops ocellatus*), red snapper (*Lutjanus campechanus*) and cobia (*Rachycentron canadum*). Aquaculture 253:731–735
12. Siddiqi S, Mansoor A, Usman S, Nasir M, Khan KM, Qamar R (2011) Characterization of Y-chromosomal short tandem repeat markers in Pakistani populations. Genet Test Mol Biomarkers 15:165–172
13. Welborn SR, Renshaw MA, Light JE (2012) Characterization of 10 polymorphic loci in the Baird's pocket gopher (*Geomys breviceps*) and cross-amplification in other gopher species. Conservat Genet Res. doi:10.1007/s12686-011-9576-3
14. Boutin-Ganache I, Raposo M, Raymond M, Deschepper CF (2001) M13-tailed primers improve the readability and usability of microsatellite analyses performed with two different allele-sizing methods. Biotechniques 31:24–28
15. GeneScan® Analysis Software Program: Version 3.1 (1998) User's Manual; The Perkin-Elmer Corporation
16. ABI Prism® Genotyper® 2.5 Software (2001) User's Manual; Applied Biosystems P/N 904648D
17. ABI Prism® 377 DNA Sequencer: For Sequencing and GeneScan® Analysis Software Applications (2000) User's Manual; Applied Biosystems P/N 4307164B
18. Toonen RJ, Hughes S (2001) Increased throughput for fragment analysis on an ABI Prism® 377 automated sequencer using a membrane comb and STRand software. Biotechniques 31:1320–1324

Chapter 14

Robust and Inexpensive SSR Markers Analyses Using LI-COR DNA Analyzer

Maria del Rosario Herrera and Marc Ghislain

Abstract

Plant genotyping is performed for different purposes which dictate to a large extent the type of molecular makers and platform to be used. The level of throughput, the technical capacity of the genotyping facility, and the availability of reagents are also part of the decision towards a particular genotyping system. SSR markers are quite popular markers because they are easily implementable in standard laboratories, can be used on manual gel electrophoresis, require inexpensive reagents, are mostly randomly distributed in the genome, can be located within genes, have a good discriminatory power, and are codominant with Mendelian inheritance. These features have made SSR the marker of choice for low-resolution genetic mapping and genetic diversity studies including genetic identity verification. The LI-COR platform offers both qualitative and quantitative improvements over the conventional assays based on agarose and polyacrylamide (PAGE) gels with DNA stained with ethidium bromide and silver or radiolabeled. A fast run coupled with an automated detection system using fluorophores makes possible to achieve routinely in our genotyping facility five runs per day using the same gel up to four times which results in 48 genotypes genotyped with ten SSR markers (two per gel electrophoresis using low-cost M13-tailed primers). This gel-base, low cost per sample and equipment, and medium throughput makes the LI-COR platform particularly useful for laboratories with intermediate skills and expectations in molecular genetics.

Key words SSR markers, Microsatellite markers, Gel electrophoresis, Genotyping, LI-COR, Genebank

1 Introduction

Simple sequence repeat markers have become most important marker for plant and animal genetics (1). Their repeat units of 1–6 nucleotides tend to mutate frequently by adding or deleting a small number of repeats generating for the same locus numerous alleles typically in the range of 10–20. This feature is unique compared to other PCR-based markers, AFLP and SNP markers, displaying only two alleles, presence or absence of the amplicon. The latter are, however, produced by high multiplex ratio assays and are markers of choice for genetic studies requiring high marker density.

Stella K. Kantartzi (ed.), *Microsatellites: Methods and Protocols*, Methods in Molecular Biology, vol. 1006,
DOI 10.1007/978-1-62703-389-3_14, © Springer Science+Business Media, LLC 2013

These are also usually used on high-throughput platforms, requiring excellent technical skills, and are significantly more expensive per assay. Whole genome sequencing techniques are certainly on the horizon and expected eventually to replace marker-based genotyping (2).

Plant genotyping are performed to identify specific genomic regions contributing to trait performance by QTL mapping or association genetics, as well as to assess genetic distance between genotypes including the characterization of their homogeneity and true-to-type identity. For the purpose of assessing genetic distances, SSR makers are generally preferred over the other PCR-based markers for their simplicity of use, information content, and locus specificity. These markers have been used to characterize and manage collections of genotypes deposited in genebanks (3). Issues such as a balanced representation of the various gene pools of a crop, the purity or diversity of an accession, and the identity preservation through conservation of the accessions are part of the routine genotyping activities of such facilities performed using diverse molecular marker systems (4). SSR markers have also been proposed to serve as reference marker for variety identification and registry into official variety catalogue (5).

Genotyping platforms have evolved tremendously over the last decade from the early days of individual sample preparation, single agarose or polyacrylamide gel electrophoresis, and gel-specific detection methods. The most advanced platforms today can operate in full automation starting with a nondestructive DNA extraction from seeds, assaying any DNA markers, and loading marker assay results into spreadsheet suitable for specific genetic studies (6). However, these setups are only cost-efficient when high throughputs are needed and part of a high-value commercial pipeline of products. In many laboratories, the research staff and the local reagent suppliers have intermediate skills and capacities. Often the size of their projects is in the range of a couple of hundred of genotypes to be characterized for genetic distances. In such situation, SSR markers and the gel-based system of LI-COR are particularly well suited (1).

Here we present a method using SSR markers using LI-COR DNA analyzer which is both robust and inexpensive. Our lab has been engaged in using DNA markers for genebank germplasm characterization for a long time for potato and sweet potato accessions stored under various forms: in vitro, as tubers, or plants in the field. We have studied the representation of this germplasm in our ex-situ collections, explored taxonomic issues, challenged hypothesis on spread of these crops outside their center of origin, and identified mislabeled accession which are quite common for clonally propagated accessions in genebanks. SSR markers were selected and tested extensively (7). Several genetic diversity studies were achieved using the LI-COR platform with changes in the standard method to increase throughput and decrease cost per sample (8).

2 Materials

Total DNA, reagents for PCR, and gel electrophoresis are the materials needed to perform this method. Reagents should be analytical lab quality but are of equal quality from various suppliers unless specified.

2.1 Total DNA

1. Plant materials: The quality of the DNA is dependent on the quality of the plant material used. Leaves or in vitro plantlets should be in their growth phase long before senescence starts and grown without environmental stress conditions.
2. Total DNA is extracted using the CTAB method (9) modified to avoid the use of liquid nitrogen. However, other methods for extracting total DNA should give similar results (see Note 1).
3. Total DNA extracts are cleaned of RNA using RNAse treatment and its quality and quantity estimated by conventional spectrophotometry measurements and an aliquot on agarose gel to verify its integrity.
4. Stock solution of total DNA is 100 ng/μL and stored at −20 °C. 20× dilutions are made to set up PCR.

2.2 Reagents and Components for SSR Marker Amplification by PCR

1. Primers for SSR markers are designed to produce amplicons in the range of 89–314 bp.
2. For each SSR marker, the forward primer is synthesized with an M13 forward primer sequence on the 5′-end (5′-CACGACGTTGTAAAACGAC-3′). Stock solutions are prepared as 1 μM M13-tailed SSR forward primer, 1 μM SSR reverse primer.
3. IRDye-labeled M13 primer was purchased with the fluorophores 700 or 800. Stock solutions are prepared as 1 μM labeled M13 forward primer (LI-COR IRDye 700 or 800).
4. Nuclease-free water.
5. 10× PCR buffer containing 1 M Tris–HCl, 200 mM $(NH_4)_2SO_4$, 25 mM $MgCl_2$.
6. 5 mM working mix of dNTP: Add 50 μL of each dNTP (100 mM dNTP set) and 800 μL of nuclease-free water. Aliquot and store at −20 °C.
7. Stop solution.
8. Adjustable volume pipette, 0.1–2.5 μL, 0.5–10 μL, 10–100 μL.
9. 96-well PCR plates.
10. 96-well thermal cycler.

2.3 Reagents and Components for PAGE on LI-COR Apparatus

1. (10 %) Ammonium persulfate: In a small tube, dissolve 0.1 g APS into 1.0 mL of deionized water. Use freshly made solution.
2. TEMED (*N,N,N′,N′*-Tetramethylethylenediamine).
3. KB Plus 6.5 % Gel Matrix (LI-COR).
4. TBE (Tris/borate/EDTA) electrophoresis buffer:

   0.5 M EDTA

   Weigh out 93.05 g of EDTA disodium salt. Dissolve in 400 mL deionized water and adjust the pH to 8.0 with NaOH. Top up the solution to a final volume of 500 mL.

   TBE 10×

   Make a concentrated (10×) stock solution of TBE by weighing 108 g Tris base and 55 g of boric acid. Dissolve both in approximately 900 mL deionized water. Add 40 mL of 0.5 M EDTA (pH 8.0) and adjust the solution to a final volume of 1 L. This solution can be stored at room temperature, but a precipitate will form in older solutions. Store the buffer in glass bottles and discard if a precipitate has formed.
5. Gloves (non-powdered), safety glasses, Kimwipes, front plate (notched), back plate (notched), 1 set of spacers (0.25 mm), comb, 1 set of rail assemblies, casting plate, casting stand, 20 cm^3 syringe.
6. LI-COR 4300 DNA Analyzer and SAGA^{GT} software.

3 Methods

Carry out PCR assembly in a separate room from PCR amplification, adding stop solution, storage, and loading on gels and using separate disposable tubes, pipette tips, and pipettes to avoid contaminations.

Maintain temperature around 20 °C in the gel pouring and electrophoresis room as well as constant electric energy supply.

3.1 SSR Marker Amplification by PCR

1. Set up PCR in 96-well plates in a final volume of 5 μL by adding 2.5 μL nuclease-free water, 1 μL 10× PCR buffer, 0.4 μL 5 mM dNTP, 0.2 μL 1 μM M13-tailed SSR forward primer, 0.3 μL 1 μM SSR reverse primer, 0.3 μL 1 μM labeled M13 forward primer, and 3 μL containing 1 unit of Taq polymerase.
2. Add 5 μL of the DNA solution equivalent to 25 ng of genomic DNA.
3. Homogenize by gently pipetting up and down.

4. Carry out the PCR amplification as follows: 4 min at 94 °C, followed by 33 cycles of 50 s at 94 °C, 50 s at annealing temperature (T°a), and 1 min at 72 °C, then 4 min at 72 °C as a final extension step.
5. Add 10 µL of stop solution to the PCR and heat it at 94 °C for 1 min to denature DNA, and chill immediately on ice.

3.2 PAGE Preparation

1. Plate assembly:
    - For LI-COR system, 18 and 25 cm gel plates are suitable for microsatellite analysis application (see Note 2). The 0.25 mm spacers should be used with either gel height.
    - Place the back plate down and place the spacers along the edges.
    - Place the front plate on top of the rear plate. Make sure that the plates are aligned at the bottom. Align the spacers with the outside edges of the plates.
    - Place the left and right rail assemblies over the plate edges. Make sure the rails fit tightly against the edges of both glass plates. The spacer must also be tight against the rail. A leak will occur if there is a gap between the rail and either plate, or between the rail and the spacer. Tighten the glass clamp knobs on each rail. Overtightening can break or distort the glass plates.
2. Gel preparation:
    - Standard runs for microsatellite analysis are made in 25 cm gels since they provide optimum resolution and have adequate run speeds. Each gel can be reused up to three times.
    - Bring 20 mL of KB Plus 6.5 % Gel Matrix to room temperature (10–15 min).
    - Add 150 µL of 10 % APS and 15 µL TEMED immediately before use.
    - Mix to homogenate and draw the gel solution into a 20 cm^3 syringe.
    - Incline the assembled electrophoresis apparatus on the casting stand to improve the flow of the gel between the plates.
    - Inject the gel evenly at a steady rate while moving from side to side across the notch. Occasionally, gently tap the front of the plates to prevent the formation of air bubbles. If the gel is injected correctly and plates clean, a smooth curve shaped gel front advance downward between the gel plates.
    - When the gel solution reaches the bottom of the plates, quickly lay the plate assembly flat on the bench to prevent the gel solution from running out the bottom.

- Any bubbles that form during gel pouring are removed using a bubble hook
- Select a shark's-tooth comb with thickness that matches the spacers. Make sure that the comb fits between the two plates and centered at the top of the gel. The comb is inserted upside down during polymerization to make a trough which forms the base of the wells.
- Place the casting plate into the grooved area in the rails normally occupied by the upper buffer tank. Tighten the two tank clamp knobs until finger tight.
- Allow the gel to polymerize for at least 1 h before use.
- After polymerization, add a small volume of water to the notched area on the front plate where the comb is inserted. This will help to maintain good well morphology.
- Carefully remove the comb from the gel. Remove the acrylamide accumulated in the edge of the notch and rinse with TBE 1× buffer and using a syringe of 20 cm^3 fitted with a 22 gauge needle.
- Additionally, using wipes, clean the back and front plates. Be sure to remove any small acrylamide fragments in each well.
- Before loading the samples, invert the comb and reinsert until the teeth just touch the gel.

3. Electrophoresis apparatus assembly:
   - Press the white rubber gasket into the recessed groove on the back of the upper buffer tank.
   - Loosen the upper clamp knob on each rail and slide the upper buffer tank into place.
   - Tighten the upper clamp knobs "finger tight." At this step, the electrophoresis apparatus is fully assembled.
   - In the LI-COR instrument, place the lower buffer tank into position at the base of the heater plate.
   - Mount the gel apparatus on the instrument against the heater plate, with the bottom of the gel sandwich inside the lower buffer tank. Check to see that the support arms holding the gel assembly on the instrument are seated evenly on the bracket.
   - For KB Plus gels, use 1× TBE running buffer (dilute 100 mL of 10× stock to 1 L with deionized water). Fill the upper buffer tank to the max fill line. Pour the remainder of the buffer into the lower buffer tank.
   - With buffer and using a 20 cm^3 syringe, rinse the wells to remove crystallized urea and air bubbles. Be careful not to release the teeth when rinsing wells around the shark's-tooth comb.

- Place the upper and lower buffer tank lids onto the tanks. Insert the power cable on the upper buffer tank and connect it to the high-voltage connector on the instrument chassis. Make sure that both connectors are fully inserted.
- Pre-electrophoresis and electrophoresis can be started using SAGA^{GT} Microsatellite Analysis Software.

4. Loading and electrophoresis:
   - Denature samples at 94 °C for 3 min. After 3 min, immediately put the samples on ice and cover to reduce exposure to light.
   - After pre-running, open the instrument door and remove the upper buffer tank lid. Remove particulate matter by rinsing the wells with running buffer using a 20 cm^3 syringe.
   - For loading samples, use adjustable pipette with 0.1 μL micropipette tips. Carefully place the tip between the glass plates and slowly release the sample into the wells
   - After sample loading, replace the upper buffer tank lid close the instrument door, and start the run using SAGA^{GT}.

3.3 Detection of Amplicons

The LI-COR System detects DNA using infrared (IR) fluorescence. The Model 4300 is a dual laser system that detects IRDye 700 and IRDye 800 at the same time without spectral overlap between detection channels. Two independent image files are created from the same gel during electrophoresis. The 4300 System includes a server software (SAGA Application Server, which administers an Oracle® database), SAGA client software (SAGA^{GT} software for microsatellite analysis), and the DNA analyzer (Model 4300). As it is mentioned in the manual, SAGA uses projects to manage user's workflow. Projects are containers having various experiments which are gels on the system. From these experiments, image files and genotype data are generated. By the use of projects, the gels are grouped based on the research interest.

When a project is created, there is a setup stage where data that describe DNA, locus, etc. are entered (see Note 3). After logging in the SAGA Client software, the Program Manager opens automatically and new projects can be created or existing projects can be opened. New projects and relevant information about the project are listed in the Project Manager. The Experimental Procedures Manager provides access to all other Manager windows in SAGA. The Molecular Weight Standards Manager permits to add molecular weight standard specifications. With the Locus Manager, the loci that will be used in the project are added. In this step, it is important to create panels of loci that match the samples that will be loaded on gels. With the DNA Source Manager, a text file with data for the individuals to be analyzed in the project can be imported.

And with the Gel Manager, gel templates containing sample information that can be copied when starting new gels can be created. After the setup, the daily operation consists of starting gels that will be part of the project.

After electrophoresis, the gel image files are transferred to the SAGA Application Server for automatic analysis. Then, it is recommended to review the lane finding, placement of desmile lines, and locus boundaries.

The SAGA software can generate several simple text report formats or more advanced report files that are intended for input into specific programs. Simple reports can be generated in Gel Manager from the Gelstab. After clicking Reports in the Gel Manager, the Brief list format is selected. The report contains results for each IRDye infrared dye. In the preview window, the data can be examined and then copied to excel. Scoring from all segments is then copied together in one file. When scoring of all alleles is completed, data is transformed to a binary code by using a simple interphase.

4 Notes

1. Total DNA is stored in $T_{10}E_1$ buffer (10 mM Tris–HCl pH 8.0, 1 mM EDTA pH 8.0) and is quantified using fluorescent reagent on a TBS-380 Mini-Fluorometer.
2. Two gel plates are available, 18 and 25 cm, for the electrophoresis. Choosing among them depends on the number of nucleotide repeats and fragment size. Dinucleotide repeats will be easier to distinguish with a longer run and using 25 cm will be desirable in such case.
3. It is important to highlight that the $SAGA^{GT}$ software has been created to analyze only diploid organisms. In order to perform analysis with polyploids, the gel is divided in sections with two alleles each. To determine the number and composition of these sections in the gel is important to observe the alleles present in the whole sample to be analyzed. For example, gels where five alleles have been detected are divided in three sections. Each section is scored independently. Thus, alleles are selected manually in the Gel editor window, and then, the gel is confirmed to permit the reanalysis (Fig. 1).

Acknowledgement

The authors are grateful to the conscious dedication of Luciano Fernandez for his technical support.

Fig. 1 Typical example of a LI-COR gel image of amplicons produced by PCR amplification of a SSR marker of 48 potato genotypes. Visual scoring allow quick assessment of quality and proper allele scoring using the Gel editor window

References

1. Wang ML, Barkley NA, Jenkins TM (2009) Microsatellite markers in plants and insects. Part I: applications of biotechnology. Genes Genomes Genomics 3(1):54–67, ISSN 1749-0383
2. Hamilton JP, Buell CR (2012) Advances in plant genome sequencing. Plant J 70: 177–190
3. Ghislain M, Spooner DM, Rodrıguez F, Villamon F, Nunez J, Vasquez C, Waugh R, Bonierbale M (2004) Selection of highly informative and user-friendly microsatellites (SSRs) for genotyping of cultivated potato. Theor Appl Genet 108:881–890
4. Spooner DM, van Treuren R, de Vicente MC (2005) Molecular markers for genebank management. IPGRI Technical Bulletin No. 10. International Plant Genetic Resources Institute, Rome, Italy
5. Reid A, Hof L, Felix G, Rucker B, Tams S, Milczynska E, Esselink D, Uenk G, Vosman B, Weitz A (2011) Construction of an integrated microsatellite and key morphological characteristic database of potato varieties on the EU common catalogue. Euphytica 182:239–249
6. Gao SB, Martinez C, Skinner DJ, Krivanek AF, Crouch JH, Xu Y (2008) Development of a seed DNA-based genotyping system for marker-assisted selection in maize. Mol Breeding 22:477–494
7. Ghislain M, Núñez J, Herrera MR, Pignataro J, Guzman F, Bonierbale M, Spooner DM (2009) Robust and highly informative microsatellite-based genetic identity kit for potato. Mol Breeding 23:377–388
8. Spooner DM, Nunez J, Trujillo G, Herrera MR, Guzman F, Ghislain M (2007) Extensive simple sequence repeat genotyping of potato landraces supports a major reevaluation of their gene pool structure and classification. Proc Natl Acad Sci U S A 104:19398–19403
9. Doyle J, Doyle JL (1987) Genomic plant DNA preparation from fresh tissue—the CTAB method. Phytochem Bull 19:11–15

Chapter 15

The Use of the MegaBACE for Sequencing and Genotype Analysis

Pamela A. Burger

Abstract

Despite the advent of next generation sequencing techniques, which provide access to an enormous amount of genomic information in a relatively short time, the conventional Sanger sequencing and microsatellite genotyping analyses present a straightforward method to answer clearly defined questions in population genetics, phylogeography, or forensics. The MegaBACE is a platform that provides both applications with equally reliable performance. In this overview, protocols for the classical techniques of Sanger sequencing and microsatellite genotyping are described. This chapter aims to supply the user of the MegaBACE with methodological tools and some "insider" knowledge of this highly sensitive apparatus.

Key words Sanger sequencing, Microsatellites, Multiplex PCR

1 Introduction

Medical and biological research throughout all disciplines, today, relies heavily on the accurate and fast reproducible knowledge of nucleotide sequences. The development of advanced DNA sequencing methods started with the use of dideoxynucleotide triphosphates (ddNTPs) as DNA chain terminators (Sanger sequencing (1)). This was soon followed by the application of fluorescently labeled ddNTPs and primers for automated, high-throughput DNA sequencing (2). The *dye-terminator sequencing* as performed in the MegaBACE utilizes the labeling of the chain terminator ddNTPs, which permits sequencing in a single reaction rather than four reactions as in the labeled-primer method. Each of the four dideoxynucleotide chain terminators is labeled with fluorescent dyes, which emit light at different wavelengths. These labeled ddNTPs are mixed together with regular nucleotides in a

Stella K. Kantartzi (ed.), *Microsatellites: Methods and Protocols*, Methods in Molecular Biology, vol. 1006,
DOI 10.1007/978-1-62703-389-3_15,

sequencing reaction that provides the termination of elongation at all positions in a template sequence. The amplified products of different lengths can then be separated with capillary electrophoresis, and the terminating base is recognized based on its emitting light by an optical system (3, 4).

It becomes more and more evident that massive parallel sequencing platforms like 454 and Illumina have revolutionized sequencing approaches today and opened completely new fields of research topics that can now be addressed in an experimental and not only in a theoretical way. Nevertheless, it should be noted that the first human genome in 2001 (5) was sequenced with the Sanger method. For specific research questions, small DNA fragments and small sample size the Sanger sequencing and capillary electrophoresis present a cost-efficient and straightforward method. The MegaBACE can also be the sequencing platform of choice, if the high PHRED quality score of Sanger sequencing is taken into consideration compared with the high error rates of massive parallel sequencing technologies including their advanced bioinformatic requirements.

Next to sequencing the second widely used application of the MegaBACE is genotyping using microsatellite loci. These nuclear markers are short sequence motifs (e.g., GT) that are tandemly repeated and distributed over the entire genome. A special mutation process, DNA-replication slippage, renders them a highly polymorphic marker making them well suited for genome scans (6). Due to their high mutations rates, microsatellites are highly informative and very useful for the characterization of recent divergence (up to hundreds of generations) or selective sweeps. Microsatellites are significantly less abundant in the genome than single nucleotide polymorphisms (SNPs) but still occur at high numbers in most species and are often applied in forensics. One drawback in the amplification of these highly repetitive loci are artifacts that occur by DNA-replication slippage during the PCR. These so-called stutter bands are shorter than the actual microsatellite allele and make an automated scoring of the alleles difficult and prone to errors (7). Compared to mitochondrial or nuclear markers, microsatellites have been shown to better resolve genetic relationships among recently diverged populations or species (8).

This methodological overview provides a selection of applications that are available for the use of the MegaBACE. However, the methods described here have been successfully applied by the author and her colleagues for several years (e.g., refs. 9 and 10). Based on the author's experience, the methods of choice should be the ones that best fit to the particular application (sequencing or genotyping) and to the particular research question.

2 Materials

2.1 Quantification of PCR Products

1. *X0.5 TBE buffer*

   For a stock solution of X5 TBE buffer, weigh 54 g Tris base and 27.5 g boric acid and dissolve both in 950 ml deionized water. Add 20 ml of 0.5 M EDTA (pH 8.0), and adjust the solution to a final volume of 1 l (1l). This solution can be stored at room temperature after autoclaving. Use a 1:10 dilution of the TBE buffer for gel electrophoresis.

2. *0.8 % Agarose gel*

   Add 0.8 g of agarose powder to 100 ml X0.5 TBE buffer, dissolve with heat, let it cool until 60 °C, add 5–7 μl of ethidium bromide, and poor it into a prepared electrophoresis gel slide bed. Wait until the gel is solid.

3. *Low DNA Mass Ladder*

   Mix 4 volumes of Low DNA Mass Ladder (Invitrogen) with 1 volume of gel loading buffer containing dye (e.g., 4 μl ladder with 1 μl dye).

2.2 Sequencing Reaction

1. *Sequencing chemistry*

   BigDye® Terminator v3.1 Cycle Sequencing Kit (Applied Biosystems): The BigDye® Terminator v3.1 Ready Reaction Mix and the sequencing buffer are ready to use and should be stored on −20 and 4 °C, respectively, and kept on ice prior to usage.

   DYEnamic ET Dye Terminator Cycle Sequencing Kit (GE Healthcare): The DYEnamic ET terminator reagent premix is ready to use and should be stored on −20 °C or kept on ice prior to usage.

2. *Sequencing primers*

   The same primers that have been used for the PCR can be applied for the sequencing reaction in a concentration of 5 pmol/μl. Alternatively, nested primers (30–50 bp inside the PCR product) can be used in the same concentration (see Note 1).

3. *Deonized, distilled water*

2.3 Purification of Sequencing Products

1. *Purification of sequencing products by gel filtration columns*

   Sephadex® G-50 (Sigma-Aldrich): With this chromatographic separation media, the sequencing products are filtered and excess terminators are removed.

   MultiScreen®-HV 96-well plates (Millipore)

   MultiScreen® Column Loader 45 μl (Millipore): For loading the Sephadex® powder into the MultiScreen®-HV plates

Centrifuge Alignment Frame (Millipore; MACF09604): For positioning the prefilled MultiScreen®-HV plate on top of the receiver plate during centrifugation (see Note 2).

2. *Purification of sequencing products by salt/ ethanol precipitation*

   3 M *sodium acetate* (Na-Acetate; NaOAc) pH 4.8–5.2: Dissolve 408.3 g of sodium acetate in 800 ml of deionized, distilled water. Adjust to pH 4.8–5.2 with acetic acid and fill up to final volume of 1 l.

   96 % ethanol

   70 % ethanol

   Deionized, distilled water

2.4 Genotyping with Microsatellite Loci

1. *Genotyping PCR reagents*

   $MgCl_2$ (25 mM), fluorescent-labeled microsatellite primers (10 pmol/μl), dNTP mix (2 mM each), X10 PCR buffer (usually provided together with the polymerase), polymerase (e.g., FIREPol® DNA Polymerase; Solis BioDyne). After defrosting, keep the reagents on ice prior to usage.

2. *Dilution of PCR product*

   Dilute the PCR product 1:10 with deionized, distilled water, i.e., to an amount of 20 μl PCR product add 180 μl deionized, distilled water directly to the PCR tubes using a multichannel pipette (see Note 3).

   MegaBACE ET400/550/900-R Size Standard (GE Healthcare).

   Depending on the size of the fragment the ET400-R, the ET550-R or the ET900-R Size Standard are used.

2.5 Running the MegaBACE

1. *MegaBACE chemistry*

   MegaBACE Long Read Matrix: Centrifuge three or six tubes of linear polyacrylamide (LPA) long read matrix (GE Healthcare) depending on the usage 48 (MegaBACE 500) or 96 capillaries (MegaBACE 1000), respectively, for less than 30 s at 376×*g* (see Note 4).

   MegaBACE X1 running buffer: 1:10 dilution of the X10 stock MegaBACE running buffer (GE Healthcare), store in fridge (4 °C).

   Buffer plate: Fill half (48 capillaries; MegaBACE 500) or a full 96-well plate (96 capillaries; MegaBACE 1000) applicable for the MegaBACE with 150–200 μl of X1 running buffer.

   Buffer tubes: Fill 3 (6) tubes with 2 ml of X1 running buffer.

2. *Deionized, distilled water*

   Fill the complete MegaBACE water tank and 3 (6) tubes with deionized, distilled water.

3 Methods

3.1 Quantification of PCR Products

Quantify the purified PCR products (200–800 bp) with a Low DNA Mass Ladder on a 0.8 % agarose gel (11). By loading 2.5 μl of the Low DNA Mass Ladder, the reference bands deliver standardized DNA amounts as presented in Table 1. The amount of DNA in the PCR product is measured based on the reference bands using any gel documentation system (Fig. 1). For an alternative method of DNA quantification, see Note 5.

3.2 Sequencing Reaction

1. *Sequencing with BigDye® Terminator v3.1 Cycle Sequencing Kit*

   Sequence the purified PCR products according to the protocol given in Table 2 (see Notes 6 and 7). The PCR parameters are 25–35 cycles of 20 s at 96 °C for denaturation, 5 s at 45–50 °C for primer annealing, and 4 min at 60 °C for extension.

2. *Sequencing with the DYEnamic ET Dye Terminator Cycle Sequencing Kit*

   The sequencing reaction is presented in Table 3. The cycling parameters are 25–30 cycles of 20 s at 95 °C for denaturation, 15 s at 45–65 °C for primer annealing, and 1 min at 60 °C for product extension (see Note 8).

3.3 Purification of Sequencing Products

3.3.1 Purification of Sequencing Products by Gel Filtration

1. Load dry Sephadex® G-50 Superfine into 48 wells of a MultiScreen®-HV 96-well plate by using the 45 μl Column Loader as follows. Add the Sephadex® to the Column Loader. Remove excess powder off the top of the Column Loader with the supplied scraper as shown in Fig. 2. Place the MultiScreen®-HV plate upside-down on top of the Column Loader and invert both. Tap lightly on top and side of the Column Loader to release the complete amount of Sephadex.

Table 1
Amount of DNA content in the Low DNA Mass Ladder (Invitrogen™) reference bands

Band	Fragment size (bp)	Amount of DNA (ng)
1	2,000	100
2	1,200	60
3	800	40
4	400	20
5	200	10
6	100	5

The volumes in the table are for ladder only, not ladder plus dye

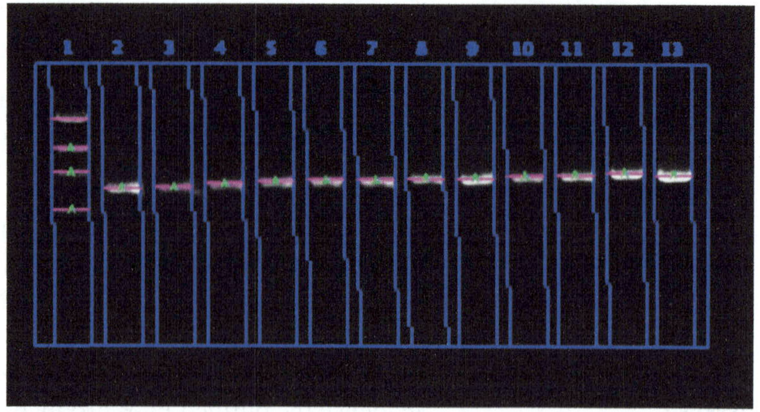

Fig. 1 Quantification of PCR products using an agarose gel. The PCR products are quantified with a mass ladder and according to the density of the band. The size and DNA content of the standardized bands are given in Table 1. *Lane 1*—Low DNA Mass Ladder (Invitrogen), *lane 2*—10 PCR products (2 μl)

Table 2
Sequencing reaction using the BigDye® Terminator v3.1 Cycle Sequencing Kit

Reaction volume	Reagent
5–30 ng	Purified PCR product depending on fragment size
1.5 μl	BigDye® Terminator v3.1 Ready Reaction Mix
1 μl	Sequencing buffer
1 μl	Primer (5 pmol/μl)
Add 10 μl	Deionized, distilled water
10	Final volume

Table 3
Sequencing reaction with the DYEnamic ET Dye Terminator Cycle Sequencing Kit

Reaction volume	Reagent
100 ng per 1 kb	Purified PCR product
2 μl	DYEnamic ET terminator reagent premix
1 μl	Sequencing buffer
1 μl	Primer (5 pmol/μl)
Add 10 μl	Deionized, distilled water
10	Final volume

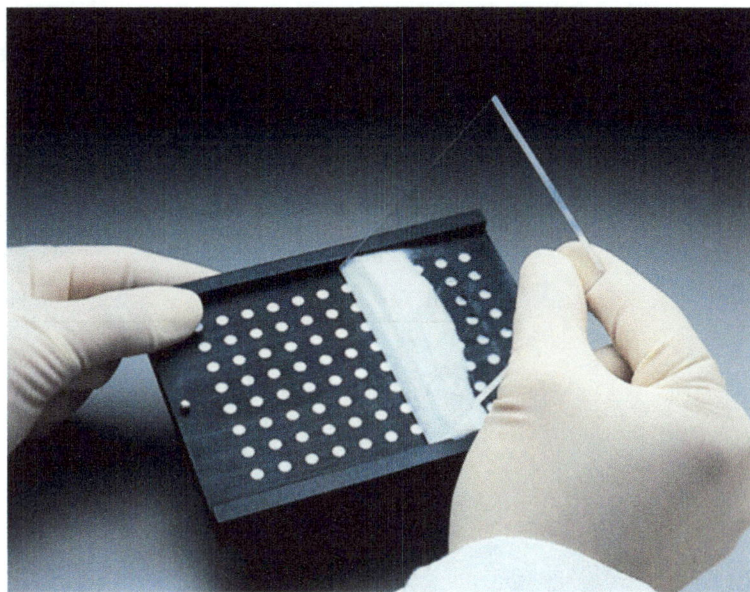

Fig. 2 MultiScreen® Column Loader. The Sephadex® powder is loaded into the MultiScreen®-HV plates by means of a Column Loader 45 μl (Millipore; Cat. No. MACL09645). Excess powder is removed with a scraper. The image was provided with the courtesy of the Millipore Corporation, which owns all copyrights

2. Place a Centrifugation Alignment Frame between a standard 96-well microplate and the MultiScreen®-HV plate. Add 300 μl of deionized, distilled water to each well to swell the resin, and incubate at room temperate for at least 3 h (see Note 9). Centrifuge at 357×*g* for 5 min to pack the mini-columns, and remove excess water. Replace the standard 96-well microplate used for the washing step by a 96-well collection plate applicable for the MegaBACE.
3. Add 7–10 μl of deionized, distilled water to the PCR products, and load the sequencing reaction product (10 μl + 7 μl H_2O) in the center of the column bed slowly. Make sure that the sample does not touch the sides of the column and that the pipette tip does not touch the column surface. Centrifuge at 1,920 rpm for 5 min (see Note 10), remove the MultiScreen® plate and the alignment frame, and seal the plate tightly. The plate is now ready to be load in the MegaBACE or to be stored at −20 °C for a later procession.

3.3.2 Purification of Sequencing Products by Salt/Ethanol Precipitation

1. Transfer the 10 μl sequencing extension product in an 1.5 ml tube, and add 10 μl of deionized, distilled water, 2 μl (1/10 volume) of 3 M Na-Acetate, pH 4.8–5.2, and 50 μl (2.5 volume) of 96 % ethanol. Vortex well and incubate for 10 min on ice.

Table 4
Dilution of the genotyping PCR product with deionized, distilled water and ET-R Size Standard

2.00 μl	PCR product (1:20 diluted)
Add 2.75 μl	Water
Add 0.25 μl	ET400 (550)-R Size Standard
5 μl	Final volume

2. Centrifuge at 78,400 × *g* for 20 min in a cooled (4 °C) centrifuge, discard the fluid carefully, and wash the pellet with 500 μl of 70 % ethanol (see Note 11).
3. Centrifuge at maximum speed for 5 min. Discard the fluid and dry the pellet in a vacuum centrifuge for 2–3 min to remove residual ethanol (see Note 12). Resuspend the pellet in 20 μl of deionized, distilled water, and transfer to a 96-well plate applicable for the MegaBACE (see Note 13). Seal the plate tightly and store at −20 °C if it is not processed immediately.

3.4 Genotyping with Microsatellite Loci

3.4.1 Multiplex PCR Reaction Using Fluorescent-Labeled Microsatellite Primers

The genotyping PCR is carried out in volume of 20 μl with final primer concentrations of 0.15–0.30 μM depending on the performance of the respective primer in the multiplex reaction. In addition, final concentrations of 1.5–2.5 nM $MgCl_2$ (see Note 14), 0.2 nM dNTPs, 1× PCR buffer, and 1.5 U of polymerase are used. The cycling parameters are 1 denaturation cycle of 5 min at 95 °C, 38 cycles of 30 s at 95 °C, 1 min at the optimal primer annealing temperature, 3 min at 65 °C, and a final extension step for 7 min at 72 °C.

3.4.2 Dilution of the PCR Product

Each PCR product dilution (1:10) is mixed with the appropriate size standard as presented in Table 4 and distributed in the 96-well MegaBACE plate (see Note 15).

3.5 Running the MegaBACE

This is the protocol for the specific MegaBACE used by the author. Please note that there might be slightly different procedures necessary for other machines as the MegaBACE is an extremely sensitive apparatus. Therefore, first-time users should seek instructions of an experienced person, who is familiar with the respective MegaBACE, and never try to start the machine alone.

3.5.1 Switch on the MegaBACE

This step is necessary, if the machine was really switched off as it was not in use for a longer time. If the MegaBACE is not used for 16–72 h, the function *Store capillaries* should be applied, where the capillaries are stored in deionized, distilled water for up to 72 h.

1. First switch on the MegaBACE machine and wait for approximately 30 s until you hear a distinct 3× peeping sound.
2. After that you can switch on the corresponding computer, open *Host Scan* in the MegaBACE Folder and wait until the command *complete init* appears.
3. Open the applications *Instrument Control Manager* and *Molecular Dynamics Instrument Control Studio* (*ICS*) in the Service Folder (see Note 16).

3.5.2 Before Starting a Run

1. Choose the correct application either for sequencing or genotyping: In the *Instrument Control Manager Window*, go to the bar menu *Configure* and choose in *Application* either *Genotyping* or *Sequencing*.
2. Check, if the correct filters are placed in the MegaBACE. The MegaBACE detection system uses emission filters and beam splitters to separate the emitted light from the four fluorescent dyes and record them in four separate channels. For sequencing, choose the BigDye or the ET Terminator Filter Set according to the chemistry (BigDye® or DYE ET Terminator) that was used for the sequencing reaction. Choose Filter Set 1 (FAM, HEX, TET, ROX) for genotyping (see Note 17).

3.5.3 Heating Up the MegaBACE

1. To use the MegaBACE after the capillaries were stored in deionized, distilled water: In the *Instrument Control Manager* window, go to *Store Capillaries*, click *Stop*, and wait until it stops blinking.
2. To start heating the MegaBACE up in order to reach the correct operation temperature of 44 °C, go to *Matrix Fill and Prerun* and press *Start*. At the step *ventp on*, click *Stop* to abort the *Matrix Fill and Prerun* (as the machine has to warm up before resuming the protocol). The MegaBACE is now heating up to 44 °C (see Note 18). This might take approximately 15 min.
3. In the meantime, the running parameters and the sample names can be defined: In the *Instrument Control Manager* window switch to *Plate setup* and click *New* to enter the different parameters for sequencing (Table 5) and genotyping (Table 6), respectively. Fill the sample sheet with the names of the samples (see Note 19). Save the run parameters and proceed with the next Subheading 3.5.4.

3.5.4 Rinse Tips

The *Rinse tips* has to be performed in the morning or whenever the MegaBACE is used for the first time of the day. It is not necessary to do it in between two or several runs. First, prepare all the necessary materials (they are also listed on the screen), a full water tank, and full water tubes.

Table 5
Electrophoresis and chemistry parameters for sequencing

Electrophoresis parameters			
Injection voltage	2 KV	Sample injection time	120 s
Run voltage	8 KV	Run time	120–180 min[a]
Chemistry parameters			
Big dye or ET terminators			
Optional parameters			
Do not change anything			

[a]The run time depends on the fragment length. 160–180 min for 700–800 bp are recommended. For up to 1,000 bp, a longer run time up to 3 h is suggested (GE Healthcare 2006)

Table 6
Electrophoresis and chemistry parameters for genotyping

Electrophoresis parameters			
Injection voltage	3 KV	Sample injection time	45 s
Run voltage	10 KV	Run time	60–120 min[a]
Chemistry parameters			
GT dye set 1 (ET-ROX-FAM-HEX-TET)			
Optional parameters			
Do not change anything			

[a]The run time depends on the ET-R Size Standard. 70–80 min are recommended for the ET400-R and 80–90 min for the ET550-R Size Standard, respectively

1. Start the procedure by clicking on *Rinse tips* and then *Start* (see Note 20). Follow the instructions on the screen and on the left and right displays next to the drawers of the MegaBACE.
2. *Load full water tank*: If the full water tank is already inside the machine, because the capillaries were stored over night, then just open and close the left drawer without changing the water tank.
3. Afterwards, the MegaBACE will ask to *insert full water tubes*. The *Rinse tips* is in now progress and will take approximately 5 min.

3.5.5 Start the MegaBACE Run

When a run is started, the MegaBACE should not be left alone until the protocol is finished. The MegaBACE waits for a prompt response to its commands by the user, and a delay of 30–60 s can lead to the problem that the MegaBACE gets stuck. It can happen that the machine does not respond anymore, which implies that the whole process has to be aborted and started anew.

1. Confirm that all the necessary material are prepared (they are also listed on the screen), i.e., matrix tubes, an old buffer plate (used in the previous run and stored overnight tightly sealed at 4 °C), a new buffer plate filled with 150 μl of fresh X1 running buffer, and fresh buffer tubes.
2. Before starting a run, check in the *Molecular Dynamics Instrument Control Studio* that the *Hi-pressure* is highlighted in yellow (not in red!) and does not go above 1,040 or below 930 and that the *Low-pressure* is normal (see Note 21). Check that the machine has reached its operation temperature of 44 °C.
3. Begin the process by clicking on *Matrix Fill and Prerun* and *Start*, and follow the different commands on the screen and on the small displays next to the left- and right-hand drawers of the MegaBACE.
4. Insert the *old buffer plate* in the left drawer (see Note 22).
5. At the command *load Matrix tubes* open the right drawer, remove the water tubes (when *Rinse tips* was done) or the buffer tubes (when a previous run was performed) and *insert the matrix tubes*. The MegaBACE uses now high pressure to inject the linear polyacrylamide matrix, and the small red light for high pressure on the MegaBACE will be turned on. This step takes 3 min followed by 1 min of *Matrix equilibration*.
6. Click *Continue* to *change the buffer tubes* and to remove the matrix tubes on the right drawer of the MegaBACE.
7. Click *Continue* to insert the new buffer plate to the left drawer. Now the *Prerun is in progress* and will take approximately 5 min. In this time, you can put the matrix tubes back to the fridge, wash the old buffer plate with water, and refill it with fresh buffer for the next run. Store the buffer plate tightly sealed at 4 °C.
8. *For genotyping only*: During the prerun, there is time to denature the DNA samples at 94 °C for 2 min. Be careful that the plate is tightly sealed and that a PCR machine with a heated lid is used to avoid evaporation of the samples. After denaturation is finished, put the plate immediately on ice and check that all wells containing samples are well surrounded by ice (see Note 23).

3.5.6 Inject Samples

During this step, the samples are injected in the capillaries. Confirm that the sample plate is ready (centrifuge, if it was de-thawed after storage at −20 °C) and kept on ice (only for genotyping). Prepare a full water tank.

1. Click *Inject samples and run* and *Start*.
2. Choose the correct file with your sample sheet and run parameters.
3. When the command *Load water tank* is visible on the left display of the MegaBACE, remove the buffer plate, but keep it within reach and load the water tank filled with fresh deionized, distilled water. A short Rinse tips step is now performed.
4. Click *Continue* and follow the command *quickly load samples* by removing the water tank and loading the samples plate (see Note 22). The samples are now injected. This step takes between 45 s (genotyping) and 2 min (sequencing).
5. After the sample injection is finished, *load the buffer plate* again. The sample run is now in progress and the small green light for *Data scan* is turned on.
6. To monitor the sample run in the menu bar go to *Options* and *Current Monitor*. The current should be close to 7 for genotyping and around 6 for sequencing runs. Also check the baseline in the *Run image* window, it should be below 1,500 for all wells.

3.5.7 Stop a MegaBACE Run

1. Never click simply on Stop!
2. In the bottom toolbar, click *Full run time*. In the opening window *Run length* you can shorten (e.g., −10) or increase (+10) the run for the given time in min. The run will end automatically.
3. If you have chosen *Sleep after this run* in the *Instrument Control Manager* window, the MegaBACE will automatically store the capillaries in buffer and decrease its temperature to 25 °C. This procedure should only be applied if the capillaries are stored up to a maximum of 16 h. For longer storage time, the capillaries should be stored in deionized, distilled water (see Subheading 3.5.8).

3.5.8 Store Capillaries (or Put the MegaBACE to Sleep)

In this step, the capillaries are stored up to 72 h (maximum!) in deionized, distilled water. The sleeping time can be adjusted to the time, when the next run on the following day is planned. The machine will automatically start with rising its temperature to 44 °C at the programmed time.

1. Prepare the water tank and 2 ml tubes with fresh deionized, distilled water.
2. Click *Store capillaries* and *Start*.
3. Follow the command *Load full water tank* on the left display of the MegaBACE and remove the buffer plate. It is now the old buffer plate for the next run and should be kept tightly sealed either at 4 °C for a longer period or at room temperature overnight.

4. At the request *Load full water tubes*, load the water tubes in the right drawer of the MegaBACE and remove the buffer tubes. Wash them with water and let them dry for refill during the next run.

If the MegaBACE is not used for a longer time (up to 1 week), the *Store Capillaries* step can be repeated several times. However, if the MegaBACE is not in use for more than two weeks, it is better to switch the machine off completely (see Subheading 3.5.9).

3.5.9 Switch Off the MegaBACE

The MegaBACE can only be switched of when the capillaries are stored in water. Therefore, the step *Store capillaries* has to precede this procedure.

1. Close the *Instrument Control Manager* and the *Molecular Dynamics Instrument Studio* windows on the computer.
2. Go to the *Host Scan* window and type *bye*. The computer closes the applications and the *Host Scan* window is closed automatically.
3. Only if the *Host Scan* window is closed, *switch off* the MegaBACE machine.
4. Only when the MegaBACE is switched off you can *shut down* the computer.

4 Notes

1. The authors always applied normal PCR primers for sequencing with good results.
2. MultiScreen HV plates can be reused up to 20 times (the author has been reused it up to 40 times). After centrifugation of the sequencing product into the MegaBACE plate, remove Sephadex by simply inverting the plate. Put the plate into a container with deionized, distilled water and let it stand for a while, then dry the plate and reuse.
3. It is advisable to make a test dilution prior to high-throughput sample processing, normally a 1:10 dilution is sufficient.
4. It is only necessary to centrifuge the tubes when they are new or if there are bubbles inside. (Don't use a higher speed than 2,000 rpm as it may damage the matrix).
5. Alternatively, the DNA concentration of the PCR product can be measured using NanoDrop (Thermo Fisher Scientific Inc., Wilmington, USA; http://www.nanodrop.com).
6. In this protocol, we reduced the amount of reaction volume and BigDye® Terminator v3.1 Ready Reaction Mix to 10 and 1.5 μl, respectively, and retrieved good results. In the original

protocol, however, a reaction volume of 20 μl and an amount of 8 μl BigDye® Terminator v3.1 Ready Reaction Mix are recommended (reference BigDye protocol).

7. For a base material, where a low DNA quantity and quality is expected (e.g., feces or hair samples), the author prefers the BigDye sequencing chemistry. Using BigDye sequencing chemistry, the amount of PCR product should not exceed 30 ng for a 700 bp fragment. For higher DNA quantity (tissue or blood samples), the ET Terminator kit works very reliable. The DNA amount depends on length of the fragment (up 100 ng per 1 kb) and should be adjusted accordingly. Higher DNA quantities may result in shorter read lengths and top-heavy data with a high background.
8. In this protocol, we reduced the amount of reaction volume and DYEnamic ET terminator reagent premix to 10 and 2 μl, respectively, and retrieved good results. In the original protocol, however, a reaction volume of 20 μl and an amount of 8 μl DYEnamic ET terminator reagent premix are recommended (GE Healthcare 2006).
9. Once the mini-columns are swollen in the MultiScreen® plates, they can be stored at 4 °C for up to 2 weeks either by tightly sealing the plate with parafilm or by storing it in a sealed plastic bag containing a moist, lint-free cloth to ensure humidity.
10. Gel filtration is an acknowledged method to remove salt, dye terminators, and other small molecule contaminants. It is faster and more reproducible than most ethanol precipitation methods.
11. While loading the 1.5 ml tubes in the centrifuge, be careful to position the lid strap outwardly, as the pellet will be placed accordingly during the spinning, and it should not be touched by the tip of the pipette in the following washing step.
12. Be careful that the ethanol is completely removed (you can smell this!) as residual ethanol may give worse sequencing results.
13. The salt–ethanol precipitation method is useful when single samples are processed. For a high throughput, the author recommends to use the faster gel filtration method.
14. Increasing the $MgCl_2$ concentration may provide better PCR results. Similarly, the addition of BSA (final concentration 10×) acts as PCR enhancer when typing difficult DNA samples.
15. It is easier to mix the ET-ROX and deionized, distilled water prior to distributing it into the 96-well plate, e.g., for 96-well plate, prepare a mix of 275 μl deionized, distilled water and 25 μl Et-ROX 400 (550). Pipette 3 μl of the ET-ROX—water mix into each well of the plate and add 2 μl of the diluted (1:10) PCR product.

16. The location of the applications might differ on other computers.
17. Only change the filters before the step "Rinse Tips" or before "Matrix Fill and Prerun," but never when a procedure is active, otherwise the laser may damage the eye.
18. This warming-up step or "awakening" after the capillaries were stored ("sleep") is a little bit tricky in the machine used by the author and might be different in other machines. However, it is critical not to start a run, before the MegaBACE reached its operation temperature of 44 °C.
19. It is possible to tell the MegaBACE not to analyze some wells by typing not used in the sample sheet. The MegaBACE will consider these wells as empty. However, as much as possible run full but not partial plates, otherwise the capillaries become worn out unevenly. The command not used should therefore be used only in exceptional situations.
20. In the MegaBACE used by the author, it is necessary to activate the Rinse tips by a right click on the mouse and by typing the password cap to overwrite or force the MegaBACE to do something. In this case, it is to start a new process as the previous process Matrix Fill and Prerun was aborted to heat up the machine.
21. If the Hi-pressure is out of range, then ask an experienced person for help to show you where to adjust the pressure.
22. If the MegaBACE 500 is used with 48 capillaries, be careful to load the 96-well plate, which contains only 48 prefilled wells, correctly—with the filled wells towards you!
23. Do not let the plate cool down slowly, otherwise the PCR products will renature; be fast!

Acknowledgment

Special thanks to Max Kauer, Brigitte Trummer and Viola Nolte, whose previous protocols and extensive experience with the MegaBACE were incorporated in this manuscript. P. Burger is recipient of an APART fellowship from the Austrian Academy of Science.

References

1. Sanger F, Nicklen S, Coulson AR (1977) DNA sequencing with chain terminating inhibitors. Proc Natl Acad Sci USA 74:5463–5467
2. Smith LM, Sanders JZ, Kaiser RJ et al (1986) Fluorescence detection in automated DNA sequence analysis. Nature 321:674–679
3. Applied Biosystems (2002) BigDye® Terminator v3.1 Cycle Sequencing Kit Protocol http://www.ibt.lt/sc/files/BDTv3.1_Protocol_04337035.pdf. Accessed 23 Apr 2012
4. GE Healthcare (2006) DYEnamic ET Dye Terminator Cycle Sequencing Kit for MegaBace

DNA Analysis Systems http://sai.unizar.es/nucleicos/doc/DYEnamic%20ET%20Dye%20Terminator%20Cycle%20Sequencing%20Kit.pdf. Accessed 23 Apr 2012

5. Venter JC, Adams MD, Myers EW et al (2001) The sequence of the human genome. Science 291:1304–1351
6. Schlötterer C (2003) Hitchhiking mapping—functional genomics from the population genetics perspective. Trends Genet 19:32–38
7. Schlötterer C (2004) The evolution of molecular marker—just a matter of fashion? Nat Rev Genet 5:63–69
8. De Leon LF, Bermingham E, Podos J, Hendry AP (2010) Divergence with gene flow as facilitated by ecological differences within-island variation in Darwin's finches. Phil Trans R Soc B Biol Sci 365:1041–1052
9. Charruau P, Fernandes C, Orozco ter-Wengel P et al (2011) Phylogeography, genetic structure and population divergence time of cheetahs in Africa and Asia: evidence for long-term geographic isolates. Mol Ecol 20:706–724
10. Silbermayr K, Orozco ter-Wengel P, Charruau P et al (2010) High mitochondrial differentiation levels between wild and domestic Bactrian camels: a basis for rapid detection of maternal hybridization. Anim Genet 41:315–318
11. Sambrook J, Russel DW (2001) Molecular cloning: a laboratory manual, 3rd edn. Cold Spring Harbor Laboratory Press, Cold Spring Harbor, NY

Chapter 16

Analyzing Microsatellites Using the QIAxcel System

Deborah A. Dean, Phillip A. Wadl, Denita Hadziabdic, Xinwang Wang, and Robert N. Trigiano

Abstract

Microsatellites are ubiquitous throughout eukaryotic genomes and are useful in analyzing populations and genetic diversity. The QIAxcel system, an automated capillary electrophoresis device, allows the user to determine the size of microsatellite fragments, to discern allelic polymorphisms among individuals, and to differentiate homozygous and heterozygous individuals. This system provides comparable base pair resolution to more expensive systems at a relatively affordable cost.

Key words Alignment marker, Amplicon, BioCalculator™, Capillary gel electrophoresis, Microsatellites, DNA size marker, Electropherogram, Polymorphisms, QIAxcel system, Simple sequence repeats

1 Introduction

Nucleic acid separation is a powerful analytical tool that can be used to size fragments of DNA/RNA and to discern the genetic constitution of an organism. Microsatellites or simple sequence repeats (SSRs) are repeated motifs (such as GT, AC, CAT) and are ubiquitous throughout eukaryotic genomes. SSRs are codominant markers, which are often used in interspecific and intraspecific diversity studies (1). Mutations can arise during DNA replication as the polymerase may undergo strand slippage, thereby producing insertions or deletions in the repeated DNA motif. Changes in the SSR sequence may also occur in the flanking and primer regions. Hence, microsatellites typically express a wealth of different length polymorphisms. The QIAxcel system automates the process of detecting and measuring the base pair (bp) size of PCR-amplified DNA fragments. Our lab has utilized the QIAxcel system to conduct the following studies involving SSRs: cross transferability among related species and genera (2), development of a genetic linkage map (3), development of patent protection and data for cultivar patents (4), and assessment of genetic diversity (5). Much

Stella K. Kantartzi (ed.), *Microsatellites: Methods and Protocols*, Methods in Molecular Biology, vol. 1006,
DOI 10.1007/978-1-62703-389-3_16, © Springer Science+Business Media, LLC 2013

of the following information has been gathered from the QIAxcel DNA Handbook (6) and the QIAxcel User Manual (7).

A typical PCR reaction produces polymers of nucleic acids (amplicons) of different weights and lengths. Gel electrophoresis is an amplicon separation technique that exploits the negative charge of the phosphate group within the DNA backbone, which can be used to analyze PCR product size. Applications of electrophoresis involving DNA include the following: DNA fingerprinting, detecting mutants, analyzing molecular markers [such as with SSRs, amplified fragment length polymorphisms (AFLPs), random amplified polymorphic DNA (RAPD)], and sequencing of DNA. Relatively inexpensive methodologies for separating strands of DNA consist of agarose and acrylamide gel electrophoresis using a box containing the matrix and DC electric current supplied by a power pack. After electrophoresis is completed, the separated amplicons are visualized using ethidium bromide (agarose), SYBR®Green (CEQ8000 or ABI sequencer) or silver stains (polyacrylamide). Shortcomings of using polyacrylamide and agarose gels include difficulty in determining amplicon size and the inability to replicate the results within and among different laboratories (3). Furthermore, gel electrophoresis directly exposes the individual to ethidium bromide (a mutagen), and the data is not qualitatively recorded or stored, which can complicate subsequent data analysis. The advent of capillary electrophoresis (CE) has ameliorated the shortcomings of gel electrophoresis (8). In various CE systems, the capillaries are filled with an acrylamide matrix, and the detection of nucleic acids occurs via a florescent molecule attached to the amplicon with a primer or by a physical chemical staining agent such as ethidium bromide. CE system units are self-contained (i.e., buffer trays, power supply, detection systems, etc.) and are automated. The QIAxcel system is one such CE system commonly used today.

Twelve individual capillaries, containing an acrylamide gel matrix impregnated with ethidium bromide, reside in each QIAxcel cartridge. The negatively charged DNA sample is injected into the capillary, and it migrates within the electric field through the gel matrix toward the positive charge area of the capillary. The nucleic acid molecules pass through a UV-light source coupled with a photomultiplier detector. The BioCalculator™ software converts the emission signal strength to an electropherogram (revealing the DNA fragment size) and a constructed "gel image" which can be viewed in real time. The QIAxcel system affords a DNA fragment size resolution, which is similar to that of the CEQ8000 and ABI 3130xl (3).

This chapter describes how the QIAxcel system can be utilized to obtain and analyze data from microsatellite fragments. Materials required, preparation of samples, and selection of alignment marker and DNA size marker are outlined. Detailed instructions to run, align, and analyze samples are also provided.

2 Materials

Our laboratory uses the QIAxcel DNA High Resolution Kit (1200) along with the BioCalculator™ software for SSR analysis using the QIAxcel system. The QIAxcel DNA High Resolution Kit (1200) provides up to a 2 bp resolution when used with the OM700 method on fragments that range 100–500 bp in size (7). A 2 bp resolution has been substantiated in a study that sequenced microsatellite loci samples and compared the size of the sequenced repeat to the QIAxcel system results (2). QIAGEN recommends using the QIAxcel DNA High Resolution Kit with method OM700 when working with PCR products that have been amplified for 30–40 PCR cycles (6, 7). Methods can be adjusted for the needs of individual labs. This kit is also recommended to be used when analyzing DNA fragments sized 15 bp to 5 kb (7). The cartridge supplied with this kit provides 100 runs and 12 samples per run (a run is equivalent to 1 row on a 96-well plate) for a total of 1,200 samples. The BioCalculator™ software can be installed from a CD or the QIAGEN website and is included with the QIAxcel system. Once installed, the BioCalculator™ software automatically launches when the QIAxcel BioCalculator™ program is opened. The QIAxcel DNA High Resolution Kit (1200) includes the following: the QIAxcel DNA High Resolution Cartridge, mineral oil, QX Separation Buffer, QX Wash Buffer, QX DNA Dilution Buffer, 12-tube strips, and QX Intensity Calibration Marker (stored at 2–8 °C). A DNA size marker is not included in the kit and must be purchased from QIAGEN or elsewhere separately. The DNA size marker is used as a reference to calculate the bp size of the samples (to ensure accuracy, sample bp size should fall within the limits of the internal DNA size (15 and 500 bp) markers). The alignment marker (stored at 2–8 °C), stand for the cartridge, and nitrogen tanks are also purchased separately. To install and calibrate each new cartridge, follow the instructions outlined in the QIAGEN DNA Handbook and the QIAGEN User Manual (6, 7). Gloves should be worn when handling cartridges for they contain ethidium bromide.

3 Methods

3.1 Selecting Alignment and DNA Size Markers

An alignment marker and a DNA size marker must also be selected and are purchased separately from QIAGEN (see the QIAxcel® DNA Handbook for recommendations for alignment markers and DNA size markers). QIAGEN recommends changing the alignment marker every 50 runs or 3 days and storing the unused portion at 2–8 °C. We have found that the alignment marker can be changed less frequently with no effect on the quality of the data. If storing, mix and centrifuge the vial (30 s at 10,000 × *g*) and allow the stored alignment marker to reach room temperature before using. The alignment marker is injected from the "Marker 1" position in the buffer tray, and co-migrates with the DNA samples. It is used to calibrate variations of migration times across all of the channels. When using the QIAxcel DNA High Resolution Kit (1200) for microsatellite analysis, the fragment size of the alignment marker should be close to the size of the amplicon samples. We use the 15 bp/500 bp alignment marker because most of our microsatellite amplicons fall within this size range.

When using the QIAxcel DNA High Resolution Kit (1200), QIAGEN suggests selecting a DNA size marker of a similar range to that of the DNA samples. The DNA size marker will be used as a reference to calculate the bp size of the samples, as the migration time of each sample is compared to that of the reference DNA size marker. Any DNA size marker can be used as long as the size of the amplicon falls within the smallest and largest fragments of the reference DNA size marker. We use the 25 bp DNA Step Ladder at a final concentration of 10 ng/μl (as suggested by QIAGEN) for size determination. Any size marker can be added to the "DNA Size Marker" folder. When preparing a 96-well plate for analysis, the DNA size marker can be placed into an arbitrary well within each row. Although the position of the marker in the row does not matter, QIAGEN recommends applying a DNA size marker to each run (row) for microsatellite genotyping in order to minimize size reading error (6).

3.2 Sample/Plate Preparation

Our typical PCR reaction consists of a volume of 10 μl and includes the following reagents: 1–4 ng of genomic DNA, 0.25 μM of forward and reverse primer mix, 2.0 mM $MgCl_2$, 0.2 mM dNTPs, 1× PCR Gold Buffer, 0.4 U AmpliTaq Gold DNA polymerase and sterile water. Although others may be used, these are typical PCR conditions: 1 cycle of 94 °C for 3 min, 35 cycles of 94 °C for 40 s, 55–58 °C for 40 s, 72 °C for 30 s, and 1 cycle of 72 °C for 4 min. The PCR products can be directly separated with the QIAxcel system, and DNA samples should be prepared with a minimum volume of 10 μl; however, the instrument injects less than 0.1 μl of the sample into the individual capillary. The low volume per injection allows the remaining sample to be used to rerun the samples if necessary or for downstream applications such as cloning or sequencing of desired PCR product. To prevent damage to the capillaries, all empty wells (those not containing samples or marker) should be filled with 10 μl QX DNA Dilution Buffer.

3.3 Instructions to Run Samples

1. Ensure the sample door is closed after loading the samples. Opening the door during the run will cause the system to pause. Open the program to launch the QIAxcel BioCalculator™ software. This opens the "Instrument Control" dialogue box with options (Fig. 1).

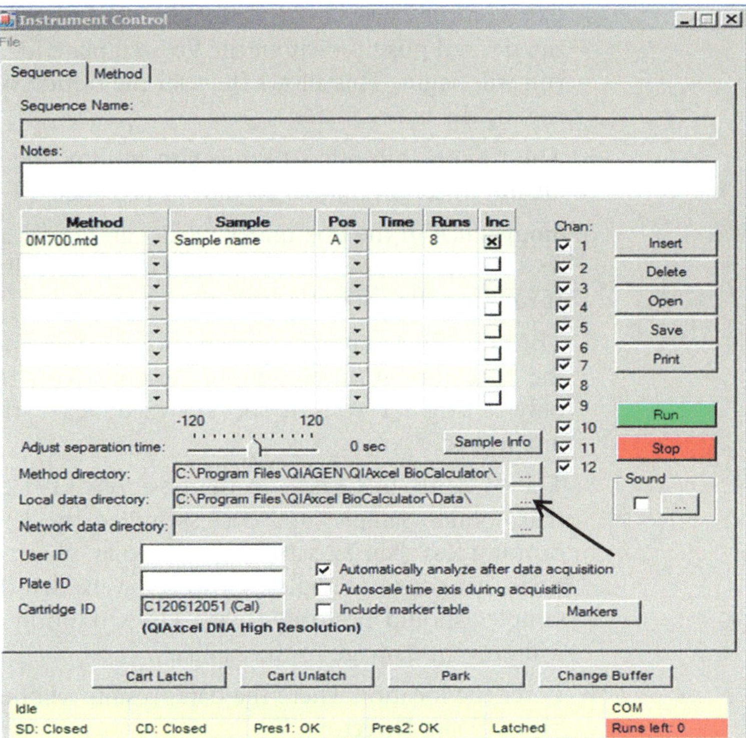

Fig. 1 To process samples using the QIAxcel system, open the Instrument Control dialogue box, then select method OM700.mtd, label sample, choose beginning position and number of runs, and check Inc if more than one run is desired. Save the data to an appropriate location using the local data directory tab (*arrow*). Prior to selecting the *green* "Run" tab, check that all instrument doors are closed, that there is adequate nitrogen pressure, and that the cartridge is calibrated and latched

2. Click on the drop-down menu under the "Method" tab to select a method from the drop-down box. The 700 series provides up to a 2 bp resolution for fragments that are 100–500 bp in length. The L, M, H designation relates to injection time and voltage options, and these recommendations are for low, medium, or high concentrations of DNA. The method OM700 is ideal when working with PCR amplicons from genomic DNA that have undergone 30–40 cycles and in which the DNA concentration is 10–100 ng/µl. For more detail, see the QIAxcel® DNA Handbook or manual (6, 7).
3. Single click in the box under "Sample" and name the sample run using a succinct identification label such as organism name and date.

4. Click on the drop-down menu under the “Pos” tab and select the desired position within the 96-well plate (A–H) where the run will begin. This allows the user the option of running one row or the entire plate.
5. The “Time” tab refers to injection time of the sample, and a default injection time of 10–20 s is provided.
6. Single click in the box under “Run” and enter a value from 1 to 8 to choose how many rows on the plate will be processed. If running the entire 96-well plate, enter 8.
7. The “Inc” box is used to process the plate incrementally. Check the “Inc” box if running more than one row (see Note 1). If only a single row is to be analyzed, leave the “Inc” box unchecked.
8. The “Chan:” box refers to the positions (1–12) in each row that contain samples and correspond to the 12 capillaries. If running less than 12 samples in the rows, check only the positions that contain samples. However, wells that do not contain samples should be loaded with DNA dilution buffer, which will prevent damage to the capillary.
9. Enter the location where the data/results will be stored under “Local Data Directory.”
10. Select “Automatically Analyze after Data Acquisition” which will analyze the data with default parameter settings.
11. Review status of the QIAxcel system. Ensure that the cartridge door and sample door are closed. The system operates with nitrogen, and each nitrogen tank is purchased separately. A “Low Pressure” message will appear at the bottom of the “Instrument Control” dialogue box when the tank needs to be replaced.
12. Click the green “Run” button to initiate the QIAxcel system and the samples will begin migration. A window will open and display the gel image and electropherogram of each sample. Once the run has started, the number of runs remaining on the cartridge will be visible.
13. When the run is finished, the “Run Complete” box will appear.

3.4 After the Run Is Complete

When the run is completed, the gel image will be displayed (see Note 2). Click onto an individual capillary channel to observe the electropherograms. Here the bands of the alignment markers and DNA samples will be present. This raw data provides information concerning the size range of the amplicons and the presence of homozygosity or heterozygosity for diploid individuals (Fig. 2). On the electropherogram, peaks corresponding to the amplicon size will appear. The peaks are the relative fluorescent units detected

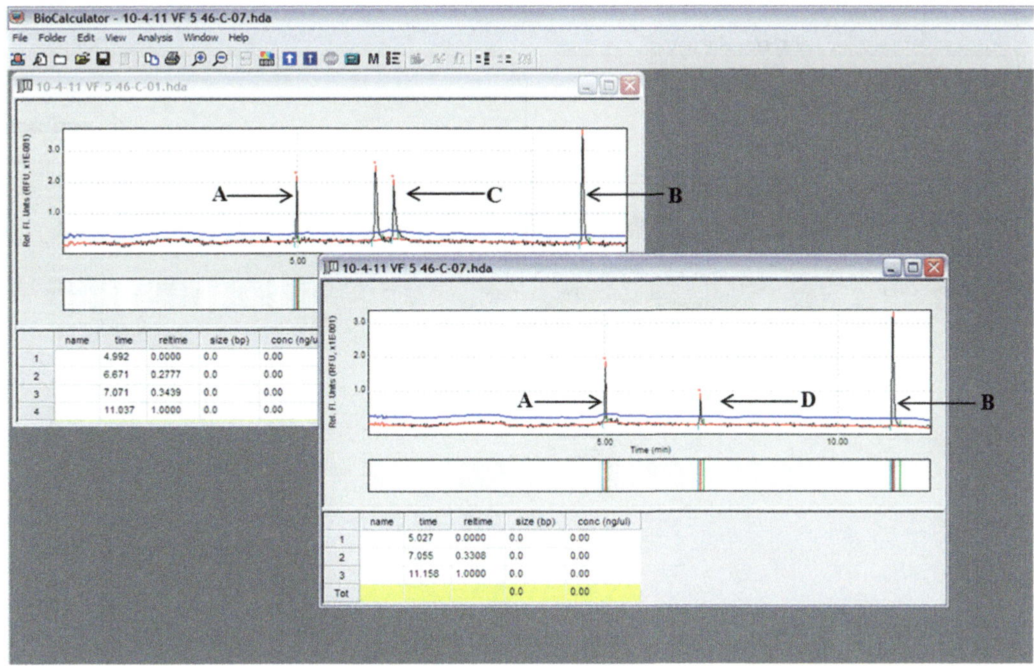

Fig. 2 When the runs are complete, a gel image and the individual electropherograms will be displayed. The beginning and end of the run is recognized and appears as alignment marker peaks *arrows* (*A* and *B*). In the following examples, heterozygous (*arrow C*) and homozygous (*arrow D*) individuals are being analyzed. Peaks above the positive threshold (the *blue line*) are recognized by the software as significant and marked with "k"

from the laser. Peaks that the system recognizes (see Note 3) are those peaks that are above the positive threshold (see QIAxcel manual for details), and these peaks are annotated with a "K" (Fig. 2).

3.5 Automatically Aligning Data (Using BioCalculator™ Software)

When beginning a run, if the "Instrument Control" panel option to "Automatically Analyze After Data Acquisition" is selected, all data will be automatically analyzed using the BioCalculator™ software default values. After the run is completed and the data is aligned, review or change the default parameters in the following manor:

1. Open "Parameters*" from either "Analysis" on the toolbar or the parameter setup icon, and the "Parameter Setup" dialogue box appears (Fig. 3).
2. In the "Parameter Setup" dialogue box, default values will appear for Baseline Filter, Pos. Threshold, Minimum Distance, Suspend Integration, and Data Smoothing Filter (pts).
3. Check "First Peak" and "Last Peak." These peaks correspond to the alignment marker peaks and the beginning (15 bp alignment marker) and last data point or other alignment markers, for example, 15-1.5kb, or 15-3.0kb (500 bp alignment marker) and serve as internal standard markers for sizing the DNA samples.

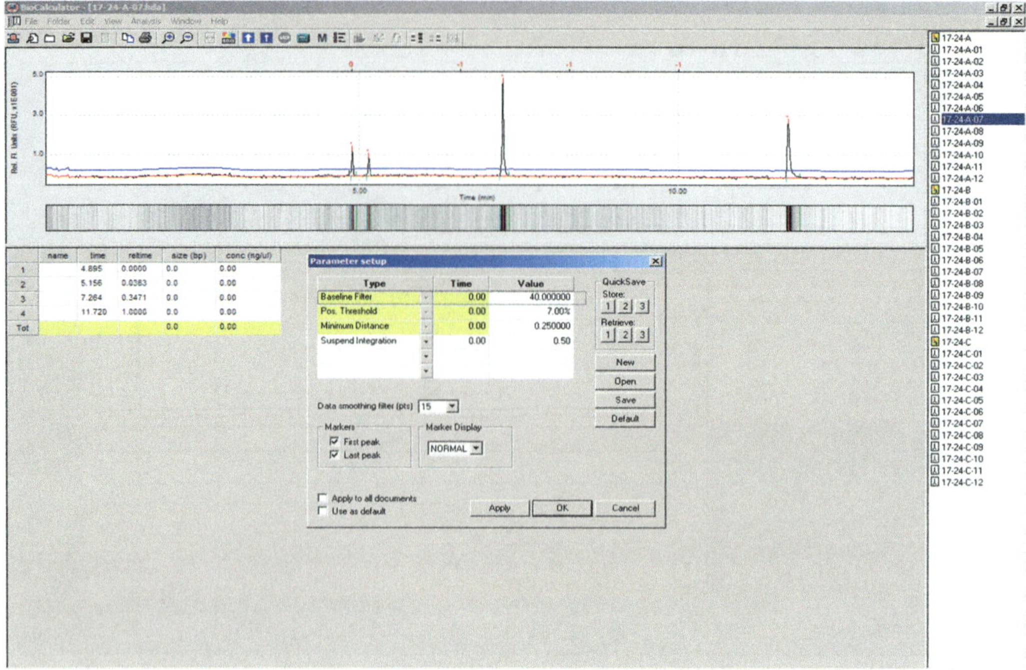

Fig. 3 To facilitate automatic alignment, the BioCalculator™ software will display and utilize a set of default parameters (highlighted) as noted in the "Parameter Setup" dialogue box

4. Select "Apply to All Documents" and then "Ok."
5. Click the wide white arrow on the BioCalculator™ toolbar (run analysis icon) on the toolbar to apply all default/selected parameters to all files within the open folder.

3.6 Manual Alignment (If Data Is Not Aligned Properly After Automatic Alignment)

1. Open the gel image folder and observe which channels are not aligned properly (Fig. 4). Open the individual channel by double clicking the colored bar at the top of the gel view. This will open the electropherogram of the channel to be aligned.
2. Look for superfluous peaks or peaks not recognized that occur before the first alignment marker or after the last alignment marker.
3. To add or delete peaks, place cursor over the peak of interest and right click. When the cursor is placed over a peak, the migration time of the selected peak is highlighted in light blue in the table below the electropherogram. A prompt box will provide the option to "Add/Delete Peak" (Fig. 4). To delete

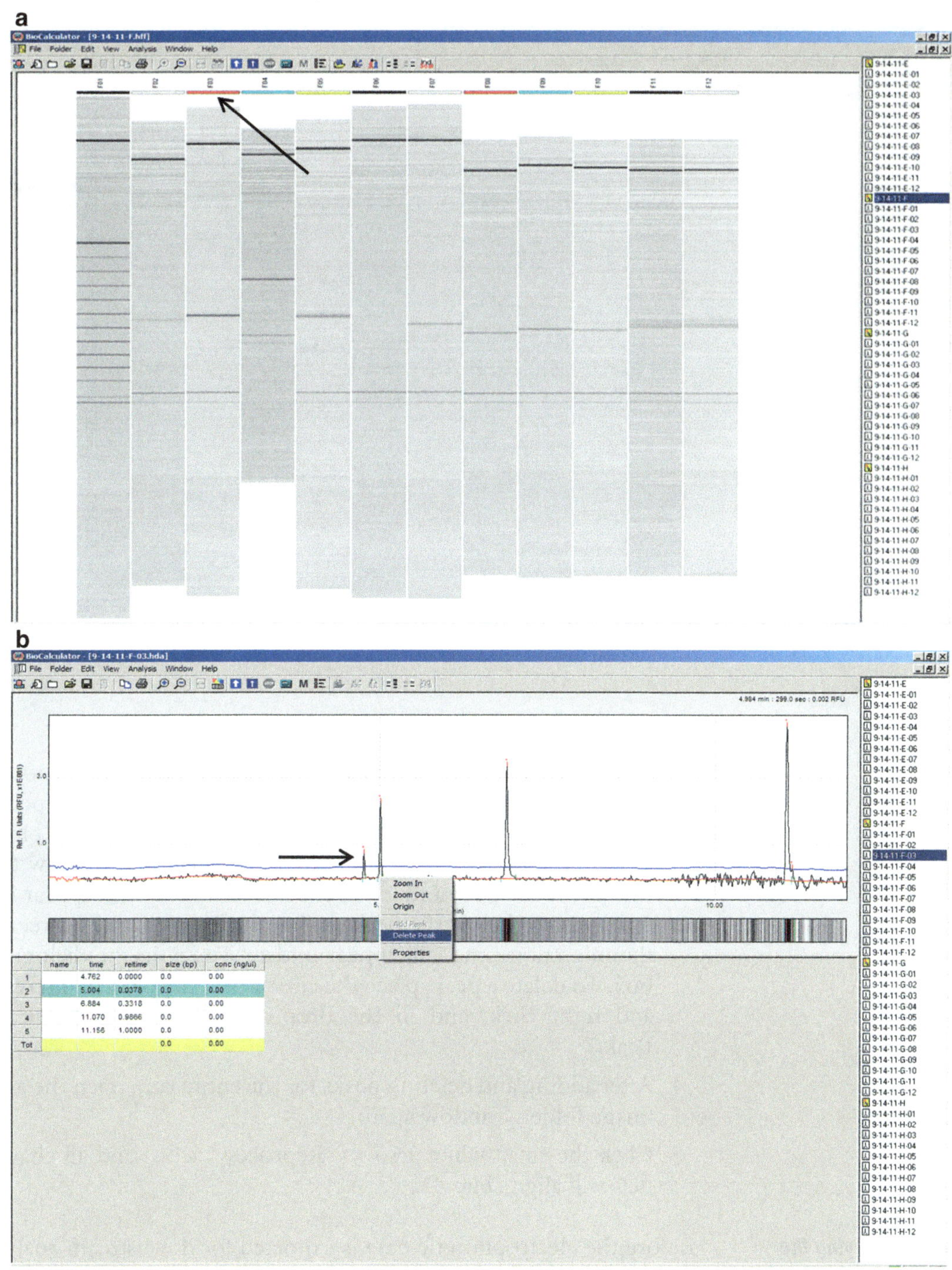

Fig. 4 Gel image of unaligned electropherogram files, which can be aligned manually. (**a**) To begin, select an individual capillary from the gel view screen (*arrow*) and that electropherogram will be displayed. (**b**) Within the electropherogram view, the alignment markers can be distinguished and erroneous peaks (*arrow*) can be deleted. Peaks can be added or deleted when aligning. Place the cursor over any individual peak, and that peak's migration time is highlighted in *blue* under "time" in the dialogue box. (**c**) In the gel view, select the *white* thin "reprocess" arrow (*arrow A*) and all of the samples properly align. The next step will be to include the DNA size marker. Selecting the size marker capillary (*arrow B*) will open the electropherogram for the DNA size marker

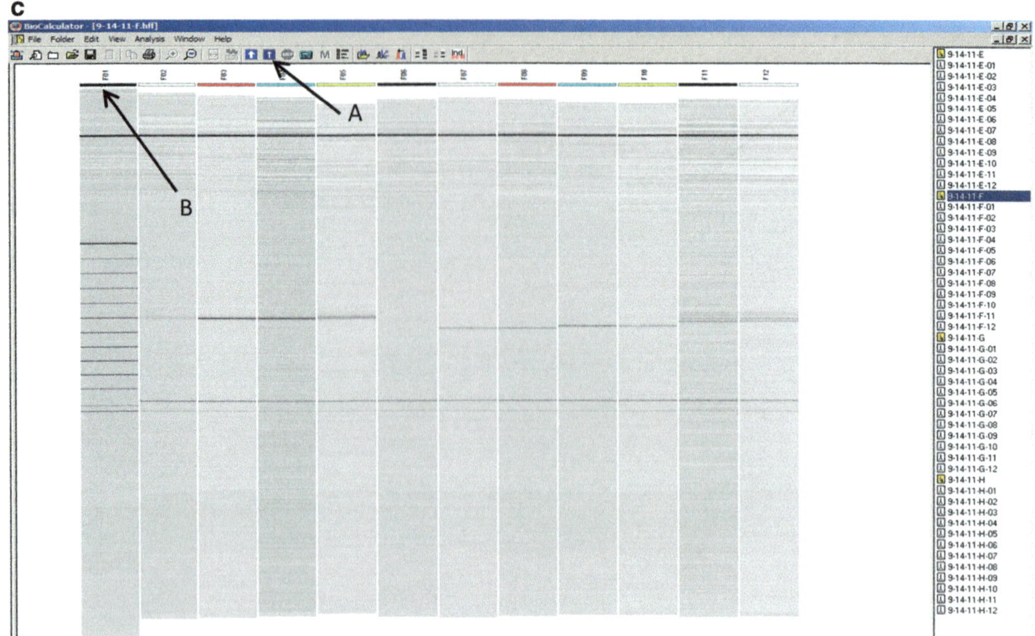

Fig. 4 (continued)

a peak, left click "Delete Peak" and a small blue arrow cursor appears. Use the blue arrow to mark the beginning of the peak by right clicking (a small green arrow will appear, and the blue arrow now marks that spot). Move the small green arrow to the end of the peak and left click. The letter "K" will appear in red above or below the peak, and at the bottom of the screen, the migration time will appear under "Time" in the dialogue box. To delete a peak, place the cursor over the peak of interest and right click, and in the drop-down box, click "Delete Peak."

4. After adding and deleting peaks for the entire run, open the gel image folder/window again.
5. Click the small white arrow ("Reprocess" icon) and all channels will align (Fig. 4).

3.7 Applying the Selected DNA Size Marker

Before the electrophoretic data is exported for downstream analyses, it must be aligned correctly and sized with an appropriate DNA size marker.

1. From the gel view image or the list of files to the right of the image, open the channel/file that contains the DNA marker by double clicking that well.
2. Select the marker icon "M" from the toolbar.

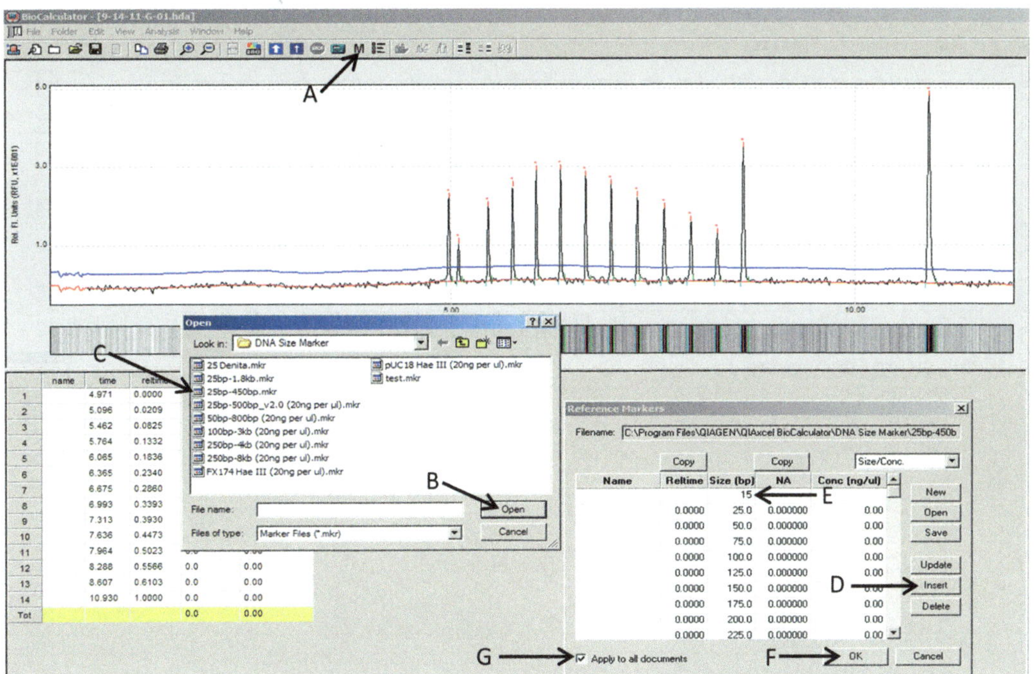

Fig. 5 To select and apply a DNA size marker, select the marker icon (*arrow A*) which opens the "Reference Markers" dialogue box. Selecting "Open" (*arrow B*) in the "Reference Markers" dialogue box will open the file that contains the various DNA size markers (*arrow C*). The proper DNA size marker can now be selected and applied to the samples. Select "Insert" (*arrow D*) and enter the first peak value of the alignment marker in the first blank box in the "Size (bp)" column (*arrow E*). Next, scroll down the same column and enter the last value of the alignment marker in the last box and select "Ok" (*arrow F*). All peaks occurring between these values will be annotated on the electropherogram. Be certain that "Apply to all documents" (*arrow G*) is selected to apply the DNA size marker to all samples in the run

3. From the "Reference Markers" dialogue box, select "Open."
4. From the "Open" dialogue box (Fig. 5), double click the appropriate marker (SSRs typically use 100–300 bp marker) and the "Reference Markers" dialogue box appears (see Note 5). In the first row, click into the first box under "Size (bp)." Next, select "Insert" and a new row appears.
5. Enter the value of the first peak of the alignment marker in the blank box under "Size (bp)" and select enter on the keyboard. Example: The QX Alignment marker is 15 bp/500 bp; therefore, enter 15 in the blank box under the "Size (bp)" box (Fig. 5).
6. Under the "Size (bp)" column, scroll down to the first open box below all of the values, and enter the value of the second peak of the alignment marker being used (500 bp in the case

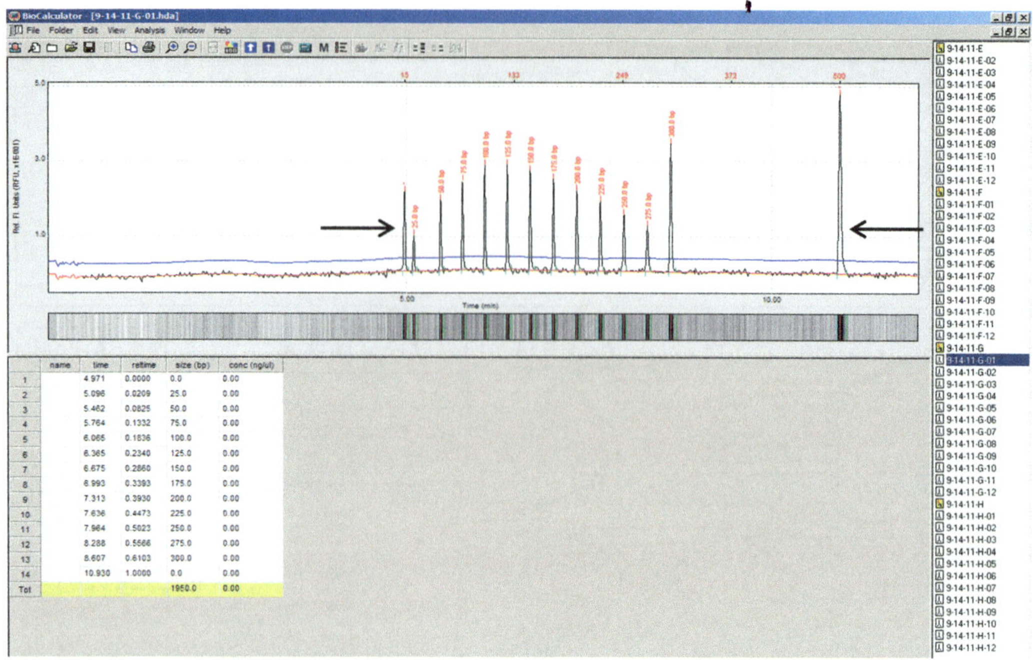

Fig. 6 Following application of the DNA size marker, the ladder with numeric base pairs size appears. The first and last peaks annotated with "K" represent the alignment marker (*arrows*)

of our example above) in the "Size (bp)" box, and hit enter on keyboard.

7. Single click each "Copy" button (located above the headings "Reltime" and "NA").
8. Check "Apply to All Documents" and click "Ok." The size marker is now applied to the gel view window and also each individual channel (Fig. 6).
9. Choosing "Apply to All" will apply the marker to that entire row. Repeat as necessary for remaining runs that require size determination.

3.8 Exporting Data

The data can be exported in both Microsoft Excel and jpg format (the gel image). Prior to exporting the data, remove any unwanted peaks (see Notes 3, 4 and 6).

1. Click "File" and select "Export" which will display the "Plate Image & Result File Creator" dialogue box (Fig. 7).
2. Use "Select" to choose which data will be exported. Ensure that the folder name of the desired data appears under "Plate Directory." Note the pathway of the data, for this is the destination of the exported Excel data file.
3. To export the gel image in jpg format, select and enter a filename in the "Image/Result File Name" and this is where the image will be stored.

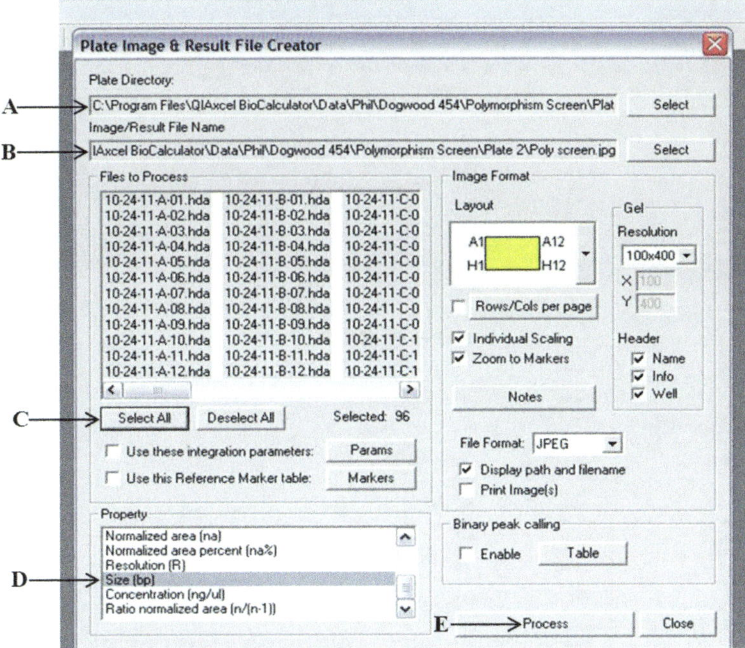

Fig. 7 After samples are processed, the data can be exported to an Excel file. In the "Plate Image & Result File Creator" dialogue box, select the "Plate Directory" to access the electropherogram files to be exported (*arrow A*). Choose the lower "Select" box and a prompt asks where to save the jpeg files (*arrow B*). Within "Files to Process," highlight which files are to be processed or choose "Select All" if all files are desired (*arrow C*). From "Property," select the parameter to be exported. "Size (bp)" (*arrow D*) is utilized when analyzing microsatellites. Finally, select "Process" (*arrow E*) and the raw data allelic sizes are processed and exported to an Excel spreadsheet

4. Under "Files to Process" select any of the 96 desired files to process, or choose "Select All" to process the entire plate.
5. In the "Property" section, highlight "Size (bp)." Although there are other export parameters available, we typically only export the size of the amplicons when analyzing SSRs.
6. Select "Process" to process and export the data to Excel format.
7. To retrieve the exported data, open the Excel file from the location selected in the "Image/Result File Name" in step 2. The Excel file contains information of the filename, sample well, repeat, sample info, and property (Fig. 8).
8. The raw allele length data can be used in a program such as Flexi Bin, an automated SSR binning program that places the raw allele size data into allelic size classes (9). Utilization of a binning program mitigates rounding errors, provides consistency, and compensates for the resolution limitation of the QIAxcel system.

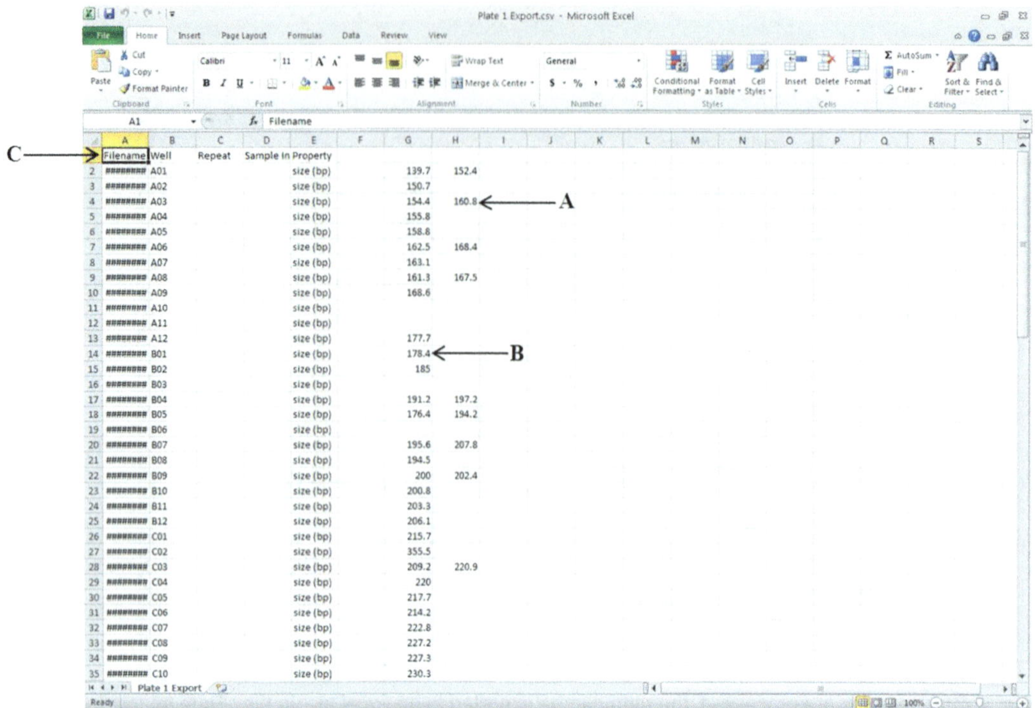

	A	B	C	D	E	F	G	H
1	Filename	Well	Repeat	Sample	In Property			
2	########	A01			size (bp)		139.7	152.4
3	########	A02			size (bp)		150.7	
4	########	A03			size (bp)		154.4	160.8
5	########	A04			size (bp)		155.8	
6	########	A05			size (bp)		158.8	
7	########	A06			size (bp)		162.5	168.4
8	########	A07			size (bp)		163.1	
9	########	A08			size (bp)		161.3	167.5
10	########	A09			size (bp)		168.6	
11	########	A10			size (bp)			
12	########	A11			size (bp)			
13	########	A12			size (bp)		177.7	
14	########	B01			size (bp)		178.4	
15	########	B02			size (bp)		185	
16	########	B03			size (bp)			
17	########	B04			size (bp)		191.2	197.2
18	########	B05			size (bp)		176.4	194.2
19	########	B06			size (bp)			
20	########	B07			size (bp)		195.6	207.8
21	########	B08			size (bp)		194.5	
22	########	B09			size (bp)		200	202.4
23	########	B10			size (bp)		200.8	
24	########	B11			size (bp)		203.3	
25	########	B12			size (bp)		206.1	
26	########	C01			size (bp)		215.7	
27	########	C02			size (bp)		355.5	
28	########	C03			size (bp)		209.2	220.9
29	########	C04			size (bp)		220	
30	########	C05			size (bp)		217.7	
31	########	C06			size (bp)		214.2	
32	########	C07			size (bp)		222.8	
33	########	C08			size (bp)		227.2	
34	########	C09			size (bp)		227.3	
35	########	C10			size (bp)		230.3	

Fig. 8 Raw data has been exported and is displayed on an Excel spreadsheet. All data peaks (hence, the resulting sizes) on the electropherogram are displayed in columns G and H. The individuals in this sample are both heterozygous (*arrow A*) and homozygous (*arrow B*). Headings (*arrow C*) and well locations can be annotated to organize the raw data. This information will be used in downstream analysis of these samples

3.9 Multiplexing Using QIAxcel System

The QIAxcel system can be utilized for analysis of multiplex polymerase chain reaction (PCR)-based assays (10, 11). Multiplexing is a simple, efficient, and cost-effective PCR amplification process that allows multiple primers to be used in a single reaction. It was originally described as a rapid method for detecting deletions in the Duchenne muscular dystrophy (DMD) gene (12). In our experiments, PCR products are separated using the OM700 method with purge time 30 s, injection 5 s, sample injection 10 s, separation time 700 s, and final purge 20 s. The QX 15–500 bp alignment marker is used as an internal standard marker. The size of each amplicon is determined using the 25 bp DNA size ladder. To reduce cost, we use Promega 25 bp DNA Step Ladder as previously described. With use of the BioCalculator™ software, 10 μl (10 ng/lane) of the 25 bp DNA size marker is sufficient to visualize each run.

The PCR amplifications of all loci are performed in 10 μl reaction mixtures (Table 1) using a touchdown PCR (13) program with the following modified cycling conditions (14): 94 °C for 3 min, 15 cycles of 94 °C for 40 s, 40 s initially at 63 °C and subsequently decreasing 0.5 °C per cycle, and 72 °C for 30 s. The next

Table 1
PCR amplification of plant and fungal DNA using one, two, three, or four primers in a single PCR reaction (singleplex, multiplex 2, multiplex 3, and multiplex 4, respectively; 4 ng/μl and 2 ng/μl of DNA template for *Cornus florida* and *Geosmithia morbida*, respectively)

PCR reaction (10 μl)	Singleplex	Multiplex 2	Multiplex 3	Multiplex 4
DNA template	1	2	2.5	2.5
10×PCR buffer II	1	1	1	1
25 mM $MgCl_2$	1	1	1	1
2 mM dNTPs (each)	1	1	1	1
2.5 μM each primer (forward and reverse combined)	1	2	3	4
5 U/μl AmpliTaq Gold® DNA polymerase	0.16	0.24	0.32	0.40
Sterile water	4.84	2.76	1.18	0.10
Total volume	10	10	10	10

20 cycles consisted of 94 °C for 40 s, 55 °C for 40 s, and 72 °C for 30 s ending with 72 °C for 4 min.

See the following figures for multiplex PCR data for the diploid tree *Cornus florida* (Fig. 9) and the haploid fungal pathogen *Geosmithia morbida* (Fig. 10) as analyzed on the QIAxcel system. Primer information is listed (Table 2).

4 Notes

1. Analysis rows/runs are repeated. When preparing the samples on the "Instrument Control" dialogue screen, a value (2–8) has been entered in the "Runs" box, but the "Inc" box has not been selected. This oversight prevents the incremental progression of each row and the first run is repeated.
2. Bands are difficult to visualize. Select the "Contrast" icon. Adjust the contrast slide (decrease from 100 %) and the image will darken to individual preference. The "Invert" icon will invert the light and dark contrast of this image, which can assist in visualization of bands on the gel image.
3. There is background "noise" (Fig. 11) that leads to extraneous peaks being called. Select "Analysis" and then "Parameters*." In the "Parameter Setup" dialogue box, double click the "Pos. Threshold" "Value" of 7.00 %. Raise this value slightly and the positive threshold increases, disregarding the background noise.

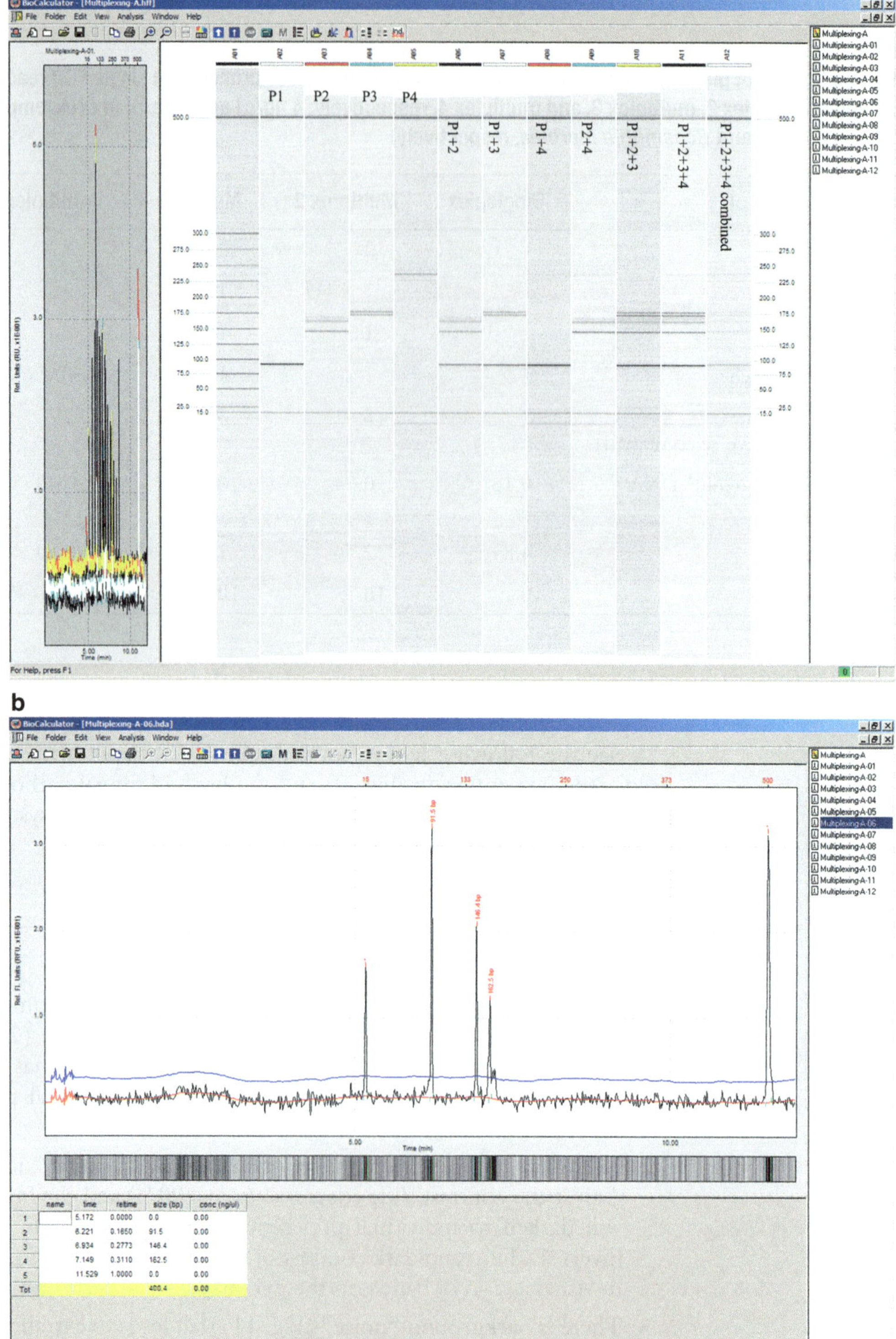

Fig. 9 (**a**) *Cornus florida* DNA sample (diploid species) amplified using one, two, three, or four *C. florida* SSRs (CF127, CF 20, CF113, and CF236 labeled as P1, P2, P3, and P4) (Table 2). The same DNA sample is used in all multiplex reactions. (**b**) Multiplexing of *C. florida* DNA sample with two primers (CF12, CF20, and P1 + 2, respectively). (**c**) Multiplexing of *C. florida* DNA sample with three primer pairs (CF127, CF20, CF113, and P1 + P2 + P3, respectively)

c

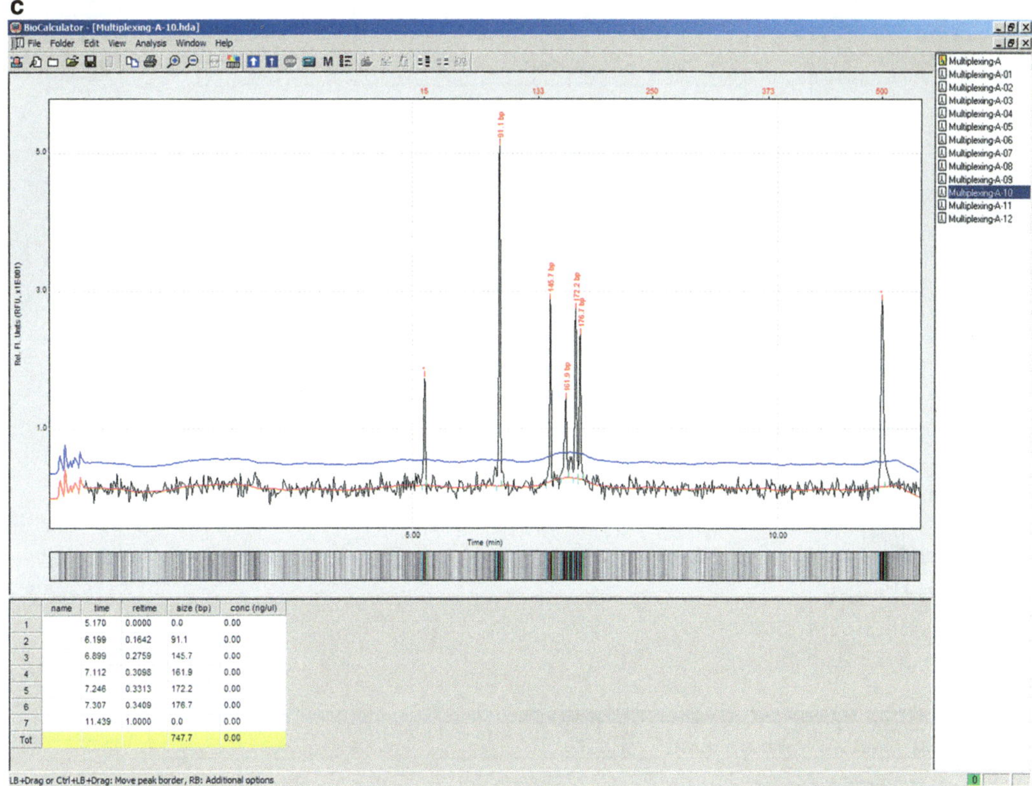

Fig. 9 (continued)

4. Preventing unwanted peaks from being included in analysis data. In the "Parameter Setup" dialogue box, double click "Suspend Integration." Change the default parameter of 0–0.5 to higher value. For example, if 0–1.0 is entered, at time 0 data collection is suspended for 1 min (Fig. 11).

5. If the incorrect DNA size marker is selected, extraneous peaks will be present or necessary peaks are not recognized on the DNA size marker electropherogram. As a result, the size marker increments will be incorrect and may not appear as whole numbers.

6. It is extremely important to be aware (and annotate in publications) what version of the BioCalculator™ software is being used for analysis of a data set. QIAGEN will update the software at times, and we have found that switching between different BioCalculator™ software versions will result in bp size discrepancies. We advise using the same software version for the entire data set, and try to use the same cartridge for analysis to minimize discrepancies in bp size.

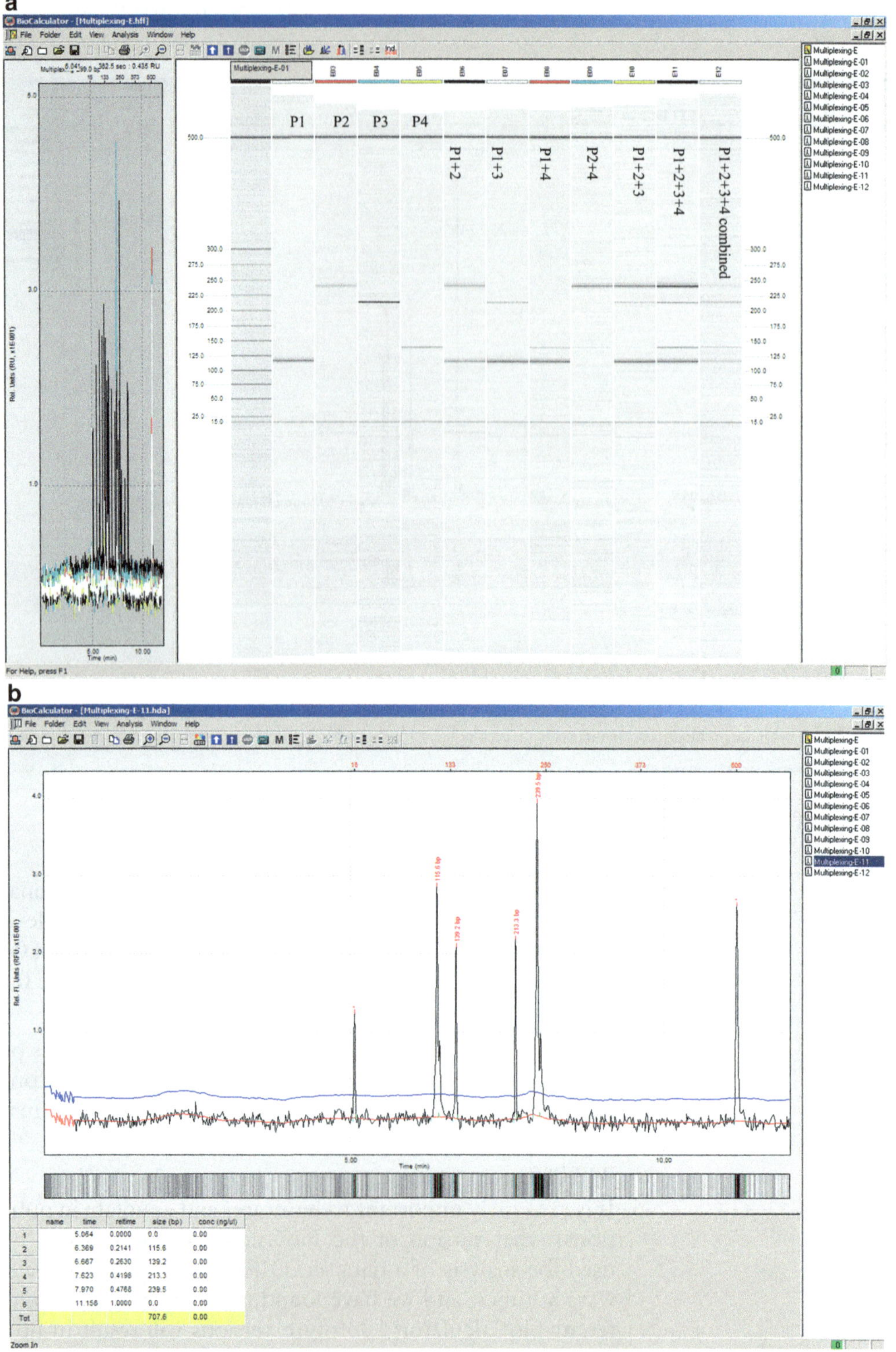

Fig. 10 (**a**) *Geosmithia morbida* DNA sample (haploid species) amplified using one, two, three, or four *G. morbida* SSRs (GS0036, GS0037, GS0060, and GS0078, labeled as P1, P2, P3, and P4) (Table 1). Lane (well) E12 contained a PCR reaction of all four primers individually amplified and combined prior to QIAxcel analysis (2 μl of each reaction from E2, E3, E4, and E5 are combined [GS0036, GS0037, GS0060, and GS0078, respectively] into E12 [P1 + 2 + 3 + 4 combined] well). The same DNA sample is used in all multiplex reactions. (**b**) Multiplexing of a *G. morbida* DNA sample with four primer pairs (GS0036, GS0037, GS0060, and GS0078) labeled P1 + 2 + 3 + 4 (E11)

Table 2
Primer information for eight microsatellite loci used to analyze *Cornus florida* and *Geosmithia morbida* samples in multiplex experiments, representing diploid and haploid organisms, respectively

Locus	Primer sequence (5′–3′)	Repeat	Observed size (bp)	GenBank accession number
Cornus florida				
CF020	F:TATGGCTTGCTTTGGCTAATTGTT R:CCAACTTATGCACACAGTGACACA	$(TC)_{22}$	146	ED651708
CF113	F:ATTTGTTGACTTTTGGTTGGAG R:CCTAATGAAGTTGTTAGGCACA	$(TG)_8(AG)_4$	177	ED651789
CF127	F:TGGATGAGAGAAGTGTTTTGTTTTGT R:CAAGAATTATTGCTCCCCATTCC	$(AG)_6$	91	ED651802
CF236	F:CCTTACCAAATGGAACACTTGTTTTT R:TGTGATGATCTAGAACCCACCTGA	$(AC)_{15}$	236	ED651892
Geosmithia morbida				
GS0036	F:CTAGGGAAAAATGGTCAGCATC R:TCGACATCTAGATCACGGAATG	$(AC)_6$	116	JN580439
GS0037	F:GTTTGCCATCTGCATTACAAAA R:TCCGTCTTATTCTTGGTGTGTG	$(CACCCA)_3...(CTCCCA)_5...(CA)_8c(CCACGT)_3c(CA)_8t(ACACATAC)_2$	240	JN580440
GS0060	F:CGAATCCTGATCTTGTCTTTCC R:CTGGACCAATAAGGTGCTGCT	$(TGC)_6...(TTG)_3(CTGTTG)_3(TTG)_2(CTG)_3t(GTG)_5(TTG)_{11}$	213	JN580445
GS0078	F:CAACTCCCCTCCAGTACACAAC	$[CAT(CAC)_2]_2CAT(CAC)_3$	139	JN580447

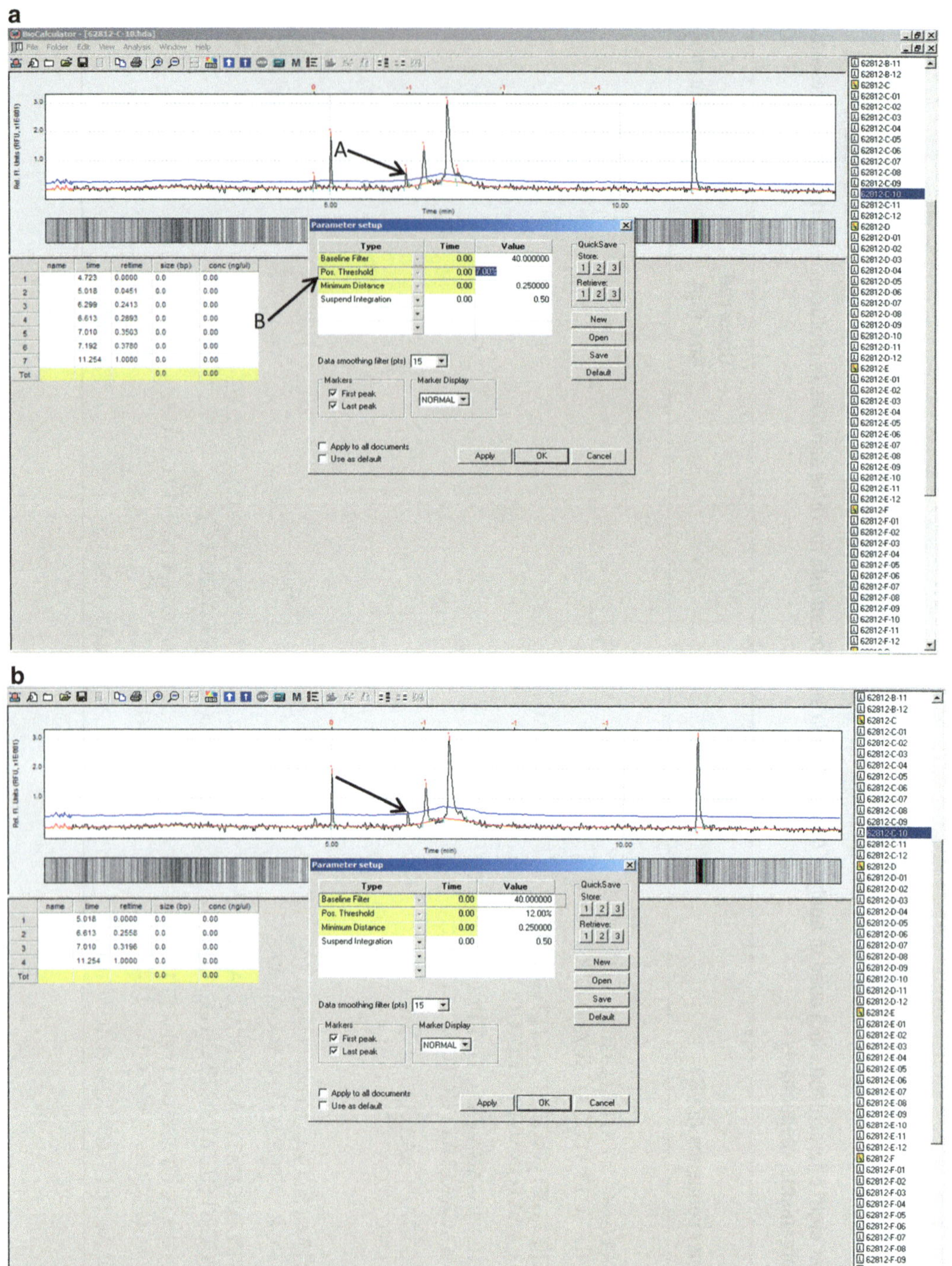

Fig. 11 (**a**) Extraneous peaks (*arrow A*), which are recognized by the BioCalculator™ software, can occur in noisy runs. Select "Pos. Threshold" (*arrow B*) and change the default value (7.00 %) to a slightly higher value. (**b**) Increasing the "Pos. Threshold" to 12 % (the blue positive threshold line will rise, and only acknowledge peaks above the *blue line*) discounts interference and assists in cleaning up the data, as the extraneous peaks are not acknowledged (*arrow*) by the software. This is a quick way to delete insignificant peaks

References

1. Gupta PK, Varshney RK (2000) The development and use of microsatellite markers for genetics and plant breeding with emphasis on bread wheat. Euphytica 113:163–185
2. Wadl PA, Wang X, Moulton JK, Hokanson SC, Skinner JA, Rinehart TA, Reed SM, Pantalone VR, Trigiano RN (2010) Transfer of *Cornus florida* and *C. kousa* simple sequence repeats to selected *Cornus* (Cornaceae) species. J Am Soc Hort Sci 135:279–288
3. Wang X, Rinehart TA, Wadl PA, Spiers JM, Hadziabdic D, Windham MT, Trigiano RN (2009) A new electrophoresis technique to separate microsatellite alleles. Afr J Biotechnol 8:2432–2436
4. Hadziabdic D, Wang X, Wadl PA, Rinehart TA, Ownley BH, Trigiano RN (2012) Genetic diversity of flowering dogwood in the Great Smoky Mountains National Park. Tree Genet Genomes. doi:10.1007/s11295-012-0471-1
5. Dean D, Wadl PA, Wang X, Klingeman WE, Ownley BH, Rinehart TA, Scheffler BE, Trigiano RN (2011) Screening and characterization of 11 novel microsatellite markers from *Viburnum dilatatum*. HortScience 46:1456–1459
6. QIAGEN (2011) QIAxcel® DNA handbook. 2nd edn. http://www.QIAGEN.com/literature/handbooks/literature.aspx?id=2000066. Accessed 03 Feb 2012
7. QIAGEN (2008) User manual. http://www.qiagen.com/products/qiaxcelsystem.aspx#Tabs=t5. Accessed 03 Feb 2012
8. Gupta V, Dorsey G, Hubbard A, Rosenthal PJ, Greenhouse B (2010) Gel versus capillary electrophoresis genotyping for categorizing treatment outcomes in two anti-malarial trials in Uganda. Malaria J. 9. http://malariajourna.com/content/9/1/19. Accessed 12 Feb 2012
9. Amos W, Hoffmann JI, Frodsham A, Zhang L, Best S, Hill AVS (2007) Automated binning of microsatellite alleles: problems and solutions. Mol Ecol Notes 7:10–14
10. Jakubauskas A, Griskevicius L (2010) KRas and BRaf mutational status analysis from formalin-fixed, paraffin-embedded tissues using multiplex polymerase chain reaction-based assay. Arch Pathol Lab Med 134:620–624
11. Salgotra RM, Millwood RJ, Agarwal S, Stewart N (2011) High-throughput functional marker assay for detection of Xa/xa and fgr genes in rice (*Oryza sativa L.*). Electrophoresis 32:2216–2222
12. Chamberlain JS, Gibbs RA, Ranier JE, Nguyen PN, Caskey CT (1988) Deletion screening of the Duchenne muscular dystrophy locus via multiplex DNA amplification. Nucleic Acids Res 16:11141–11156
13. Korbie DJ, Mattick JS (2008) Touchdown PCR for increased specificity and sensitivity in PCR amplification. Nat Protoc 3:1452–1456
14. Trigiano RN, Wadl PA, Dean D, Hadziabdic D, Scheffler BE, Runge F, Telle S, Thines M, Ristaino J, Spring O (2012) Ten polymorphic microsatellite loci identified from a small insert genomic library for *Peronospora tabacina*. Mycologia. doi:10.3852/11-288

Part IV

Scoring and Data Analysis

Chapter 17

Microsatellite Analysis of Malaria Parasites

Pamela Orjuela-Sánchez, Michelle C. Brandi, and Marcelo U. Ferreira

Abstract

Microsatellites have been increasingly used to investigate the population structure of malaria parasites, to map genetic loci contributing to phenotypes such as drug resistance and virulence in laboratory crosses and genome-wide association studies, and to distinguish between treatment failures and new infections in clinical trials. Here, we provide optimized protocols for genotyping highly polymorphic microsatellites sampled from across the genomes of the human malaria parasites *Plasmodium falciparum* and *P. vivax* that have been extensively used in research laboratories worldwide.

Key words Malaria, Microsatellites, *Plasmodium falciparum*, *Plasmodium vivax*, Genotyping

1 Introduction

Despite the continuous development of novel molecular typing methods and high-throughput platforms, microsatellites (tandem repeats of motifs of one to six nucleotides) remain among the most popular and informative markers currently used by population geneticists (1). The extensive variation found in most microsatellite-type sequences, which typically results in multiple alleles per locus, is mostly generated by strand-slippage events during DNA replication. Observed mutation rates (10^{-3} to 10^{-4} per locus per generation) result from the interplay between strand-slippage events and mismatch repair, which counteracts DNA slippage during replication (2).

Microsatellite-type repeats are highly abundant in the malaria parasite *Plasmodium falciparum*; an average of one microsatellite locus is found every 2–3 kb of genome sequence (3). The first reported use of microsatellites in malaria research aimed at mapping inherited traits in the progeny of a genetic cross (4, 5). Subsequently, selected trinucleotide repeats were used to compare the population structure of parasites from different endemic settings (6–10). Only 160 microsatellites have been found in the genome of another

Stella K. Kantartzi (ed.), *Microsatellites: Methods and Protocols*, Methods in Molecular Biology, vol. 1006,
DOI 10.1007/978-1-62703-389-3_17, © Springer Science+Business Media, LLC 2013

major human malaria parasite, *P. vivax* (11). Despite this limitation, microsatellite-based studies have provided valuable information on the genetic diversity of *P. vivax* worldwide, suggesting that a spectrum of population structures also exists for this species (12–21). Microsatellite typing has also been standardized for the human malaria parasite *P. malariae* (22), the rodent species *P. yoelii* (23, 24), and the lizard species *P. mexicanum* (25).

Here, we provide detailed protocols for microsatellite genotyping of patient-derived isolates of *P. falciparum* and *P. vivax*. These single-copy microsatellites consist of tandem repeats of tri- or tetranucleotide motifs and have been extensively used in laboratories across the world for studying the population structure of malaria parasites. They have also been used: (a) to detect naturally occurring multiple-clone infections (10, 26), (b) to distinguish between recrudescences and new infections in clinical trials (27–29), (c) to distinguish between *P. vivax* relapses and new infections in population-based cohort studies (14, 17, 30–32) (see Note 1), (d) to compare levels of genetic diversity in sympatric populations of different species (9, 14, 33), and (e) to track the geographic origin of infections (19).

Although highly sensitive nested and hemi-nested PCR protocols are available to amplify microsatellite loci of both *P. falciparum* and *P. vivax* (34, 35), here we provide single-stage PCR protocols that are less likely to distort the relative proportion of alleles originally found in multiple-clone infections (see Note 2). When limited amounts of template DNA are available, we suggest the use of multiple displacement whole-genome amplification (36) as a means to generate enough starting material for single-stage PCR (see Note 3). Since identical cycling parameters are used for all loci of each species, multiplex PCR assays may be standardized from the protocols described here (see Note 4).

2 Materials

2.1 Equipment

1. Single-channel pipettes with dispensing volume ranges of 0.1–2, 1–10, 20–200, and 200–1,000 μL.
2. Bench-top microcentrifuge.
3. Heating block.
4. Thermal cycler. Amplification protocols have been optimized using GeneAmp PCR System 9700 and 2400 equipment (Applied Biosystems, Foster City, CA); further standardization may be needed when different equipment is used.
5. Automated DNA sequencer. We have used ABI 377, ABI 310, ABI 3100, ABI 3500, and ABI 3700 sequencers (Applied Biosystems); the three-color primer labeling strategy, combined with ROX-500 internal size standard, is compatible with all of them.

2.2 Consumables

1. 1 Aerosol-barrier pipette tips.
2. 1.5-mL microtubes.
3. 200-μL thin-walled PCR tubes.
4. 0.5-mL sample tubes (catalog number 401957) and septas (catalog number 401956) (Applied Biosystems).

2.3 PCR

1. 1 Sterile, deionized, and distilled water (ddH_2O).
2. Recombinant *Taq* DNA polymerase 5 U/μL (catalog number EP0402; Fermentas, Burlington, Canada).
3. 10× *Taq* buffer with $(NH_4)_2SO_4$ (catalog number EP0404) (Fermentas). This buffer contains 750 mM Tris–HCl (pH 8.8), 200 mM $(NH_4)_2SO_4$, and 0.1 % (v/v) Tween 20.
4. 25 mM $MgCl_2$ (Fermentas).
5. dNTP mix (dATP, dCTP, dGTP, and dTTP in ddH_2O, 10 mM each).

2.4 Fragment Analysis

1. ROX-500 internal size standard (Applied Biosystems catalog number 401734).
2. Highly deionized formamide (Hi-Di Formamide; Applied Biosystems catalog number 4311320).

3 Methods

3.1 P. falciparum Microsatellites

3.1.1 Oligonucleotides

1. The set of oligonucleotides used to type 11 single-copy trinucleotide microsatellites of *P. falciparum* (4, 34) is described in Table 1. Three microsatellites (Poly∝, TA42, and TA109) display complex mutation patterns that may be inconsistent with a standard stepwise mutation model (see Note 5). The labeled oligonucleotides are supplied lyophilized by Applied Biosystems. We routinely use three labels, the fluorescent dyes 6-FAM (6-carboxyfluorescein, "blue" label), VIC (Applied Biosystems proprietary "green" fluorescent dye), and NED (Applied Biosystems proprietary "yellow" fluorescent dye); the internal size standard is ROX-500 ("red"). However, when ABI 3500 and ABI 3700 sequencers (Applied Biosystems) are used for fragment analysis, primers may be labeled with four dyes (6-FAM, VIC, NED, and PET), combined with LIZ 500 or LIZ 600 as internal size standards.
2. The stock solution is prepared by dissolving oligonucleotide primers in ddH_2O to a final concentration of 100 μM. The stock solution is further diluted to a final working concentration of 20 μM. All labeled oligonucleotide solutions should be protected from light.

Table 1
Sequences of oligonucleotide primers used to amplify 11 microsatellite loci of *Plasmodium falciparum*

Name	5′ Label	Sequence 5′-3′	Chromosome
Polyα-F		AAAATATAGACGAACAGA	4
Polyα-R	VIC	ATCAGATAATTGTTGGTA	
TA60-F	VIC	CTCAAAGAAAAATAATTCA	13
TA60-R		AAAAAGGAGGATAAATACAT	
ARA2-F	NED	GTACATATGAATCACCAA	11
ARA2-R		GCTTTGAGTATTATTAATA	
PfG377-F		GATCTCAACGGAAATTAT	12
PfG377-R	NED	TTATGTTGGTACCGTGT	
PfPK2-F		CTTTCATCGATACTACGA	12
PfPK2-R	NED	CCTCAGACTGAAATGCAT	
TA87-F	VIC	ATGGGTTAAATGAGGTACA	6
TA87-R		ACATGTTCATATTACTCAC	
TA109-F	FAM	TAGGGAACATCATAAGGAT	6
TA109-R		CCTATACCAAACATGCTAAA	
C2M3-F	NED	GGTTAATATGATCACAAAATG	2
C2M3-R		ATTGTTGATTCATGAAATGCA	
TAA81-F	FAM	GAAGAAATAAGGGAAGGT	5
TAA81-R		TTTCACACAACACAGGATT	
TAA42-F	VIC	ACAAAAGGGTGGTGATTCT	5
TAA42-R		GTATTATTACTACTACTAAAG	
2490-F		TTCTAAATAGATCCAAAG	10
2490-R	FAM	ATGATGTGCAGATGACGA	

Note: *F* forward, *R* reverse. The fluorescent dyes used to label primers were 6-FAM (6-carboxyfluorescein, "blue" label), VIC (Applied Biosystems proprietary "green" fluorescent dye), and NED (Applied Biosystems proprietary "yellow" fluorescent dye)

3.1.2 PCR Master Mix

Reagents are stored at −20 °C and thawed shortly before use.

1. Gently vortex and briefly centrifuge all solutions after thawing.
2. Prepare sufficient master mix, on ice, for the number of reactions plus one extra.

3. Add the reagents in the following order:
   - ddH_2O to bring total volume to 13 μL
   - 10× *Taq* buffer with $(NH_4)_2SO_4$, 1.5 μL
   - $MgCl_2$ solution (25 mM), 1.2 μL
   - dNTP mix solution (10 mM of each), 0.6 μL
   - Oligonucleotide primers (20-μM working solution), 1.5 μL of each (forward and reverse)
   - *Taq* DNA polymerase (5 U/μL), 0.2 μL
4. Aliquot 13 μL of PCR master mix into individual thin-walled PCR tubes.
5. Add 2 μL of the template DNA solution or water (for the negative controls) and mix well by pipetting. To avoid contamination of reagents with other templates and amplicons, set up the PCR mixture in an area that has not been used for DNA extraction, thermal cycling, or fragment analysis.
6. Pulse the tubes in the microcentrifuge, place them in the thermal cycler, and run the cycling program.

3.1.3 PCR Cycling Parameters

The same cycling parameters are used for all oligonucleotide primer pairs: (1) 94 °C for 2 min; (2) 94 °C for 30 s; (3) 42 °C for 30 s; (4) 40 °C for 30 s; (5) 65 °C for 30 s; (6) repeat steps 2–5 for 40 cycles; (7) 65 °C for 5 min; (8) hold at 15 °C.

3.1.4 Fragment Analysis

This protocol has been used on an ABI 310 capillary sequencer and may require further optimization when different equipment is used.

1. Prepare sufficient loading buffer, on ice, for the number of reactions plus one extra. The loading buffer consists of 14.75 μL of Hi-Di formamide and 0.25 μL of ROX-500 size standard (total volume, 15 μL per sample).
2. Mix well by pipetting and aliquot 15 μL of loading buffer into individual 0.5-mL sample tubes (catalog number 401957, Applied Biosystems).
3. Dilute the PCR products (1:10 v/v) with ddH_2O. Briefly centrifuge the PCR tubes before opening them. The dilution factor for PCR products may vary according to the PCR yield, type of label, and DNA sequencer used.
4. Add 1 μL of each amplicon into individual sample tubes containing 15 μL of loading buffer. Always include a negative control (formamide plus ROX) to be sure that your reagents are not contaminated with labeled PCR products. For capillary electrophoresis, amplicons are pooled as follows: TA60 + ARA2, PfG377 + TA87, PfPK2 + TA109, TA81 + TA42, and Ploy∝ + 2490 + C2M3.
5. Cap the sample tubes with the septas (catalog number 401956, Applied Biosystems). The septas can be reused after washing and drying at room temperature.

6. Denature the samples in a heating block for 5 min at 94 °C.
7. Cool down the samples on ice for 5 min and load them immediately in the automated DNA sequencer.

3.2 *P. vivax* Microsatellites

3.2.1 Oligonucleotides

The set of oligonucleotides used to type 14 single-copy microsatellites of *P. vivax* (37) is described in Table 2. These loci contain either tetra- (MS2) or trinucleotide repeats (all other loci). Six of them (MS1, MS3, MS4, MS7, MS9, and MS15) contain perfect repeats, while the others have degenerate repeats that may be inconsistent with a standard stepwise mutation model (see Note 5). A CTGTCTT tail (lowercase letters in Table 2) was added to the 5′ end of the reverse primers to promote nontemplate-directed nucleotide addition (+A) to amplicons in a reproducible way (see Note 6).

The oligonucleotides are supplied lyophilized by Applied Biosystems; stock (100 μM) and working (40 μM) solutions are prepared as described for *P. falciparum* microsatellites. We routinely label the primers with the fluorescent dyes 6-FAM (6-carboxyfluorescein, "blue" label), VIC (Applied Biosystems proprietary "green" fluorescent dye), or NED (Applied Biosystems proprietary "yellow" fluorescent dye), but four labels may also be used, combined with LIZ 500 or LIZ 600 size standards (37). Labeled oligonucleotides should be protected from light.

3.2.2 PCR Master Mix

Reagents are stored at −20 °C and thawed shortly before use.

1. Gently vortex and briefly centrifuge all solutions after thawing.
2. Prepare sufficient master mix, on ice, for the number of reactions plus one extra.
3. Add the reagents in the following order:
   - ddH_2O to bring total volume to 12 μL
   - 10× *Taq* buffer with $(NH_4)_2SO_4$, 1.5 μL
   - $MgCl_2$ solution (25 mM), 1.2 μL
   - dNTP mix solution (10 mM of each), 0.6 μL
   - Oligonucleotide primers (40 μM working solution), 0.7 μL of each (forward and reverse)
   - *Taq* DNA polymerase (5 U/μL), 0.25 μL
4. Aliquot 12 μL of PCR master mix into individual thin-walled PCR tubes.
5. Add 3 μL of the template DNA solution or water (for the negative controls) and mix well by pipetting. Follow standard precautions to prevent PCR contamination.
6. Pulse the tubes in the microcentrifuge, place them in the thermal cycler, and run the cycling program.

Table 2
Sequences of oligonucleotide primers used to amplify 14 microsatellite loci of *Plasmodium vivax*

Name	5′ Label	Sequence 5′-3′	Chromosome
MS1-F	FAM	TCAACTGTTGGAAGGGCAAT	3
MS1-R		ctgtcttTTGCTGCGTTTTTGTTTCTG	
MS2-F	VIC	GAGCTAGCCAAAGGTTCAACA	6
MS2-R		ctgtcttTGGGGAGAGACTCCCTTTTC	
MS3-F	NED	GAAGATCCTGTGGAGGAGCA	4
MS3-R		ctgtcttCTCCTTCGCTCCTTTCCTTT	
MS4-F	FAM	CGATTTACTGTTGACGCTGAA	6
MS4-R		ctgtcttCAAAGGAACATGCTCGATGA	
MS5-F	NED	CGTCCTCTATCGCGTACACA	6
MS5-R		ctgtcttAAAGGGAGAGGAGCGAAAAC	
MS6-F	VIC	GGTTCTTCGGTGATCTCTGC	11
MS6-R		ctgtcttGGAGGAÇATCAACGGGATT	
MS7-F	FAM	TTGCAGAAAATGCAGAGAGC	12
MS7-R		ctgtcttAGGGTCTTCAGCGTGTTGTT	
MS8-F	NED	AGAGGAGGCAGAAATGCAGA	12
MS8-R		ctgtcttAGCCCCTTTGCGTTCTTTAT	
MS9-F	FAM	AGATGCCTACACGTTGACGA	8
MS9-R		ctgtcttGAAGCTGCCCATGTGGTAAT	
MS10-F	NED	TTATCCCTGCTGGATGTGAA	13
MS10-R		ctgtcttTCCTTCAGGTGGGACTTGTT	
MS12-F	FAM	AATGCGCATCCTATGTCTCC	5
MS12-R		ctgtcttCTGCTGTTGTTGTTGCTGCT	
MS15-F	FAM	TGTTTGCAAAGGAATCCACA	5
MS15-R		ctgtcttCGGCCAGATGAAAAGGATAA	
MS16-F	NED	TGTTGTGGTTGTTGATGGTGA	9
MS16-R		ctgtcttGTCGGGGAGAACAACAACAT	
MS20-F	VIC	GCACAACAAATGCAAGATCC	10
MS20-R		ctgtcttGTGGCAGTGGCTCATCTTCT	

Note: *F* forward, *R* reverse. The fluorescent dyes used to label primers were 6-FAM (6-carboxyfluorescein, "blue" label), VIC (Applied Biosystems proprietary "green" fluorescent dye), and NED (Applied Biosystems proprietary "yellow" fluorescent dye). A CTGTCTT tail (lowercase letters) was added to the 5′ end of the reverse primers to promote nontemplate-directed nucleotide addition to amplicons in a reproducible way (see Note 6)

3.2.3 PCR Cycling Parameters

The same cycling parameters are used for all oligonucleotide primer pairs: (1) 94 °C for 2 min; (2) 94 °C for 30 s; (3) 58 °C for 40 s; (4) 72 °C for 50 s; (5) repeat steps 2–4 for 35 cycles; (6) 72 °C for 5 min; (7) hold at 15 °C.

This PCR protocol was optimized with recombinant *Taq* DNA polymerase supplied by Fermentas (Burlington, Canada) on a GeneAmp PCR System 9700 thermal cycler (Applied Biosystems). Similar (although not identical) cycling parameters have been used with Platinum Taq DNA polymerase (Invitrogen) on a PTC-200 thermal cycler (MJ Research) (37) and with HotStarTaq Plus (QIAGEN) on a PTC-100 thermal cycler (MJ Research) (20).

3.2.4 Fragment Analysis

This protocol has been used on an ABI 310 capillary sequencer and may require further optimization when different equipment is used.

1. Prepare sufficient loading buffer, on ice, for the number of reactions plus one extra. The loading buffer consists of 14.75 μL of Hi-Di formamide and 0.25 μL of ROX-500 size standard (total volume, 15 μL per sample).
2. Mix well by pipetting and aliquot the loading buffer into individual 0.5-mL sample tubes (catalog number 401957, Applied Biosystems).
3. Dilute the PCR products (1:10 v/v) with ddH$_2$O. Briefly centrifuge the PCR tubes before opening them. The dilution factor for PCR products varies according to the PCR yield, type of label, and DNA sequencer used. VIC-labeled amplicons, for example, are often run at a 1:20 dilution on our ABI 310 DNA sequencer. Amplicons must be diluted to 1:50 or even 1:100 for analysis on ABI 3500 or 3700 sequencers. Amplicons may also be frozen and analyzed up to 1 week after PCR amplification.
4. Add 1 μL of each diluted amplicon into individual sample tubes containing 15 μL of loading buffer. Always include a negative control. For capillary electrophoresis, amplicons are pooled as follows: MS1 + MS3 + MS9, MS2 + MS5, MS4 + MS6 + MS10, MS7 + MS15, MS8 + MS12, and MS16 + MS20.
5. Cap the sample tubes with the septas (catalog number 401956, Applied Biosystems).
6. Denature the samples in a heating block for 5 min at 94 °C.
7. Cool down the samples on ice for 5 min, and load them immediately in the automated DNA sequencer.

3.3 Data Analysis

After electrophoresis, fragment sizes are scored using either commercially available software (such as GeneMapper 4.1, Applied Biosystems) or free software (such as STRand version 2.3.79, available at http://www.vgl.ucdavis.edu/informatics/strand.php). Because all microsatellite loci used here are single-copy genes, the presence of two or more alleles at one or more loci indicates a naturally occurring mixed-clone infection.

The relative abundance of alleles is inferred from peak heights in electropherograms (measured in arbitrary fluorescence units). We score two alleles at a locus when the minor peak was more than one-third the height of the predominant peak (38, 39). Infections are considered to contain multiple clones if one or more loci show more than one allele.

Multilocus haplotypes, which characterize parasite lineages, are defined as unique combinations of alleles at each locus analyzed; only the predominant alleles are considered for haplotype assignment in multiple-clone infections (34).

4 Notes

1. Reappearance of parasitemia after drug treatment can result from either recrudescence of surviving asexual blood-stage parasites, relapse from dormant liver stages known as hypnozoites (that only exist, among human malaria parasites, for *P. vivax* and *P. ovale*), or new infections with unrelated parasites. Molecular genotyping of paired parasite samples usually makes a distinction between recrudescences (with the same genotype as the initial infection) and new infections (with a different genotype) (29). Until recently, *P. vivax* relapses were thought to be caused by hypnozoites that are genetically identical to the blood-stage parasites found in primary infections, but this view has been challenged by the finding of different parasite genotypes in primary infections and relapses most *P. vivax*-infected patients from Thailand, India, and Myanmar who provided paired blood samples for multilocus analysis (30). The current consensus is that relapses may originate from reactivation of either the same parasite clone found in the primary bloodstream infection (homologous hypnozoites) or another, genetically different clone (heterologous hypnozoites).
2. We find that, when the plateau phase of amplification is reached, the first amplification step in nested PCR protocols may reduce differences between relative proportions of alleles that were present in the original DNA template obtained from multiple-clone infections. Because it is impossible to predict whether the amplification plateau will be reached when using template DNA from patient-derived samples with widely variable parasitemias, cycling parameters cannot be tailored for specific samples. As a consequence, the predominant haplotypes in multiple-clone infections may be inaccurately assigned. For further discussion on challenges for microsatellite haplotype assignment in multiple-clone *P. falciparum* and *P. vivax* infections, see ref. 38 and 39.

3. The use of isothermal whole-genome amplification (WGA) of template DNA, with multiple displacement technology, prior to PCR-based microsatellite typing has been validated for both *P. falciparum* (40) and *P. vivax* (15, 37). Although a report suggests that WGA prior to PCR may result in the preferential detection of some alleles of a human gene (41), we found no evidence for such biases when typing *P. vivax* microsatellite markers in mixtures of DNA templates with different proportions of each allele (39).
4. Many specialists recommend the use of specialized multiplex PCR buffers, such as QIAGEN PCR multiplex kit, to standardize multiplex amplification protocols (1).
5. The stepwise mutation model assumes that a single mutational event is more likely to add or subtract one than two or more repeats. Under this model, a continuum of similarities (same size, similar size, very different size) may potentially be defined. Therefore, size differences convey additional information about the relationships between alleles. However, complex mutation events that result in the addition or deletion of multiple repeat units are often seen in degenerate repeats, violating the stepwise mutation model.
6. The tendency of *Taq* polymerase to add a nontemplated nucleotide (usually A) to the 3′ end of the amplicon (usually referred to as A+) is a potential source of inaccuracy when determining microsatellite allele size (42). Adding a 5′ tail to the nonlabeled primer, such as GCTTCT (43) or CTGTCTT (44), may favor the nontemplated addition and drive the reaction to A+ amplicons in a consistent way.

Acknowledgment

This work was supported by funds from the National Institutes of Health (NIH) grant RO1 AI 075416-01, the Conselho Nacional de Desenvolvimento Científico e Tecnológico (CNPq) grant 470570/2006-7, and the Fundação de Amparo à Pesquisa do Estado de São Paulo (FAPESP) grants 05/51988-0 and 07/51199-0. POS and MUF receive or received scholarships from CNPq.

References

1. Guichoux E, Lagache L, Wagner S, Chumeil P, Léger P, Lepais O, Lepoittevin C, Malausa T, Revardel E, Salin F, Petit RJ (2011) Current trends in microsatellite genotyping. Mol Ecol Resour 11:591–611
2. Schlötterer C (1998) Genome evolution: are microsatellites really simple sequences? Curr Biol 8:R132–R134
3. Su X, Wellems TE (1996) Toward a high-resolution *Plasmodium falciparum* linkage map: polymorphic markers from hundreds of simple sequence repeats. Genomics 33: 430–444
4. Su X, Ferdig MT, Huang Y, Huynh CQ, Liu A, You J, Wootton JC, Wellems TE (1999) A genetic map and recombination parameters of

the human malaria parasite *Plasmodium falciparum*. Science 286:1351–1353

5. Su XZ, Hayton K, Wellems TE (2007) Genetic linkage and association analyses for trait mapping in *Plasmodium falciparum*. Nat Rev Genet 8:497–506
6. Anderson TJC, Haubold B, Williams JT, Estrada-Franco JG, Richardson L, Mollinedo R, Bockarie M, Mokili J, Mharakurwa S, French N, Whitworth J, Velez ID, Brockman AH, Nosten F, Ferreira MU, Day KP (2000) Microsatellite markers reveal a spectrum of population structures in the malaria parasite *Plasmodium falciparum*. Mol Biol Evol 17:1467–1482
7. Machado RLD, Póvoa MM, Calvosa VSP, Ferreira MU, Rossit ARB, dos Santos EJM, Conway DJ (2004) Genetic structure of *Plasmodium falciparum* populations in the Brazilian Amazon region. J Infect Dis 190:1547–1555
8. Anthony TG, Conway DJ, Cox-Singh J, Matusop A, Ratnam S, Shamsul S, Singh B (2005) Fragmented population structure of *Plasmodium falciparum* in a region of declining endemicity. J Infect Dis 191:1558–1564
9. Orjuela-Sánchez P, da Silva-Nunes M, da Silva NS, Scopel KK, Gonçalves RM, Malafronte RS, Ferreira MU (2009) Population dynamics of genetically diverse *Plasmodium falciparum* lineages: community-based prospective study in rural Amazonia. Parasitology 136:1097–1105
10. Conway DJ (2007) Molecular epidemiology of malaria. Clin Microbiol Rev 20:188–204
11. Carlton JM, Adams JH, Silva JC, Bidwell SL, Lorenzi H, Caler E, Crabtree J, Angiuoli SV, Merino EF, Amedeo P, Cheng Q, Coulson RM, Crabb BS, del Portillo HA, Essien K, Feldblyum TV, Fernandez-Becerra C, Gilson PR, Gueye AH, Guo X, Kang'a S, Kooij TW, Korsinczky M, Meyer EV, Nene V, Paulsen I, White O, Ralph SA, Ren Q, Sargeant TJ, Salzberg SL, Stoeckert CJ, Sullivan SA, Yamamoto MM, Hoffman SL, Wortman JR, Gardner MJ, Galinski MR, Barnwell JW, Fraser-Liggett CM (2008) Comparative genomics of the neglected human malaria parasite *Plasmodium vivax*. Nature 455:757–763
12. Imwong M, Sudimack D, Pukrittayakamee S, Osório L, Carlton JM, Day NPJ, White NJ, Anderson TJC (2006) Microsatellite variation, repeat array length and population history of *Plasmodium vivax*. Mol Biol Evol 23: 1016–1018
13. Imwong M, Nair S, Pukrittayakamee S, Sudimack D, Williams JT, Mayxay M, Newton PN, Kim JR, Nandy A, Osorio L, Carlton JM, White NJ, Day NPJ, Anderson TJ (2007) Contrasting genetic structure in *Plasmodium vivax* populations from Asia and South America. Int J Parasitol 37:1013–1022
14. Ferreira MU, Karunaweera ND, da Silva-Nunes M, da Silva NS, Wirth DF, Hartl DL (2007) Population structure and transmission dynamics of *Plasmodium vivax* in rural Amazonia. J Infect Dis 195:1218–1226
15. Karunaweera ND, Ferreira MU, Munasinghe A, Barnwell JW, Collins WE, King CL, Kawamoto F, Hartl DL, Wirth DF (2008) Extensive microsatellite diversity in the human malaria parasite *Plasmodium vivax*. Gene 410:105–112
16. Joy DA, Gonzalez-Ceron L, Carlton JM, Gueye A, Fay M, McCutchan TF, Su XZ (2008) Local adaptation and vector-mediated population structure in *Plasmodium vivax* malaria. Mol Biol Evol 25:1245–1252
17. Orjuela-Sánchez P, da Silva NS, da Silva-Nunes M, Ferreira MU (2009) Recurrent parasitemias and population dynamics of *Plasmodium vivax* polymorphisms in rural Amazonia. Am J Trop Med Hyg 81:961–968
18. Rezende AM, Tarazona-Santos E, Fontes CJ, Souza JM, Couto AD, Carvalho LH, Brito CF (2010) Microsatellite loci: determining the genetic variability of *Plasmodium vivax*. Trop Med Int Health 15:718–726
19. Gunawardena S, Karunaweera ND, Ferreira MU, Phone-Kyaw M, Pollack RJ, Alifrangis M, Rajakaruna RS, Konradsen F, Amerasinghe PH, Schousboe ML, Galappaththy GN, Abeyasinghe RR, Hartl DL, Wirth DF (2010) Geographic structure of *Plasmodium vivax*: microsatellite analysis of parasite populations from Sri Lanka, Myanmar, and Ethiopia. Am J Trop Med Hyg 82:235–242
20. van den Eede P, Erhart A, van der Auwera G, van Overmeir C, Thang ND, le Hung X, Anné J, D'Alessandro U (2010) High complexity of *Plasmodium vivax* infections in symptomatic patients from a rural community in central Vietnam detected by microsatellite genotyping. Am J Trop Med Hyg 82:223–227
21. van den Eede P, van der Auwera G, Delgado C, Huyse T, Soto-Calle VE, Gamboa D, Grande T, Rodríguez H, Llanos A, Anné J, Erhart A, D'Alessandro U (2010) Multilocus genotyping reveals high heterogeneity and strong local population structure of the *Plasmodium vivax* population in the Peruvian Amazon. Malar J 9:151
22. Bruce MC, Macheso A, Galinski MR, Barnwell JW (2007) Characterization and application of multiple genetic markers for *Plasmodium malariae*. Parasitology 134:637–650
23. Li J, Zhang Y, Sullivan M, Hong L, Huang L, Lu F, McCutchan TF, Su XZ (2007) Typing

Plasmodium yoelii microsatellites using a simple and affordable florescent labeling method. Mol Biochem Parasitol 155:94–102

24. Li J, Zhang Y, Liu S, Hong L, Sullivan M, McCutchan TF, Carlton JM, Su XZ (2009) Hundreds of microsatellites for genotyping *Plasmodium yoelii* parasites. Mol Biochem Parasitol 166:153–158
25. Schall JJ, Vardo AM (2007) Identification of microsatellite markers in *Plasmodium mexicanum,* a lizard malaria parasite that infects nucleated erythrocytes. Mol Ecol Notes 7:227–229
26. Havryliuk T, Ferreira MU (2009) A closer look at multiple-clone *Plasmodium vivax* infections: detection methods, prevalence and consequences. Mem Inst Oswaldo Cruz 104: 67–73
27. Nyachieo A, van Overmeir C, Laurent T, Dujardin JC, D'Alessandro U (2005) *Plasmodium falciparum* genotyping by microsatellites as a method to distinguish between recrudescent and new infections. Am J Trop Med Hyg 73:210–213
28. Mwangi JM, Omar SA, Ranford-Cartwright LC (2006) Comparison of microsatellite and antigen-coding loci for differentiating recrudescing *Plasmodium falciparum* infections from reinfections in Kenya. Int J Parasitol 36:329–336
29. Juliano JJ, Gadalla N, Sutherland CJ, Meshnick SR (2010) The perils of PCR: can we accurately "correct" antimalarial trials? Trends Parasitol 26:119–124
30. Imwong M, Snounou G, Pukrittayakamee S, Tanomsing N, Kim JR, Nandy A, Guthmann JP, Nosten F, Carlton J, Looareesuwan S, Nair S, Sudimack D, Day NP, Anderson TJ, White NJ (2007) Relapses of *Plasmodium vivax* infection usually result from activation of heterologous hypnozoites. J Infect Dis 195:927–933
31. van den Eede P, Soto-Calle VE, Delgado C, Gamboa D, Grande T, Rodríguez H, Llanos-Cuentas A, Anné J, D'Alessandro U, Erhart A (2011) *Plasmodium vivax* sub-patent infections after radical treatment are common in Peruvian patients: results of a 1-year prospective cohort study. PLoS One 6:e16257
32. Restrepo E, Imwong M, Rojas W, Carmona-Fonseca J, Maestre A (2011) High genetic polymorphism of relapsing *P. vivax* isolates in northwestern Colombia. Acta Trop 119: 23–29
33. Bruce MC, Macheso A, McConnachie A, Molineux ME (2011) Comparative population structure of *Plasmodium malariae* and *Plasmodium falciparum* under different transmission settings in Malawi. Malar J 10:38
34. Anderson TJC, Su XZ, Bockarie M, Lagog M, Day KP (1999) Twelve microsatellite markers for characterization of *Plasmodium falciparum* from finger-prick blood samples. Parasitology 119:113–125
35. Koepfli C, Mueller I, Marfurt J, Goroti M, Sie A, Oa O, Genton B, Beck HP, Felger I (2009) Evaluation of *Plasmodium vivax* genotyping markers for molecular monitoring in clinical trials. J Infect Dis 199:1074–1080
36. Dean FB, Hosono S, Fang L, Wu X, Farugi AF, Bray-Ward P, Sun Z, Zong Q, Du Y, Du J, Driscoll M, Song W, Kingsmore SF, Egholm M, Lasken RS (2002) Comprehensive human genome amplification using multiple displacement amplification. Proc Natl Acad Sci USA 99:5261–5266
37. Karunaweera ND, Ferreira MU, Hartl DL, Wirth DF (2007) Fourteen polymorphic microsatellite DNA markers for the human malaria parasite *Plasmodium vivax.* Mol Ecol Notes 7:172–175
38. Greenhouse B, Myrick A, Dokomajilar C, Woo JM, Carlson EJ, Rosenthal PJ, Dorsey G (2006) Validation of microsatellite markers for use in genotyping polyclonal *Plasmodium falciparum* infections. Am J Trop Med Hyg 75:836–842
39. Havryliuk T, Orjuela-Sánchez P, Ferreira MU (2008) *Plasmodium vivax*: microsatellite analysis of multiple-clone infections. Exp Parasitol 120:330–336
40. Wang Y, Nair S, Anderson TJC (2009) Multiple displacement amplification of malaria parasite DNA. J Parasitol 95:253–255
41. Murthy KK, Mahboubi VS, Santiago A, Barragan MT, Knoll R, Schultheiss HP, O'Connor DT, Schork NJ, Rana BK (2005) Assessment of multiple displacement amplification for polymorphism discovery and haplotype determination at a highly polymorphic locus, *MC1R*. Hum Mutat 26:145–152
42. Ballard LW, Adams PS, Bao Y, Bartley D, Bintzler D, Kasch L, Petukova L, Rosato C (2002) Strategies for genotyping: effectiveness of tailing primers to increase accuracy in short tandem repeat determinations. J Biomol Tech 13:20–29
43. Brownstein MJ, Carpten JD, Smith JR (1996) Modulation of non-templated nucleotide addition by *Taq* DNA polymerase: primer modifications that facilitate genotyping. Biotechniques 20:1004–1010
44. Raby BA, Silverman EK, Lazarus R, Lange C, Kwiatkowiski DJ, Weiss ST (2003) Chromosome 12q harbors multiple genetic loci related to asthma and asthma-related phenotypes. Hum Mol Genet 12:1973–1979

Chapter 18

Informativeness of Microsatellite Markers

M. Humberto Reyes-Valdés

Abstract

Simple sequence repeats (SSR) are extensively used as genetic markers for studies of diversity, genetic mapping, and cultivar discrimination. The informativeness of a given SSR locus or a loci group depends on the number of alleles, their frequency distribution, as well as the kind of application. Here I describe several methods for calculating marker informativeness, all of them suitable for SSR polymorphisms, proposed by several authors and synthesized in an Information Theory framework. Additionally, free access software resources are described as well as their application through worked examples.

Key words Marker informativeness, Microsatellites, Information theory, PIC, Coancestry, Cultivar discrimination, QTL mapping, Software

1 Introduction

1.1 Microsatellites

Simple sequence repeats, also called microsatellites, are ubiquitous on eukaryotic genomes. They are usually composed by di- or three-nucleotide sequences, repeated around ten times. Their sequence patterns induce hypervariability in the number of repeats across any given locus, due to phenomena related to DNA replication and recombination. This high variation in length has proven to be highly useful for genetic marking, scored through amplification by the polymerase chain reaction (PCR). As it has been the case with other genetic markers, microsatellite polymorphisms have been successfully applied in areas such as the study of genetic diversity, genetic mapping, and cultivar identification.

The informativeness of SSR markers varies across loci and populations. It depends mainly on the number of alleles and their frequencies. Furthermore, their informativeness depends on the type and strategy of application. Thus, it is important to identify informative SSR markers and quantitatively evaluate their informativeness in order to delineate optimum strategies for their use, in terms of maximum efficiency and minimum cost.

Stella K. Kantartzi (ed.), *Microsatellites: Methods and Protocols*, Methods in Molecular Biology, vol. 1006, DOI 10.1007/978-1-62703-389-3_18, © Springer Science+Business Media, LLC 2013

1.2 Information Theory

I will base the general approach to informativeness calculation for SSR markers in the framework of information theory, a branch of mathematics dedicated to the storage, transmission, recovering, and measuring of information. The pioneer work in this subject was made by Claude Shannon (1), while he was working for the Bell Laboratories. His theory was based on the so-called information channel, which comprises a source of information, an encoder, a noisy channel, a decoder, and a destination. A key concept in information theory is the Shannon entropy, a measure of uncertainty. For a discrete variable, the Shannon entropy of the variable M is given by the following equation:

$$H(M) = -\sum_{i=1}^{g} p_i \log_2 p_i,$$

where $p_1, p_2, \ldots, p_g$ are probabilities assigned to the possible values of $m_1, m_2, \ldots, m_g$ of a random variable M. For g possible values of a discrete random variable, the maximal value of the Shannon entropy is $\log_2(g)$, occurring when $p_1 = p_2 = \ldots = p_g$, whereas the minimum is 0 for any $p_i = 1$. In the previous equation, the expression $0 \log_2(0)$ equals 0 by definition. Based on the entropy concept, the mutual information between two variables X and M is defined as the average reduction in the uncertainty about X given knowledge of the value of M, in accordance with the following expression:

$$I(X;M) = H(X) - H(X \mid M) = H(M) - H(M \mid X),$$

where $H(X|M)$ is the average entropy or uncertainty in X, given knowledge of the value of the variable M. Information is symmetrically defined in terms of entropies; in fact, the expression for $I(X;M)$ can also be defined as the information conveyed about a variable M by the variable X, and it can be also written as $I(M;X)$. The Shannon entropy has been applied in several situations involving genetic markers, for example, in the measurement of linkage disequilibrium (2), inference of ancestry (3), SNP selection for association studies (4, 5), statistics for association (6), information for QTL mapping (7), and transcriptome analysis (8). The entropy concept can be used as a general, firmly mathematically founded framework for calculating information provided by genetic markers for several applications.

1.3 Informativeness for Genetic Markers

The so-called Polymorphism Information Content or PIC (9) is a statistic defined to one particular type of human pedigree: one parent is affected by a rare dominant disease and is heterozygous at the disease locus, whereas the other parent is unaffected by the disease. This locus is associated with a marker with several codominant alleles. In this context, an offspring is said to be informative if we can infer from his genotype which marker allele is co-inherited with the disease allele. Thus, PIC is defined as the expected fraction of informative offspring from this type of pedigree (10). The expression for this statistic is assuming Hardy–Weinberg equilibrium:

$$\mathrm{PIC} = 1 - \sum_{i=1}^{a} p_i^2 - \sum_{i=1}^{a-1} \sum_{j=i+1}^{a} 2(p_i p_j)^2,$$

where p_i is the frequency of the i-th marker allele and a is the number of different alleles. Since PIC is the proportion of completely informative offspring, and each informative offspring allows the choice between two possible alleles as the co-inherited one, thus producing a mutual information of 1, it can also be considered as average mutual information in accordance with the Shannon theory. Alternatively, for the same type of application, heterozygosity can be used and it is estimated as follows:

$$\mathrm{HET} = 1 - \sum_{i=1}^{a} p_i^2.$$

The PIC statistic will always be equal or lower than heterozygosity, both measures being strongly correlated.

An informativeness expression, often called PIC too (11), has been used as a part of a strategy for the choice of parents in the construction of linkage maps with RFLP markers, based on the concept of gene diversity (12):

$$\mathrm{GS} = 1 - \sum_{i=1}^{g} p_i^2,$$

where p_i is the frequency of the i-th RFLP pattern or any given marker with g genotypes. Since this expression is identical to the Gini–Simpson index (13, 14), originally applied to diversity analysis in ecology, I denote this index as GS. This expression with marker genotype frequencies has been used in several works for cultivar discrimination. It is useful because it estimates, for a large sample, the probability that two random chosen individuals or lines from a population have different banding patterns (15).

An appealing alternative to calculate information for homogeneous cultivar discrimination with marker data is the direct use of Shannon entropy, with p_i being the frequency of the i-th single locus or multilocus marker genotype. In fact, if cultivars are homogeneous, e.g., lines, hybrids, or clones, the conditional entropy of genetic markers given cultivars, say $H(M|X)$, becomes 0; thus, $I(X;M)$ becomes $H(M)$, i.e., the entropy of the distribution of marker frequencies. The following properties are fulfilled by this application of the Shannon entropy to N cultivars: (i) the minimum value is 0, and it is reached when the frequency of any marker genotype equals 1; (ii) the maximum value is $\log_2(N)$, and it occurs only when the marker genotypes allow distinction of all cultivars; (iii) for g marker genotypes, the maximum value, $\log_2(g)$, is attained when all of them have the same frequency; and (iv) the simultaneous mutual information provided by a set of independent markers is the sum of the individual marker information contents. Thus, the value of the Shannon entropy gives the information that the same number of independent, fully informative, binary loci would theoretically provide, or shortly, it is the ***effective number of binary loci***. From information theory, it turns out that the number of bits required to distinguish each unit among a set of N equiprobable cultivars is $\log_2(N)$.

Marker informativeness for inference of coancestry has been proposed with an information theory basis (3). The methodology was developed mainly for genetic mapping in humans, with the key parameter being informativeness for assignment for a given locus (I_n):

$$I_n(Q;J)=\sum_{j=1}^{a}\left(-p_j\log p_j+\sum_{i=1}^{N}\frac{p_{ij}}{N}\log p_{ij}\right),$$

where p_j is the average frequency of the allele j across N populations and p_{ij} is the frequency of the allele j in population i. This is the mutual information between the population Q and an individual allele J. For a given set of populations, the minimum value of I_n occurs when all alleles have the same frequency across populations, and the maximal value $\log(N)$ occurs when $a \geq N$ and no allele is found in more than one population.

The entropy-based founder informativeness was developed for QTL analysis (7). The goal of this statistic is to measure the amount of information about the putative QTL genotype in a given genome site in a linkage map. Assume that for a given locus in a mapping population, there are f putative QTL genotypes, e.g., QQ, Qq, and qq, with probabilities p_1, p_2, …, p_f. The entropy-based founder informativeness, based on marker information, at the map location m in a given member of a mapping population is

$$\mathrm{EFI}(m)=\mathrm{Max}(H)-\sum_{i=1}^{f}p_i\log_2 p_i,$$

where $\mathrm{Max}(H)$ is the maximum entropy of the ensemble of putative QTL genotypes, calculated without marker information and assuming Mendelian segregation. The same paper (7) provides a table for $\mathrm{Max}(H)$ in several mapping populations, whereas probabilities of putative genotypes are calculated in most QTL analysis approaches. The $\mathrm{EFI}(m)$ values averaged for population members at regular intervals across a linkage map allow drawing an information content map.

The approaches described herein can be applied to several types of genetic markers, and all of them are suitable for SSR polymorphisms.

2 Software

I briefly describe the software that can be used to perform the above calculations. However, this list does not discard other alternatives.

2.1 R

R (16) is free software for statistical computing and graphics. It runs in a wide variety of Unix versions, as well as Windows and MacOSX. It can be downloaded from http://www.r-project.org/ and it has a wide availability of packages for diverse applications.

2.2 R/qtl

R/qtl (17) is an R package for QTL analysis in experimental crosses that allows importing data from different standard formats. It uses several methods for QTL analysis, like maximum likelihood and linear regression. Also, it allows numerical calculation of statistical thresholds through permutation tests. Documentation and several tutorials can be downloaded from the R site http://www.r-project.org/.

2.3 Infocalc

The *infocalc* application (18) is a small Perl script, developed by Noah Rosenberg, for calculating statistics for ancestry information content of genetic markers (3). It can be downloaded at the site http://www.stanford.edu/group/rosenberglab/infocalc.html. The instructions are inside the script.

3 Methods

3.1 Polymorphism Information Content

To calculate PIC (9), the following R function can be used with a vector of allele frequencies as argument:

```
pic<-function(x){1-sum(x^2)-sum(x^2)^2+sum(x^4)}
```

Suppose that we have the following set of allele frequencies for a random mating Mendelian population: 0.1, 0.5, 0.2, 0.2. Paste the function on the R console, and after the > prompt execute the `pic` function with its arguments:

```
> pic(c(0.1,0.5,0.2,0.2))
[1] 0.6102
```

In this way we calculate the PIC value of 0.6102. To calculate the maximum PIC for *a* alleles, use the following function:

```
mPIC<-function(a){(a-1)^2*(a+1)/a^3}
```

Heterozygosity, often called PIC, can be calculated for a vector x of allele frequencies with the following R function:

```
het<-function(x){1-sum(x^2)}
```

For a sample size of n alleles, i.e., $2N$ diploid individuals, an unbiased estimation of heterozygosity is

```
het.unbiased<-function(x,n){het(x)*n/(n-1)}
```

Example: For a set of estimated allele frequencies: 0.1, 0.5, 0.2, 0.2, from a sample of 50 individuals, proceed as follows:

Paste in the R console both functions het and het.unbiased, and press <Enter>

For the basic heterozygosity estimation, write

```
> het(c(0.1,0.5,0.2,0.2))
[1] 0.66
```

To get an unbiased estimation of heterozygosity, type

```
> het.unbiased(c(0.1,0.5,0.2,0.2),100)
[1] 0.6666667
```

There is a web calculator of the PIC statistic and biased heterozygosity as the one given by the het function, designed by Steve Kemp (19).

3.2 Gini–Simpson Index for Genotypic Frequencies

The Gini–Simpson index, often used for cultivar diversity or informativeness for cultivar discrimination, can be calculated with the het function, applied on genotypic frequencies.

Example: Consider the following set of frequencies of cultivar marker genotypes: 0.4, 0.1, 0.2, 0.15, 0.15.

Paste and execute the het function in the R console, and type

```
> het(c(0.4,0.1,0.2,0.15,0.15))
[1] 0.745
```

3.3 Mutual Information for Cultivar Discrimination

If we have a set of homogeneous cultivars, e.g., lines, hybrids, or clones, we can estimate the mutual information between one or more marker loci and cultivar identity, thus providing a measure of the discrimination ability of the marker set. The raw material for calculation is the set of frequencies of marker genotypes, which is in turn used to calculate the Shannon entropy. The following R functions allow the necessary calculations:

```
MyLog2p<-function(x){if(x==0)   0   else   x*log(x,2)}
#Defining x logx
entropy<-function(x){-sum(sapply(x,MyLog2p))}
```

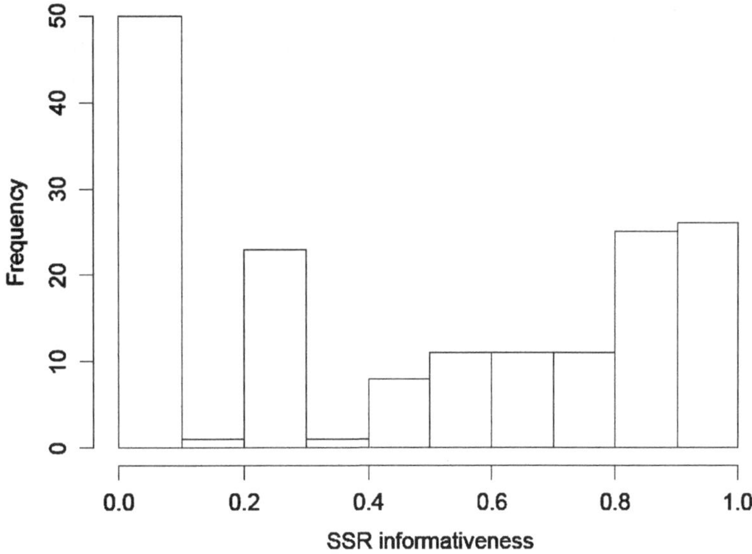

Fig. 1 Entropy-based SSR informativeness for 167 biallelic loci in soybean. Data provided by Stella Kantartzi

Assume a set of cultivars, with marker frequencies 0.091, 0.008, 0.005, 0.022, 0.086, 0.029, 0.090, 0.036, 0.047, 0.040, 0.012, 0.011, 0.087, 0.030, 0.034, 0.059, 0.042, 0.189, 0.013, 0.069. Proceed as follows:

```
> entropy(c(0.091,0.008,0.005,0.022,0.086,0.029,0.090,
0.036,
0.047,0.040,0.012,0.011,0.087,0.030,0.034,0.059,0.042,
0.189,
0.013, 0.069))
[1] 3.86779
```

This means that the marker information available for cultivar discrimination is 3.86779 bits, equivalent to the same number of fully informative independent binary markers, and enough to discriminate among $2^{3.86779} = 14.6$ cultivars. Obviously, there cannot be 14.6 cultivars, but this pictures the availability of information.

As a further example, I analyze homozygous SSR data on soybean lines from a biparental cross, kindly provided by Dr. Stella Kantartzi. In this case, information ranges from 0 to a maximum of 1, given that the biparental origin allows a maximum of two alleles. An informativeness of 0 is obtained for the SSR loci with the same genotype across lines, whereas the value of 1 is calculated for loci with a 50 % frequency of each genotype. In Fig. 1, we can appreciate the informativeness distribution of 167 SSR loci.

3.4 Marker Informativeness for Inference of Coancestry

The software *infocalc* (18) is used to calculate marker informativeness for coancestry, with one of the main parameters being informativeness for assignment (I_n). The used instructions are in the respective Perl script. The data file follows the STRUCTURE format, whose first line denotes the names of marker loci. The following lines include the genotype data of individuals, with the first five columns being individual identifiers, followed by the allele code for each locus. Each individual genotype is represented by two lines, with the order of the two alleles being irrelevant. Missing data are marked with a particular value, –9. The following five lines, taken from *infocalc*, represent codification for two individuals:

```
D9S1779   D9S1825   D7S2477   D17S784   D16S403   D3S1262
D10S189
854 86 Maya Mexico AMERICA 124 129 152 -9 138 112 186
854 86 Maya Mexico AMERICA 142 135 156 -9 140 124 186
855 86 Maya Mexico AMERICA 124 129 156 230 138 112 186
855 86 Maya Mexico AMERICA 124 129 164 234 140 112 186
```

The first line contains the names of seven marker loci. For each of the subsequent lines, the first five columns are individual code, population code, population name, country, and geographical region. The numbers that follow are either allele codes or the code –9 for missing data. Thus, for the Mayan individual coded with 854, the marker genotype for locus D9S1779 is the set of alleles 124 and 142, whose order does not indicate phase, thus being interchangeable. Weights can be defined, so a nonuniform prior for the populations can be accommodated.

If you use Unix, Linux, or MacOS X, Perl is most likely already installed. To get information about your Perl version, type perl –v at a command prompt. For Windows operating systems, the current standard Perl distribution is ActivePerl, from ActiveState, at http://www.activestate.com/ActivePerl/.

Example: I use the dataset provided by the *infocalc* site, mksp.stru, for data on four human populations: Maya, Karitiana, Suri, and Pima. For an unweighted analysis, proceed as follows:

Make the directory containing the dataset your home folder. Then type

```
./infocalc -column 3 -numpops 4 -input mksp.stru -output mksp.stru.out.txt <Enter>
```

The option `-column 3` states the population identifier column, –numpops 4 is the number of populations, `-input mksp.stru` is for the input file, and `-output mksp.stru.out.txt` is for the output file. The results are displayed as follows:

```
Locus   I_n      I_a       ORCA[1-allele]  ORCA[2-allele]
D10S189 0.761877 0.130766  0.61756         0.727457
D16S403 0.854949 0.167937  0.745536        0.87205
D17S784 0.342555 0.0572477 0.4625          0.599987
D3S1262 0.23707  0.0472184 0.460417        0.552591
```

```
D7S2477 0.332763 0.0614692 0.494048      0.607001
D9S1779 0.259119 0.0530722 0.416667      0.537326
D9S1825 0.0531744 0.0111935 0.327083     0.377795
Command: infocalc -column 3 -numpops 4 -input mksp.stru
-output mksp.stru.out.txt -weightfile [none]
PriorWeights: Karitiana 0.25 Maya 0.25 Pima 0.25 Surui
0.25
```

Besides calculating informativeness for assignment (I_n) for each locus, *infocalc* performs calculation for I_a, the informativeness for ancestry coefficients in the admixture model, and ORCA, the optimal rate of correct assignment (3). The last two lines recapitulate the options and the weights.

3.5 Information for QTL Mapping

Informativeness maps for QTL analysis can be drawn across linkage maps through entropy-based founder informativeness "EFI" (7). Since calculation requires probabilities of putative QTL genotypes across the linkage map, the R/qtl package may be used for common mapping populations. The trick is to extract those probabilities. In the example below, I use an anonymous recombinant inbred line (RIL) dataset and show how to perform this calculation. Since there are only two possible QTL genotypes, one from each parent, the maximum entropy is 1; thus, the entropy of the distribution of the putative QTL genotypes must be subtracted from 1 across the linkage map. I will not extend on details of how to use R/qtl, since the subject is extense and excellently covered by several manuals and one book (20).

Once the map file and the genotype file are saved on the working directory, the following script is executed in an R console:

```
> #Drawing a QTL information map
> dat<-
+
read.cross("mm",file="genotypes.raw",mapfile="mapR.map")
+      #Retrieving data in a MapMaker format
> class(dat)[1]<-"riself" #Declaring RILs
+   jittermap(dat) #Rectify markers located at the
+   same position
> dat <- calc.genoprob(dat, step=1,
+   error.prob=0.01,map.function="kosambi")#Calculate
+   probabilities, assuming a genotyping error rate of
    0.01
> attach(dat$geno[[3]])#Attaching   probabilities   to
  linkage
+   group 3
> dim(prob)#Check dimensions of the probability data
> MyLog2p<-function(x){if(x==0)   0   else   x*log(x,2)}
  #Define
+   function plog2p
> entropy<-function(x){-sum(sapply(x,MyLog2p))}
+   a<-NULL;length(a)<-dim(prob)[2];for(i in 1:dim(prob)
+   [2])a[i]<-1-mean(apply(prob[,i,1:2],1,entropy))
```

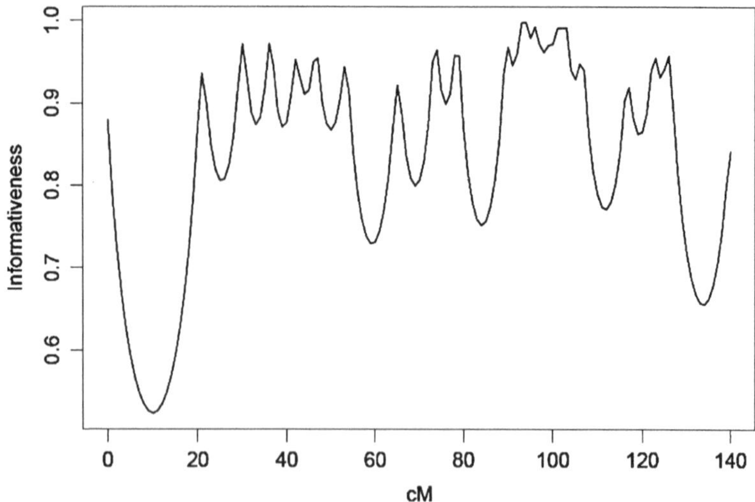

Fig. 2 Informativeness for QTL mapping along a 140 cM linkage group. Peaks correspond to SSR marker positions

```
> plot(cbind(c(0:140),a),type="l",xlab="cM",
+  ylab="Informativeness+   ")#Information   map   for
   linkage
+  group 3
```

The informativeness map for linkage group 3 is depicted in Fig. 2. The upward peaks on the plot represent local maxima for average EFI, and those points are usually located at the marker loci. One may note that even at the marker sites, EFI has not its maximum attainable of 1. There are two reasons: the first one is that we are setting a genotyping error rate of 0.01 and the second one is the presence of missing data records.

4 Notes

4.1 Recommendations on the use of R

The R object names are case sensitive; thus, one must be very careful when writing down commands and variables. One of the main problems with analyzing data is to have a correctly structured dataset in the correct directory and with a fairly simple name. In general, for file names one must keep in mind the following recommendations: (i) consider that file names are case sensitive in several systems; (ii) you can use upper and lower case letters, dots, numbers, and underscore symbol; (iii) it is better to avoid blank spaces; (iv) avoid the following characters in file names: "/," "&," "|," ":," ">," and "<." The character "/" is reserved as a directory and file name separator in a pathname; (iv) start your names with a letter or a number; and (v) make your names short but not cryptic. Datasets saved in tab separated plain text files, and in comma separated (csv) files work very well, and can be exported from Excel, OpenOffice, and LibreOffice, among others.

Sometimes things go wrong in R, because the columns in a data frame are not in the desired format. Therefore, one must make sure that we are dealing either with a factor or a numeric vector, etc., by using the class command. For example, if we are interested in analyzing a numerical vector *x*, and the command `class(x)` gives `factor` as the output, then we need to convert the variable. The following instruction has worked fine on my experience: `x<-as.numeric(as.vector(x))`.

Acknowledgements

I am thankful to Stella Kantarzi, who provided soybean SSR data to be used in one of the examples; to Noah Rosenberg, who reviewed the material related to his developments in marker informativeness; and to José Reyes, who checked my R scripts.

The R functions used in this book chapter can be accessed through the following link: http://www.uaaan.mx/~mhreyes/FunctionsChapterSSR.html.

References

1. Shannon CE (1948) A mathematical theory of communication. Bell Syst Tech J 27(379–423):623–656
2. Nothnagel M, Fürst R, Rhode K (2002) Entropy as a measure for linkage disequilibrium over multilocus haplotype blocks. Hum Hered 54:186–198
3. Rosenberg NA, Li LM, Ward R et al (2003) Informativeness of genetic markers for inference of ancestry. Am J Hum Genet 73:1402–1422
4. Hampe J, Schreiber S, Krawczak M (2003) Entropy-based SNP selection for genetic association studies. Hum Genet 114:36–43
5. Butler JM, Bishop DT, Barrett JH (2005) Strategies for selecting subsets of single-nucleotide polymorphisms to genotype in association studies. BMC Genet. doi:10.1186/1471-2156-6-S1-S72
6. Zhao J, Boerwinkle E, Xiong M (2005) An entropy-based statistic for genomewide association studies. Am J Hum Genet 77:27–40
7. Reyes-Valdés MH, Williams CG (2005) An entropy-based measure of founder informativeness. Genet Res 85:81–88
8. Martínez O, Reyes-Valdés MH (2008) Defining diversity, specialization, and gene specificity in transcriptomes through information theory. Proc Natl Acad Sci USA 105: 9709–9714
9. Botstein D, White RL, Skolnick M et al (1980) Construction of a genetic linkage map in man using restriction fragment length polymorphisms. Am J Hum Genet 32: 314–331
10. Hildebrand CE, Torney DC, Wagner RP (1992) Informativeness of polymorphic DNA markers. Los Alamos Sci 20:100–102
11. Anderson JA, Churchill GA, Autrique JE et al (1993) Optimizing parental selection for genetic linkage maps. Genome 36:181–186
12. Weir BS (1990) Genetic data analysis. Methods for discrete genetic data. Sinauer Associates, Inc., Sunderland, MA
13. Simpson EH (1949) Measurement of diversity. Nature 163:688
14. Jost L, Baños T (2006) Entropy and diversity. Oikos 113:373–375
15. Tessier C, David J, This P et al (1999) Optimization of the choice of molecular markers for varietal identification in *Vitis vinifera* L. Theor Appl Genet 98:171–177
16. R Development Core Team (2012) R: a language and environment for statistical comput-

ing. R Foundation for Statistical Computing. http://www.R-project.org

17. Broman KW, Wu H, Sen S et al (2003) R/qtl: QTL mapping in experimental crosses. Bioinformatics 19:889–890
18. Rosenberg N (2006) Infocalc—a program for calculating marker informativeness statistics. Version 1.1. http://www.stanford.edu/group/rosenberglab/infocalc.html
19. Kemp S (2002) PIC calculator. http://www.stanford.edu/group/rosenberglab/infocalc.html
20. Broman KW, Sen S (2009) A guide to QTL mapping with R/qtl. Springer, New York

Chapter 19

Microsatellite Data Analysis for Population Genetics

Kyung Seok Kim and Thomas W. Sappington

Abstract

Theories and analytical tools of population genetics have been widely applied for addressing various questions in the fields of ecological genetics, conservation biology, and any context where the role of dispersal or gene flow is important. Underlying much of population genetics is the analysis of variation at selectively neutral marker loci, and microsatellites continue to be a popular choice of marker. In recent decades, software programs to estimate population genetics parameters have been developed at an increasing pace as computational science and theoretical knowledge advance. Numerous population genetics software programs are presently available to analyze microsatellite genotype data, but only a handful are commonly employed for calculating parameters such as genetic variation, genetic structure, patterns of spatial and temporal gene flow, population demography, individual population assignment, and genetic relationships within and between populations. In this chapter, we introduce statistical analyses and relevant population genetic software programs that are commonly employed in the field of population genetics and molecular ecology.

Key words Population genetics, Genetics software, Genetic variation, Genetic structure, Gene flow, Microsatellites

1 Introduction

Population genetics is the study of the frequency and interaction of alleles and genes in populations. It has revolutionized many fields of evolutionary biology over the last 30 years and represents the essence of the modern evolutionary synthesis. Allele frequency in populations can change spatially and temporally under the influence of various evolutionary processes, particularly natural selection, genetic drift, mutation, gene flow, and mating system. Comparative analyses of spatial and temporal patterns in allele frequency provide an important entry point to identify the evolutionary forces that gave rise to them. In large part, however, the fundamental power and premise behind population genetics is that one can compare allele frequencies at selectively neutral marker loci to estimate gene flow, under reasonable assumptions about the rate of drift, mutation, and mating system. Gene flow is a parameter of critical importance in studies related to wildlife conservation, fisheries management,

Stella K. Kantartzi (ed.), *Microsatellites: Methods and Protocols*, Methods in Molecular Biology, vol. 1006,
DOI 10.1007/978-1-62703-389-3_19,

invasion biology and routes of invasion, insect resistance management, pest management, pest eradication programs, population and metapopulation dynamics, phylogeography, biosystematics, and many others. Gene flow is tied in a fundamental way to effective dispersal of the individuals carrying the genes, and information obtained about either for a given species informs understanding of the other. Thus, population genetics analyses have been widely used for examining patterns and magnitude of animal dispersal over both geographic (e.g., 1–5) and temporal dimensions (6).

The ability of population genetics to deliver on its promise of elucidating gene flow has relied on development of suitable molecular markers and population genetics theory for making robust inferences from observed variation in marker loci. Advances in population genetics theory and adoption of new types of molecular markers have been accompanied over the decades by parallel creation and improvement of analytical software programs to effectively calculate population genetics parameters. Although numerous software programs are available, only a handful have been used routinely in studies of natural populations. Microsatellites are a very popular marker for population genetics studies, in part because the abundance of alleles per locus and the ability to distinguish heterozygotes enhance their information content over many other types of markers. This chapter introduces the basic operating procedures of the software programs most commonly used to analyze microsatellite genotype data, including an overview of data formats and parameters for each. We also introduce statistical tests routinely used in the field of population genetics and molecular ecology. Issues commonly encountered and to watch for when conducting genetic analyses based on genotypic data from microsatellite markers are discussed, along with suggestions for troubleshooting.

2 Materials

Some population genetics software programs are designed for comprehensive statistical analyses, but many were specifically produced to calculate particular parameters. The best options for conducting certain types of analyses will often depend on the specific nature of the user's project and the models that the user is assuming. Many programs generate the same or similar population genetics parameters, so the choice of a particular program will depend on personal preference or availability in the user's lab. Nevertheless, basic genetic analyses in empirical population genetics and molecular ecology employ a similar framework across users.

Population genetics analyses based on a microsatellite genotype dataset can be categorized into three sequential phases: (1) Initial data manipulation, including error-checking of the raw genotype dataset and generating correctly formatted input files for other programs; (2) basic genetic analyses for obtaining summary statistics of common population genetics parameters; and (3) advanced genetic analyses for addressing specific questions or hypothesis testing.

Most of the population genetics software programs in this chapter can be downloaded free of charge from the websites listed in Table 1. Some software programs, e.g., GenAlEx and Microsatellite Toolkit, operate within Microsoft Excel (on Macs and PCs). However, other software such as Arlequin, BOTTLENECK, FSTAT, Genepop, GeneClass, Micro-Checker, and STRUCTURE operate in their own user-friendly platform environments, e.g., Dos and Java. It is important to look over the websites on a regular basis because they are often revised or updated by their curators.

2.1 Formatting and Data Manipulation

Micro-Checker (7) and Microsatellite Toolkit (8) are software programs that can be used for the beginning step of data manipulation. One of the purposes of these software programs is to detect scoring errors and to confirm that the genotype file is correct. Since accurate genotypes are critical for generating reliable results in further statistical analyses, proper and efficient use of software in phase 1 is very important. Microsatellite Toolkit offers additional functions, including the generation of an input file for other programs as an export data option. GeneAlEx (9) also generates input files for many other programs. This capacity to create correctly formatted input files is very useful, because downstream analyses in other programs then become largely a matter of strategy and interpretation of output. That is why much of this chapter, after the first phase of data manipulation is described, is concerned mainly with analytic strategies to obtain desired population genetics output from a microsatellite genotype dataset.

2.2 Basic Population Genetic Analyses

With a correctly formatted genotype file, one can proceed with the basic population genetics analyses of phase 2. Software programs such as Arlequin (10), Cervus (11), FSTAT (12), GeneAlEx (9), Genepop (13), and others provide options for calculating genetic diversity, genetic differentiation, gene flow, partitioning of genetic variation, and so on.

2.3 Advanced Population Genetic Analyses

If the user requires additional population analyses to test specific hypotheses such as population demography, individual/population genetic relationships, isolation by distance, genetic structuring, relatedness, or individual assignment/exclusion, one can use advanced programs such as BOTTLENECK (14), STRUCTURE (15), GeneClass (16), and GeneAlEx (9). Although not every population genetics study requires all such analyses, most studies can benefit from one or more of them. They are useful options to have in one's analytical toolbox, and we present the most commonly used.

Table 1
Characteristics and website information for downloading free software for population genetics studies

Software	Main use	Work environment	Website for download	References
Micro-Checker	Checks for microsatellite null alleles and scoring errors	Unique platform, Windows	http://www.microchecker.hull.ac.uk/	(7)
Microsatellite Toolkit	Error-checking Basic parameters such as diversity measures Input file for other programs	Unique platform in MS Excel, Windows	http://animalgenomics.ucd.ie/sdepark/ms-toolkit/	(8)
GenAlEx	Population genetics software package including import, management, and export of data	Unique platform in MS Excel, Windows	http://www.anu.edu.au/BoZo/GenAlEx/	(9)
Genepop	Population genetics software package, export of data	Unique platform, Windows or Mac	http://genepop.curtin.edu.au/	(13)
FSTAT	Population genetics software package to estimate and test gene diversities and differentiation statistics	Unique platform, Windows	http://www2.unil.ch/popgen/softwares/fstat.htm	(12)
BOTTLENECK	Detects recent effective population size reductions	Unique platform, Windows	http://www.montpellier.inra.fr/URLB/bottleneck/bottleneck.html	(14)
Arlequin	Population genetics software package	Unique platform, Windows or Mac	http://cmpg.unibe.ch/software/arlequin35/	(10)
STRUCTURE	Investigates population structure	Unique platform, Windows	http://pritch.bsd.uchicago.edu/structure.html	(15)
GeneClass	Selects or excludes populations as origins of individuals	Unique platform, Windows	http://www.montpellier.inra.fr/URLB/index.html	(16)

FreeNA	Estimates null allele frequencies Estimates unbiased F_{ST}	Unique platform, Windows	http://www.montpellier.inra.fr/URLB/index.html	(22)
Microsat	Estimates distance measures and diversity indices by population or individual	Unique platform, Windows	http://hpgl.stanford.edu/projects/microsat/	(19)
AGARST	A program for calculating allele frequencies, Gst and Rst from microsatellite data, plus a number of other population genetic estimates	Unique platform, Windows	Requires personal contact with the authors	(17)
DISPAN	Estimates genetic diversity, genetic distance Conducts phylogenetic analysis	Unique platform, Windows	http://www.softsea.com/review/DISPAN-Genetic-Distance-and-Phylogenetic-Analysis.html	(18)
Cervus	Provides statistical method for diversity indices and parentage analysis	Unique platform, Windows	http://www.fieldgenetics.com/pages/home.jsp	(11)
Migrate	Estimates effective population size and past migration rates between populations	Unique platform, Windows or Mac	http://popgen.sc.fsu.edu/Migrate/Migrate-n.html	(20)
RSTCALC	Calculates unbiased estimates of genetic differentiation (Rst, analogous to Fst), with their significance, for microsatellite data	Unique platform, Windows	http://www.biology.ed.ac.uk/archive/software/rst/rst.html	(21)

3 Methods

Manipulation of the genotype dataset and generation of correct input files for analytical software programs are the important initial steps in population genetics analyses. We do not describe detailed population genetics theories and assumptions underlying the specific genetic analyses in the software programs. For this information, it is highly recommended that the user read the information file included in each program's website (Table 1) and the papers it cites.

3.1 Input File and Correct File Extension for Each Program

Figures 1–12 illustrate correctly formatted input files for most of the population genetics software programs described in this chapter. Instructions on formatting are provided on the programs' respective websites (Table 1). Each input file contains the same genotype data for a total of ten individuals from four populations (two individuals for popA, three individuals for popB, two individuals for popC, three individuals for popD) at five microsatellite loci.

- Micro-Checker (7) (Fig. 1): Genepop format with a 3-digit number. Generated from Microsatellite Toolkit.
- Microsatellite Toolkit (8) (Fig. 2): Requires genotype file in Excel.
- GenAlEx (9) (Fig. 3): Requires genotype file in Excel.
- Arlequin (10) (Fig. 4): Requires special format with ".arp" extension. Generated from GenAlEx or Microsatellite Toolkit.
- Genepop (13) (Fig. 1): Requires that files have no extension. Generated from Microsatellite Toolkit or GenAlEx.
- FSTAT (12) (Fig. 5): Requires a ".dat" extension. Generated from Genepop or Microsatellite Toolkit.
- BOTTLENECK (Fig. 1) (14): Requires Genepop or FSTAT format.
- STRUCTURE (15) (Fig. 6): Requires a 3-digit genotype format. Generated from GenAlEx.
- GeneClass (Fig. 1) (16): Requires Genepop or FSTAT format.
- AGARST (17) (Fig. 7): Requires a 3-digit genotype format made manually.
- DISPAN (18) (Fig. 8): Requires special format. Generated from Microsatellite Toolkit.
- Cervus (11) (Fig. 9): Requires special format with ".csv" extension. Generated from GenAlEx.
- Microsat (19) (Fig. 10): Requires special format. Generated from Microsatellite Toolkit.

```
Title line:"3-digit GenePop"
LOC1
LOC2
LOC3
LOC4
LOC5
POP
popA , 155157 212218 253253 196196 225231
popA , 155155 212220 253263 178196 231231
POP
popB , 155155 212212 263263 196196 225231
popB , 157157 212218 259263 196196 225231
popB , 155157 220220 253253 178196 225225
POP
popC , 155157 212220 253259 196196 225225
popC , 155159 220220 259263 178196 225231
POP
popD , 157157 212212 245245 196196 225231
popD , 157157 212220 253259 196196 225225
popD , 155157 212212 245253 196196 225225
```

Fig. 1 The 3-digit Genepop input file format. FreeNA, BOTTLENECK, GeneClass, and Micro-Checker all use this format, and all look the same

	LOC1		LOC2		LOC3		LOC4		LOC5	
popA1	155	157	212	218	253	253	196	196	225	231
popA2	155	155	212	220	253	263	178	196	231	231
popB1	155	155	212	212	263	263	196	196	225	231
popB2	157	157	212	218	259	263	196	196	225	231
popB3	155	157	220	220	253	253	178	196	225	225
popC1	155	157	212	220	253	259	196	196	225	225
popC2	155	159	220	220	259	263	178	196	225	231
popD1	157	157	212	212	245	245	196	196	225	231
popD2	157	157	212	220	253	259	196	196	225	225
popD3	155	157	212	212	245	253	196	196	225	225

Fig. 2 The Microsatellite Toolkit input file format

5	10	4	2	3	2	3					
			popA	popB	popC	popD					
		LOC1		LOC2		LOC3		LOC4		LOC5	
popA1	popA	155	157	212	218	253	253	196	196	225	231
popA2	popA	155	155	212	220	253	263	178	196	231	231
popB1	popB	155	155	212	212	263	263	196	196	225	231
popB2	popB	157	157	212	218	259	263	196	196	225	231
popB3	popB	155	157	220	220	253	253	178	196	225	225
popC1	popC	155	157	212	220	253	259	196	196	225	225
popC2	popC	155	159	220	220	259	263	178	196	225	231
popD1	popD	157	157	212	212	245	245	196	196	225	231
popD2	popD	157	157	212	220	253	259	196	196	225	225
popD3	popD	155	157	212	212	245	253	196	196	225	225

Fig. 3 The GenAlEx input file format

```
[Profile]
   Title="Arlequin format"
   NbSamples=4
   GenotypicData=1
   GameticPhase=0
   RecessiveData=0
   DataType=STANDARD
   LocusSeparator=WHITESPACE
   MissingData='?'
   CompDistMatrix=1
[Data]
 [[Samples]]  #Data for 5Loci: LOC1 LOC2 LOC3 LOC4 LOC5
   SampleName="popA"
   SampleSize=2
   SampleData=    {
      popA1     1  155 212 253 196 225
                   157 218 253 196 231
      popA2     1  155 212 253 178 231
                   155 220 263 196 231
   }
   SampleName="popB"
   SampleSize=3
   SampleData=    {
      popB1     1  155 212 263 196 225
                   155 212 263 196 231
      popB2     1  157 212 259 196 225
                   157 218 263 196 231
      popB3     1  155 220 253 178 225
                   157 220 253 196 225
   }
   SampleName="popC"
   SampleSize=2
   SampleData=    {
      popC1     1  155 212 253 196 225
                   157 220 259 196 225
      popC2     1  155 220 259 178 225
                   159 220 263 196 231
   }
   SampleName="popD"
   SampleSize=3
    SampleData=    {
      popD1     1  157 212 245 196 225
                   157 212 245 196 231
      popD2     1  157 212 253 196 225
                   157 220 259 196 225

      popD3     1  155 212 245 196 225

                   157 212 253 196 225

    }
 [[Structure]]

   StructureName=" Structure"
   NbGroups=1
   IndividualLevel=0
   Group= {


            "popA"
            "popB"
            "popC"
            "popD"

  }
```

Fig. 4 The Arlequin input file format

```
4 5 263 3
LOC1
LOC2
LOC3
LOC4
LOC5
1  155157  212218  253253  196196  225231
1  155155  212220  253263  178196  231231
2  155155  212212  263263  196196  225231
2  157157  212218  259263  196196  225231
2  155157  220220  253253  178196  225225
3  155157  212220  253259  196196  225225
3  155159  220220  259263  178196  225231
4  157157  212212  245245  196196  225231
4  157157  212220  253259  196196  225225
4  155157  212212  245253  196196  225225
```

Fig. 5 The FSTAT input file format

```
LOC1 LOC2 LOC3 LOC4 LOC5
popA1 1 0 155 157 212 218 253 253 196 196 225 231
popA2 1 0 155 155 212 220 253 263 178 196 231 231
popB1 2 0 155 155 212 212 263 263 196 196 225 231
popB2 2 0 157 157 212 218 259 263 196 196 225 231
popB3 2 0 155 157 220 220 253 253 178 196 225 225
popC1 3 0 155 157 212 220 253 259 196 196 225 225
popC2 3 0 155 159 220 220 259 263 178 196 225 231
popD1 4 0 157 157 212 212 245 245 196 196 225 231
popD2 4 0 157 157 212 220 253 259 196 196 225 225
popD3 4 0 155 157 212 212 245 253 196 196 225 225
```

Fig. 6 The STRUCTURE input file format

```
Test data table Five loci, Four pps

population A
a1          155 157 212 218 253 253 196 196 225 231
a2          155 155 212 220 253 263 178 196 231 231
population B
b1          155 155 212 212 263 263 196 196 225 231
b2          157 157 212 218 259 263 196 196 225 231
b3          155 157 220 220 253 253 178 196 225 225
population C
c1          155 157 212 220 253 259 196 196 225 225
c1          155 159 220 220 259 263 178 196 225 231
population D
d1          157 157 212 212 245 245 196 196 225 231
d2          157 157 212 220 253 259 196 196 225 225
d3          155 157 212 212 245 253 196 196 225 225
```

Fig. 7 The AGARST input file format. *Important:* Do not use the words "populations", "population", or "pop" in the title or as part of the population ID

- Migrate (20) (Fig. 11): Requires special format. Generated from AGARST.
- RSTCALC (21) (Fig. 12): Requires special format. Generated from AGARST.
- FreeNA (22) (Fig. 1): Requires Genepop format.

```
#Populations = (popA,popB,popC,popD)
#Monomorphic loci = 0

@Locus 1: LOC1
#Allele = ( 155,    157,    159   )
 0.7500   0.2500   0.0000   4 popA
 0.5000   0.5000   0.0000   6 popB
 0.5000   0.2500   0.2500   4 popC
 0.1667   0.8333   0.0000   6 popD

@Locus 2: LOC2
#Allele = ( 212,    218,    220   )
 0.5000   0.2500   0.2500   4
 0.5000   0.1667   0.3333   6
 0.2500   0.0000   0.7500   4
 0.8333   0.0000   0.1667   6

@Locus 3: LOC3
#Allele = ( 245,    253,    259,    263   )
 0.0000   0.7500   0.0000   0.2500   4
 0.0000   0.3333   0.1667   0.5000   6
 0.0000   0.2500   0.5000   0.2500   4
 0.5000   0.3333   0.1667   0.0000   6

@Locus 4: LOC4
#Allele = ( 178,    196   )
 0.2500   0.7500   4
 0.1667   0.8333   6
 0.2500   0.7500   4
 0.0000   1.0000   6

@Locus 5: LOC5
#Allele = ( 225,    231   )
 0.2500   0.7500   4
 0.6667   0.3333   6
 0.7500   0.2500   4
 0.8333   0.1667   6
```

Fig. 8 The DISPAN input file format

Sample	Sex	LOC1A	LOC1B	LOC2A	LOC2B	LOC3A	LOC3B	LOC4A	LOC4B	LOC5A	LOC5B
popA1	popA	155	157	212	218	253	253	196	196	225	231
popA2	popA	155	155	212	220	253	263	178	196	231	231
popB1	popB	155	155	212	212	263	263	196	196	225	231
popB2	popB	157	157	212	218	259	263	196	196	225	231
popB3	popB	155	157	220	220	253	253	178	196	225	225
popC1	popC	155	157	212	220	253	259	196	196	225	225
popC2	popC	155	159	220	220	259	263	178	196	225	231
popD1	popD	157	157	212	212	245	245	196	196	225	231
popD2	popD	157	157	212	220	253	259	196	196	225	225
popD3	popD	155	157	212	212	245	253	196	196	225	225

Fig. 9 The Cervus input file format

3.2 Converting Genotype File in Excel to Txt File Format

Several programs can generate input files for other programs. Input files generated by MS Excel-based programs must be changed to Txt file format for use in other programs. Genotype data in Excel contains the tab character, which should be eliminated using the following procedures:

```
% individual format

popA1  MS1  155
popA1  MS1  157
popA1  MS2  212
popA1  MS2  218
popA1  MS3  253
popA1  MS3  253
popA1  MS4  196
popA1  MS4  196
popA1  MS5  225
popA1  MS5  231
popA2  MS1  155
popA2  MS1  155
popA2  MS2  212
popA2  MS2  220
popA2  MS3  253
popA2  MS3  263
popA2  MS4  178
popA2  MS4  196
popA2  MS5  231
popA2  MS5  231
popB1  MS1  155
popB1  MS1  155
popB1  MS2  212
popB1  MS2  212
popB1  MS3  263
popB1  MS3  263
popB1  MS4  196
popB1  MS4  196
popB1  MS5  225
popB1  MS5  231
popB2  MS1  157
popB2  MS1  157
popB2  MS2  212
popB2  MS2  218
popB2  MS3  259
popB2  MS3  263
popB2  MS4  196
popB2  MS4  196
popB2  MS5  225
popB2  MS5  231
popB3  MS1  155
popB3  MS1  157
popB3  MS2  220
popB3  MS2  220
popB3  MS3  253
popB3  MS3  253
popB3  MS4  178
popB3  MS4  196
popB3  MS5  225
popB3  MS5  225
popC1  MS1  155
popC1  MS1  157
popC1  MS2  212
popC1  MS2  220
popC1  MS3  253
popC1  MS3  259
popC1  MS4  196
popC1  MS4  196
```

Fig. 10 The Microsat input file format

```
 4  5  . Agarst
 2     population  1
Indiv 1   2.4  2.8  10.10  20.20  2.8
Indiv 2   2.2  2.10  10.20  2.20  8.8
 3     population  2
Indiv 1   2.2  2.2  20.20  20.20  2.8
Indiv 2   4.4  2.8  16.20  20.20  2.8
Indiv 3   2.4  10.10  10.10  2.20  2.2
 2     population  3
Indiv 1   2.4  2.10  10.16  20.20  2.2
Indiv 2   2.6  10.10  16.20  2.20  2.8
 3     population  4
Indiv 1   4.4  2.2  2.2  20.20  2.8
Indiv 2   4.4  2.10  10.16  20.20  2.2
Indiv 3   2.4  2.2  2.10  20.20  2.2
```

Fig. 11 The Migrate input file format

```
Title
 5
 3
 4
 2
 3
 2
 3
Locus1
 2
 154
Locus2
 2
 211
Locus3
 2
 244
Locus4
 18
 177
Locus5
 6
 224
Pop1
 1 2  1 2  2 2  2 2  1 2
 1 1  1 3  2 4  1 2  2 2
Pop2
 1 1  1 1  4 4  2 2  1 2
 2 2  1 2  3 4  2 2  1 2
 1 2  3 3  2 2  1 2  1 1
Pop3
 1 2  1 3  2 3  2 2  1 1
 1 3  3 3  3 4  1 2  1 2
Pop4
 2 2  1 1  1 1  2 2  1 2
 2 2  1 3  2 3  2 2  1 1
 1 2  1 1  1 2  2 2  1 1
```

Fig. 12 The RSTCALC input file format

1. Create input file using Excel-based software, e.g., Microsatellite Toolkit.
2. Copy all of the data on the Excel worksheet and paste into MS word as unformatted text using the "paste special" command.

3. Eliminate all unnecessary keystrokes (all tabs ^t), as some programs are very sensitive in that regard.
4. In general, select "All" to copy and paste the data set into Notepad and save as a *.dat file (or file without extension) to import into the prescribed software program.

3.3 Formatting and Data Manipulation

We list three possible programs for initial data manipulation. These have the advantages of providing options for generating input files for other downstream analysis programs (see Subheading 3.1) and of procedures to check for errors in the dataset. Corrections to data in the genotype file require the process to start from the program selected for error-checking and verification of genotypes. Microsatellite Toolkit and Micro-Checker are commonly used to check for errors in the genotype data. Microsatellite Toolkit and GenAlEx generate input data files for other downstream software programs.

3.3.1 Microsatellite Toolkit

Features:

- Detects invalid alleles, incompletely typed samples (for diploid data), and invalid sample/population names recognized as duplicated or genetically identical samples.
- Calculates allele frequencies per population or locus, heterozygosity, allelic diversity, and individual relationship based on shared allele frequency.
- Creates input files for population genetics analysis programs such as Arlequin, Genepop, FSTAT, DISPAN, and Microsat.

Use:

1. Open **MS_tools.xla** (Excel add-in tools for microsatellite data) and conduct further analysis after selecting "Macro included".
2. Select Diploid one-column format or two-column format in Input data format after selecting "Microsatellite Toolkit" at additional function (two-column format is illustrated in Fig. 2). Toolkit automatically recognizes the number of samples and loci.
3. Click "OK" by selecting "Check data for errors" and by default setting in "Data checking parameters". Then 1 Col data format is created and if click "OK!", "Format options" window will pop up.
4. "Format options" allows data conversion for Arlequin, Genepop, Microsat, FSTAT, and DISPAN and provides summary statistics including allele frequencies and diversity statistics.

3.3.2 Micro-Checker

Features:

- Detects mistyped allele sizes and typographic errors and deviations from a regular repeat motif (suggesting indels or typos).

- Detects evidence of null alleles (one or more alleles fail to amplify during PCR) (see Note 1), stuttering (slight changes occur in the allele sizes during PCR), and large allele dropout (large alleles do not amplify as efficiently as small alleles).

Use:

1. Open **StartMicroChecker.exe** then open the "Data" file. On the lower toolbar, select each locus and identify the repeat motif for each locus. Unless a locus has a size greater than 350 bp, accept the default parameters.
2. For each locus, select "Check" for unusual observations.
3. This will open a display window to the right of the data file window with any unusual observations identified for each locus.
4. Record these unusual genotypes. You will need to return to the "original" worksheet to correct an unusual observation or accept it if you believe it is correct after verification.

3.3.3 GenAlEx

Features:

- The GenAlEx package can be used to generate input files for other useful population genetic software programs including Arlequin, Cervus, GeneClass, Genepop, and STRUCTURE.
- The program also has options to carry out various genetic analyses in MS Excel. Once its own input file is made, the program is a user-friendly package that can perform population genetics analyses including summary statistics such as diversity measures, tests of Hardy–Weinberg equilibrium, as well as advanced statistics such as genetic distance, Analysis of molecular variance (AMOVA), Mantel tests, Principal Coordinates Analysis of multivariate genetic data, estimation of pairwise relatedness among individuals, population assignment, and many more.

Use:

1. Open **GenAlEx6.1.xla** (Excel add-in tools for microsatellite and DNA sequence data) and select "Macro included".
2. Make correct input file. One can start this using the input file for Microsatellite Toolkit because it is regarded as the most basic data format.
3. Make column 2 for population ID using the column inserting option.
4. Make rows for basic information of dataset by selecting "Insert Header Rows" in Parameters.
5. Insert total number of loci by selecting "No. Codominant Loci" in Parameters.
6. Insert information for each population by selecting "Pops from Col2" in Parameters.

7. You can continue to use various options for both basic and advanced population genetic analyses or generate input files for advanced population genetic analyses using the "Export data" option. GenAlEx generates input files in the correct format for other programs.

3.4 Basic Population Genetic Analyses

3.4.1 Indices of Genetic Diversity

Estimation of genetic diversity is an essential component of population genetics analyses of natural organisms. Within-population indices of genetic diversity include the numbers of different alleles per locus, allelic richness, and expected (H_E) and observed (H_O) heterozygosity. The measures of heterozygosity are highly correlated, but expected (H_E) is considered a better estimator of the genetic variability present in a population (23). Since genetic diversity information is the most basic approach in empirical population genetics, numerous software programs are designed to provide such information. Indices of genetic diversity can be calculated using the programs AGARST, Arlequin, Cervus, DISPAN, FSTAT, GenAlEx, Genepop, and Microsatellite Toolkit (see Note 2).

3.4.2 Test of Hardy–Weinberg Equilibrium

A test of Hardy–Weinberg equilibrium (HWE) should be carried out as an initial step of population genetics analyses. Under the Hardy–Weinberg principle, frequencies of alleles remain constant in a population in the absence of selection, mutation, migration, and genetic drift. Thus, tests of Hardy–Weinberg equilibrium interrogate the stability of allele frequencies over time. The Hardy–Weinberg principle concerns the effects of a single generation of random mating where genotype frequencies can be predicted from the allele frequencies. HWE is expected for populations in which mating is random, and such a population should show no significant difference between observed and expected heterozygosity. Excessive deviation from HWE indicates violation of one of the assumptions of population genetics analyses through such processes as nonrandom mating or a lack of selective neutrality. However, significant deviation from HWE can also arise from physical error during genotyping, e.g., null alleles, and data must be interpreted with caution (see Note 1). Tests of HWE and its significance (see Note 3) can be carried out using the programs Arlequin, FSTAT, GenAlEx, and Genepop.

3.4.3 Test for Genotypic Linkage Disequilibrium

Population genetic parameters are calculated from genetic data across multiple loci which are assumed to assort independently of one another during meiosis. If two loci are located too close together on a chromosome, they are considered linked, resulting in genotypic linkage disequilibrium. Tests for genotypic linkage disequilibrium test the null hypothesis that genotypes at one locus are independent from genotypes at the other locus. A test of genotypic linkage disequilibrium and significance should be conducted during the initial step of marker selection or genetic analyses (24).

In the case of significant disequilibrium, the best course of action is to exclude one of the two makers from further population genetic analyses. Tests of genotypic linkage disequilibrium can be carried out using the programs Arlequin, FSTAT, and Genepop.

3.4.4 Measure of Fixation Indices and Genetic Differentiation Between Populations

Genetic differentiation can be measured by difference in frequency distribution of alleles between populations. Information of fixation indices such as *F*-statistics, F_{IS}, F_{IT}, and F_{ST} (25), per locus across all populations, should be investigated at the initial step of population genetics analyses. A significant difference between observed and expected heterozygosity results in a significant F_{IS} value and may indicate the presence of null alleles (see Note 1), the Wahlund effect, or some other anomaly. F_{ST} estimates are potentially in the range 0–1 and are a measure of how genetically different two populations are at selectively neutral loci, with an F_{ST} of 0 indicating that no genetic differentiation has occurred and a value of 1 indicating that the two populations share no genotypes in common. Extent of genetic differentiation between populations using F_{ST} (an estimate of population subdivision under the infinite allele model) and other *F*-statistics (25) per locus across all populations and their respective *p*-values (see Note 3) can be calculated by the programs Arlequin, FSTAT, GenAlEx, Genepop, and many more. AGARST, FSTAT, and RSTCALC can calculate R_{ST} (an estimate of population subdivision for stepwise mutation processes, ref. 26). FSTAT can provide an adjusted *p*-value to derive significance levels for analyses involving multiple comparisons (see Note 4).

3.4.5 Gene Flow Measures

Patterns and extent of gene flow provide important information on dispersal pattern and capacity of the study species. Indirect estimates of gene flow between populations can be measured with different approaches. First, one can calculate population genetic structure-based gene flow according to the relationship $Nem = (1 - F_{ST})/4\ F_{ST}$ (27), where *Nem* is the effective number of migrants per generation, *Ne* is the effective population size of each population, and *m* is the immigration rate. This classical measure of gene flow is based on equilibrium between the forces of immigration and genetic drift under the assumptions of the island model, i.e., that migration occurs among populations of equal size with symmetrical migration rates. Pairwise estimates of genetic differentiation among subpopulations and their significance can be quantified by F_{ST} (25) and R_{ST} (26) using the program FSTAT and RSTCALC, respectively (see Note 5). Second, maximum likelihood estimates of gene flow can be calculated using the coalescent-based Markov Chain Monte Carlo (MCMC) simulation approach, which takes into account the genealogical relationship of the samples and asymmetry in gene flow (20, 28). The necessary migration parameters, such as $4Ne\mu$, where μ is the mutation rate per generation at a locus and M $(=m/\mu)$, can be calculated using the program Migrate (20) (see Note 6).

Rate of migration can also be calculated from the frequency of private alleles. A private or rare allele is defined as an allele found in only one subpopulation, but not found in other subpopulations. Estimating gene flow using private alleles was developed by Slatkin (29) based on the following equation: $\ln[p(1)] \approx a \ln(Nm) + b$, where $a = -0.505$ and $b = -2.440$, and $p(1)$ is the frequencies of private alleles. Therefore, the logarithm of expected frequency of a private allele ($p(1)$) is approximately a decreasing linear function of the logarithm of Nm with a slope of −0.505 (Fig. 1 in ref. 29). Simulation showed that this method is relatively insensitive to changes in parameters of the model other than Nm and the number of individuals sampled per population, and the author provided a rough way to correct for differences in sample size (29). Therefore, one can use the value of $p(1)$ to estimate Nm using the program Genepop.

3.4.6 Analysis of Molecular Variance Test

An AMOVA estimates the proportion of genetic diversity within and between populations, or among groups of populations that the user categorizes based on criteria such as region. The AMOVA test is therefore used to evaluate the level of genetic differentiation within and among populations, regions, or other specified hierarchical categories. The partitioning of population genetic variance in such a hierarchical AMOVA can be conducted using the program GenAlEx or Arlequin. The significance of differentiation within and among populations within regions can be determined by permutations of samples, e.g., 1,000 replicates. The AMOVA is calculated based on Euclidean distances between individuals in GenAlEx and the closest model of evolution in Arlequin.

3.5 Advanced Population Genetic Analyses

3.5.1 Bottleneck Tests

Bottleneck tests are commonly used to examine population demography in recent time for evidence of a severe reduction in population size sufficient to leave a genetic signature. Evidence of recent population bottlenecks can be assessed using three different approaches. Three tests, including the Wilcoxon test which produces the most reliable results, are available in the program BOTTLENECK to determine whether deviations of observed heterozygosity (designated H_e in software documentation or H_o in (14)) relative to that expected at drift–mutation equilibrium (designated H_{eq} in software documentation or H_1 in (14)) are significant ($\alpha = 0.05$). Both a strict stepwise mutation model (SMM) (30) and a two-phase model (TPM) (31) with 1,000 iterations can be applied. For the TPM, a generalized stepwise mutation model (GSM), in which a proportion of SMM is set to 0 with a variance in mutation lengths of 0.36 (32), can be applied. Secondly, one can look for a mode-shift in allele frequency distribution from the L-shaped distribution expected under mutation–drift equilibrium, which can be used as a qualitative indicator of population bottlenecks (33). Third, the M value of Garza and Williamson (34) and its variance across loci are calculated using the program AGARST. *M* is the mean ratio of the number of alleles to the range of allele size. This test is useful for detecting a bottleneck experienced further in the past. After a bottleneck, the *M* statistic will display persistently low values for about 100 generations. When compared to the results of the other two tests, the *M* test can distinguish populations that have been reduced in size recently from those which have been small for a long time (34).

3.5.2 Genetic Relationships Between Samples

Construction of a genetic relationship tree or of a scatter diagram from principal component analysis (PCA) or principal coordinate analysis (PCoA) of a multivariate dataset is performed to visualize pairwise differentiation between individuals or populations. Genetic divergence between populations based on allele frequencies can be calculated as genetic distance (D_A) (35) using the DISPAN computer program. Phylogenetic trees are constructed by neighbor-joining (NJ) clustering (36) or by the unweighted pair group method with the arithmetic mean (UPGMA) (37) using DA distance. Bootstrap resampling ($n = 1,000$) is applied to test the robustness of dendrogram topologies. A principal component analysis (PCA) is applied to a covariance matrix of allele frequencies across all variable loci using the program PCAGEN (38). Principal coordinate analysis (PCoA) can be conducted using the program GenAlEx. The geometric relationship among populations is visualized with a scattergram of the factor score data along the two PC axes that account for the most variation. To visualize genetic relationships among individuals, inter-individual genetic distances can be calculated based on the proportion of shared alleles using the Microsat computer program. These distance values are used to construct a UPGMA tree as implemented in the NEIGHBOR module of the PHYLIP software package (39).

3.5.3 Inferring Real-Time Migration Rate

Temporal analyses, the estimation of effective population size (Ne) and the migration rate (*m*) from samples collected over time, provide a way of measuring real-time migration regardless of population history (40–42). They also provide the most robust estimates possible of effective population size and migration rate (43). Temporal analysis is less sensitive to drift–migration equilibrium than population genetic structure-based gene flow (43), making it useful for estimating gene flow in invasive species or species that have undergone a recent range expansion, where estimates based on spatial data from geographic samples is problematic (see Note 5). The computer program MLNE allows estimation of *m* and *Ne* simultaneously using a maximum likelihood strategy (43). This method uses a temporal approach that compares allele frequencies from at least two generations. Simulation studies show that it performs better than other temporal methods (43).

3.5.4 Identification of Migrants in Current Generation

The Monte Carlo simulation approach of Paetkau et al. (42) enables the identification of immigrant individuals in the current generation, allowing an estimate of gene flow among populations at a much narrower time scale. The premise of this approach is based on resampling gametes rather than alleles to preserve linkage disequilibrium in recent immigrants. The analysis can be conducted using the "Detection of first generation migrants" criterion implemented in the program GeneClass, which assigns each potential immigrant to the most likely source population at a specified confidence level (42). First generation (F0) migrants are defined as individuals that traveled from site A to site B in year X (or the current generation) or individuals born in year X to a gravid female that moved from site A to B in year $X-1$ (or the previous generation). Two test statistics (the ratio $L_{\mathrm{home}}/L_{\mathrm{max}}$ and L_{home}) can be used to compute the likelihood of migrant detection (L) (42). In cases where it is unclear whether all potential source populations for immigrants have been sampled, L_{home} is the more appropriate test statistic but has reduced power to identify immigrants (42).

3.5.5 Assignment/Exclusion Tests

To compute the probability of each individual's belonging to a set of reference (current or potential source) populations, assignment/exclusion tests using the direct and simulation approaches can be conducted using options implemented in the program GeneClass. The direct assignment tests allocate an individual to one of the reference populations without probability computation, thereby simply calculating the proportion of correctly assigned individuals to the most likely population of origin, even if the true population of origin is not among the reference populations. In contrast, the exclusion method uses a simulation approach. This method computes the likelihood of a genotype occurring in the population by simulating multilocus genotypes based on allele frequencies of each reference population and compares the likelihood of the genotype of an individual to the distribution of likelihoods of simulated genotypes for each reference population. If the individual genotype likelihood is below a given threshold (e.g., $\alpha = 0.01$), the population is excluded as a possible origin of the individual (40). Unlike the direct assignment method, the exclusion method does not assume that the true population of origin has been sampled, because each population is treated independently (40). The Bayesian statistical approach of Rannala and Mountain (44), which has proven to be more accurate than frequency and distance based methods (40), is used for both assignment and exclusion tests. Frequency probabilities of multilocus genotypes in each reference population are determined in the exclusion test using Monte Carlo simulations of 10,000 independent individuals for the population (42). The assignment likelihoods of individuals from a geographic population to putative source populations can be further calculated and averaged using the Bayesian statistical method (44). The statistic $L_{i\ \mathrm{to}\ j}$, the mean individual assignment likelihoods of individuals collected in population *i* and assigned to population *j*, provides valuable asymmetric information for the origin of the newly introduced population under the assumptions that it is new in the location and that the putative source population was sampled.

GenAlEx can be used to visualize the relative position of individuals from spatial population data between locations by plotting the log likelihood of an individual's genotype arising in the populations using the frequency-based assignment test (42, 45). If the individual's allele is absent from one of the represented populations, the value can be set to 0.01 and the "leave one out" option (46) is applied for the assignment test.

3.5.6 Inferring the Number of Distinct Genetic Populations

The program STRUCTURE uses a model-based Bayesian clustering method to infer the number of distinct populations (K) from which samples have been drawn and to infer the genetic ancestry of the individuals sampled, based on microsatellite genotypes at multiple loci. This approach provides an independent assessment of these parameters, free of the prior assumption that each sample location constitutes a population. Thus, the results complement those of the genetic tree (Subheading 3.5.2), population structuring (Subheadings 3.4.4 and 3.4.6), and population assignment tests (Subheading 3.5.5) described above. The program is used to estimate $\Pr(X/K)$, the probability of the observed set of genotypes (X), conditional on a given K. The program can be run using different replications for both burn-in and the consequent resampling. An initial burn-in of 100,000 iterations followed by 1,000,000 iterations is common. An admixture model of individual ancestry and correlated allele frequencies among populations are appropriate for most natural populations. Multiple runs are required to test performance for each value of K to verify that estimates of $\Pr(K/X)$ were consistent between runs. The posterior probabilities of K, $\Pr(K/X)$, are calculated according to Pritchard et al. (14). The "real" value of K (number of unique populations represented by the genotypes within the sample) is estimated from the ln $\Pr(K/X)$ values output for each replicate of K using the $m(|L''(K)|)/s[L(K)]$ statistic described by Evanno et al. (47). In brief, the "real" value of K within the dataset is determined as the ln $\Pr(K/X)$ that maximizes the value of $\Delta K = m(|L''(K)|)/s[L(K)]$.

3.5.7 Genetic Isolation by Geographic Distance

A special, but common, problem is to examine gene flow within a continuously distributed population. In such cases, one would expect genetic differentiation between locations within the large continuous population to increase with distance alone. A pattern of isolation by distance (IBD) can be examined through regression of the genetic distance on geographic distance among locations. Slatkin (48) suggested that a pattern of isolation by distance should be detectable when a population is at or near equilibrium under its current patterns of dispersal. The absence of isolation by distance pattern suggests that the population either is far from equilibrium, and that genetic structuring may reflect a recent range expansion rather than current levels of gene flow, or that the spatial scale sampled was too small relative to normal dispersal distances. IBD (49) is inferred from the relationship between $F_{ST}/(1-F_{ST})$ (a measure of genetic distance) and the geographic distance between all pairs of sampled locations. It is recommended that untransformed distance (km) be used for a one-dimensional (i.e., linear) sampling scheme and the logarithm of distance be used for two-dimensional sampling schemes in the regression (50). Regression of $F_{ST}/(1-F_{ST})$ on geographic distance between all pairs of sampling locations and the probability that there is no relationship based on permutations of samples can be calculated using the Matrix Comparison option in Arlequin, FSTAT, Genepop, and GenAlEx (see Note 7).

4 Notes

1. A null allele is caused when nucleotide variation in the flanking region of the microsatellite locus prevents primer binding and PCR amplification, making the locus appear homozygous for the one allele that does amplify (51) or resulting in no amplification at all at the locus if both alleles are null. This functionally recessive behavior leads to a decrease in genotyping accuracy, which in turn can result in a number of artifacts including heterozygote deficiency, inaccurate allele frequency estimates, and inflated F_{IS}, F_{ST}, and genetic distance estimates (22, 51, 52). The extent to which null alleles tend to overestimate the true population differentiation has not been investigated (22) but can lead to overestimates of population differentiation due to effects on subsequent calculations of F_{ST} and genetic distances (53, 54). Therefore, in any population genetics study using microsatellites, the potential for null alleles must be addressed (52, 55).

   Microsatellite loci that deviate significantly from HWE show evidence of null alleles according to the distribution of homozygote-size classes. The program Micro-Checker is used to estimate the frequency of null alleles and other genotyping errors such as stuttering and allele drop out. Null alleles are suspected for a given locus when the Micro-Checker program rejects Hardy–Weinberg equilibrium (HWE) among genotypes and if the excess homozygote genotypes are evenly distributed among allele size classes. In the case of alleles harboring the potential null alleles, corrected pairwise F_{ST} estimates are calculated for all populations by applying the ENA correction in the FreeNA package.

2. Adjusted allelic diversity to account for variation in sample sizes can be calculated using both bootstrapping and jackknifing techniques implemented in the program AGARST or using allelic richness in the program FSTAT.
3. Determining the significance of differences in multiple comparisons requires a correction to avoid inflated type I error rates. Calculation of population genetics parameters such as genotypic linkage disequilibrium, pairwise F_{ST}, and HWE test often requires multiple tests since multiple populations from different sampling sites are used for calculations in a single table. One of the most popular methods for correcting for such multiple tests is the sequential Bonferroni correction, which provides adjusted *p*-values to maintain the intended α level of significance (56).
4. The calculations underlying the Bonferroni correction (see Note 3) are appropriate only for multiple independent tests. To account for the presence of multiple dependent tests within pairwise F_{ST} estimates, we suggest a correction to the significance thresholds for the critical value according to the B–Y method of Benjamini and Yekutieli (57).
5. An underlying assumption of spatial pairwise F_{ST} estimates is that the populations are in migration–drift equilibrium. This assumption is most often violated in the case of an invasive species or in a region where a species has undergone a recent range expansion. After a range expansion, F_{ST} values are often low and nonsignificant because of genetic founder effects, even though dispersal and gene flow may be limited. In such cases, estimates of gene flow are best obtained by analyzing temporal genetic data, i.e., data collected over time at the same locations (see Subheading 3.5.3).
6. Both the traditional gene flow measures based on allele frequency distributions and the coalescent-based maximum likelihood estimation of gene flow mainly reflect relatively long-term gene flow and thus may not accurately represent current levels. There are other methods available to determine whether each individual is a resident in the population in which it was sampled or an immigrant, and to estimate the number of immigrant individuals present in the current generation. These are covered in Subheadings 3.5.4 and 3.5.5.
7. Because the pairwise F_{ST} estimates are not independent data, a simple linear regression is not appropriate, and the permutation method is required.

Acknowledgements

This work was supported by the Korea Science and Engineering Foundation (KOSEF) grant funded by the Korean government (MEST) (No. 2009-0080227). Mention of trade names or commercial products in this publication is solely for the purpose of providing specific information and does not imply recommendation or endorsement by the US Department of Agriculture. USDA is an equal opportunity provider and employer.

References

1. Kim KS, Sappington TW (2006) Molecular genetic variation of boll weevil populations in North America estimated with microsatellites: implications for patterns of dispersal. Genetica 127:143–161
2. Jiang X-F, Luo L-Z, Zhang L (2007) Amplified fragment length polymorphism analysis of *Mythimna separata* (Lepidoptera: Noctuidae) geographic and melanic laboratory populations in China. J Econ Entomol 100: 1525–2532
3. Jiang X-F, Cao W-J, Zhang L, Luo L-Z (2010) Beet webworm (Lepidoptera: Pyralidae) migration in China: evidence from genetic markers. Environ Entomol 39:232–242
4. Nagoshi RN, Fleischer S, Meagher RL (2009) Texas is the overwintering source of fall armyworm in central Pennsylvania: implications for migration into the northeastern United States. Environ Entomol 38:1546–1554
5. Kim KS, Coates BS, Bagley MJ, Hellmich RL, Sappington TW (2011) Genetic structure and gene flow among European corn borer (Lepidoptera: Crambidae) populations from the Great Plains to the Appalachians of North America. Agric For Entomol 13:383–393
6. Kim KS, Bagley MJ, Coates BS, Hellmich RL, Sappington TW (2009) Spatial and temporal genetic analyses show high gene flow among European corn borer (Lepidoptera: Crambidae) populations across the central U.S. Corn Belt. Environ Entomol 38:1312–1323
7. Van Oosterhout C, Hutchinson W, Wills D, Shipley P (2004) Micro-Checker: software for identifying and correcting genotyping errors in microsatellite data. Mol Ecol Resour 4:535–538
8. Park SDE (2001) Trypanotolerance in West African cattle and the population genetic effects of selection. Ph.D. thesis, University of Dublin
9. Peakall R, Smouse PE (2006) GENALEX 6: genetic analysis in Excel. Population genetic software for teaching and research. Mol Ecol Notes 6:288–295
10. Excoffier L, Lischer HEL (2010) Arlequin suite ver 3.5: a new series of programs to perform population genetics analyses under Linux and Windows. Mol Ecol Resour 10:564–567
11. Kalinowski ST, Taper ML, Marshall TC (2007) Revising how the computer program Cervus accommodates genotyping error increases success in paternity assignment. Mol Ecol 16:1099–1106
12. Goudet J (1995) Fstat version 1.2: a computer program to calculate F statistics (version 2.9.03). J Hered 86:485–486
13. Raymond M, Rousset F (1995) GENEPOP (version 1.2): population genetics software for exact tests and ecumenicism. Heredity 86:248–249
14. Cornuet J, Luikart G (1996) Description and power analysis of two tests for detecting recent population bottlenecks from allele frequency data. Genetics 144:2001–2014
15. Pritchard JK, Stephens M, Donnelly P (2000) Inference of population structure using multilocus genotype data. Genetics 155:945–959
16. Piry S, Alapetite A, Cornuet JM, Paetkau D, Baudouin L, Estoup A (2004) GeneClass2: a software for genetic assignment and first-generation migrant detection. Heredity 95:536–539
17. Harley EH (2001) AGARst. A programme for calculating allele frequencies, G_{ST} and R_{ST} from microsatellite data, version 2. University of Cape Town, Cape Town, South Africa

18. Ota T (1993) DISPAN: genetic distance and phylogenetic analysis. Pennsylvania State University, University Park, PA
19. Minch E (1998) MICROSAT version 1.5b. University of Stanford, Stanford, CA
20. Beerli P, Felsenstein J (1999) Maximum-likelihood estimation of migration rates and effective population numbers in two populations using a coalescent approach. Genetics 152:763–773
21. Goodman SJ (1997) Rst Calc: a collection of computer programs for calculating estimates of genetic differentiation from microsatellite data and determining their significance. Mol Ecol 6:881–885
22. Chapuis M-P, Estoup A (2007) Microsatellite null alleles and estimation of population differentiation. Mol Biol Evol 24:621–631
23. Nei M (1987) Molecular evolutionary genetics. Columbia University Press, New York
24. Kim KS, Stolz U, Miller NJ, Waits ER, Guillemaud T, Sumerford DV, Sappington TW (2008) A core set of microsatellite markers for western corn rootworm (Coleoptera: Chrysomelidae) population genetics studies. Environ Entomol 37:293–300
25. Weir BS, Cockerham CC (1984) Estimating F-statistics for the analysis of population structure. Evolution 38:1358–1370
26. Slatkin M (1985) Gene flow in natural populations. Annu Rev Ecol Syst 16:393–430
27. Wright S (1931) Evolution in Mendelian populations. Genetics 16:97–159
28. Beerli P, Felsenstein J (2001) Maximum likelihood estimation of a migration matrix and effective population sizes in *n* subpopulations by using a coalescent approach. Proc Natl Acad Sci USA 98:4563–4568
29. Slatkin M (1985) Rare alleles as indicators of gene flow. Evolution 39:53–65
30. Kimura M, Ohta T (1978) Stepwise mutation model and distribution of allelic frequencies in a finite population. Proc Natl Acad Sci USA 75:2868–2872
31. Di Rienzo A, Peterson AC, Garza JC, Valdes AM, Slatkin M, Freimer NB (1994) Mutational processes of simple-sequence repeat loci in human populations. Proc Natl Acad Sci USA 91:3166–3170
32. Estoup A, Wilson IJ, Sullivan C, Cornuet JM, Moritz C (2001) Inferring population history from microsatellite and enzyme data in serially introduced cane toads, *Bufo marinus*. Genetics 159:1671–1687
33. Luikart G, Allendorf FW, Cornuet JM, Sherwin B (1998) Distortion of allele frequency distributions provides a test for recent population bottlenecks. J Hered 89:238–247
34. Garza JC, Williamson EG (2001) Detection of reduction of population size using data from microsatellite loci. Mol Ecol 10:305–318
35. Nei M, Tajima F, Tateno Y (1983) Accuracy of estimated phylogenetic trees from molecular data. J Mol Evol 19:153–170
36. Saitou N, Nei M (1987) The neighbor-joining method: a new method for reconstructing phylogenetic trees. Mol Biol Evol 4: 406–425
37. Sneath PHA, Sokal RR (1973) Numerical taxonomy. W.H. Freedman and Co., San Francisco
38. Goudet J (1999) PCAGEN version 1.2. Population genetics laboratory, University of Lausanne, Lausanne, Switzerland
39. Felsenstein J (1993) PHYLIP-phylogenetic inference package, version 3.5c. University of Washington, Seattle, WA
40. Cornuet JM, Piry S, Luikart G, Estoup A, Solignac M (1999) New methods employing multilocus genotypes to select or exclude populations as origins of individuals. Genetics 153:1989–2000
41. Wilson GA, Rannala B (2003) Bayesian inference of recent migration rates using multilocus genotypes. Genetics 163:1177–1191
42. Paetkau D, Slade R, Burdens M, Estoup A (2004) Genetic assignment methods for the direct, real-time estimation of migration rate: a simulation based exploration of accuracy and power. Mol Ecol 13:55–65
43. Wang J, Whitlock MC (2003) Estimating effective population size and migration rates from genetic samples over space and time. Genetics 163:429–446
44. Rannala B, Mountain JL (1997) Detecting immigration by using multilocus genotypes. Proc Natl Acad Sci USA 94:9197–9201
45. Paetkau D, Calvert W, Stirling I, Strobeck C (1995) Microsatellite analysis of population structure in Canadian polar bears. Mol Ecol 4:347–354
46. Efron B (1983) Estimating the error rate of a prediction rule: improvement on cross-validation. J Am Stat Assoc 78:316–331
47. Evanno G, Regnaut S, Goudet J (2005) Detecting the number of clusters of individuals using the software structure: a simulation study. Mol Ecol 14:2611–2620

48. Slatkin M (1993) Isolation by distance in equilibrium and nonequilibrium populations. Evolution 47:264–279
49. Wright S (1943) Isolation by distance. Genetics 28:114–138
50. Rousset F (1997) Genetic differentiation and estimation of gene flow from *F*-statistics under isolation by distance. Genetics 145:1219–1228
51. de Sousa SN, Finkeldey R, Gailing O (2005) Experimental verification of microsatellite null alleles in Norway spruce (*Picea abies* [L.] Karst.): implications for population genetic studies. Plant Mol Biol Rep 23:113–119
52. Girard P, Angers B (2008) Assessment of power and accuracy of methods for detection and frequency-estimation of null alleles. Genetica 134:187–197
53. Slatkin M (1995) Hitchhiking and associative overdominance at a microsatellite locus. Mol Biol Evol 12:473–480
54. Paetkau D, Waits IP, Clarkson PL, Craighead I, Strobeck C (1997) An empirical evaluation of genetic distance statistics using microsatellite data from bear (Ursidae) populations. Genetics 147:1943–1957
55. Pemberton JM, Slate J, Bancroft DR, Barrett JA (1995) Nonamplifying alleles at microsatellite loci: a caution for parentage and population studies. Mol Ecol 4:249–252
56. Rice WR (1989) Analysing tables of statistical tests. Evolution 43:223–225
57. Benjamini Y, Yekutieli D (2001) The control of false discovery rate under dependency. Ann Stat 29:1165–1188

Chapter 20

Molecular Mapping and Breeding with Microsatellite Markers

David A. Lightfoot and Muhammad J. Iqbal

Abstract

In genetics databases for crop plant species across the world, there are thousands of mapped loci that underlie quantitative traits, oligogenic traits, and simple traits recognized by association mapping in populations. The number of loci will increase as new phenotypes are measured in more diverse genotypes and genetic maps based on saturating numbers of markers are developed. A period of locus reevaluation will decrease the number of important loci as those underlying mega-environmental effects are recognized. A second wave of reevaluation of loci will follow from developmental series analysis, especially for harvest traits like seed yield and composition. Breeding methods to properly use the accurate maps of QTL are being developed. New methods to map, fine map, and isolate the genes underlying the loci will be critical to future advances in crop biotechnology. Microsatellite markers are the most useful tool for breeders. They are codominant, abundant in all genomes, highly polymorphic so useful in many populations, and both economical and technically easy to use. The selective genotyping approaches, including genotype ranking (indexing) based on partial phenotype data combined with favorable allele data and bulked segregation event (segregant) analysis (BSA), will be increasingly important uses for microsatellites. Examples of the methods for developing and using microsatellites derived from genomic sequences are presented for monogenic, oligogenic, and polygenic traits. Examples of successful mapping, fine mapping, and gene isolation are given. When combined with high-throughput methods for genotyping and a genome sequence, the use of association mapping with microsatellite markers will provide critical advances in the analysis of crop traits.

Key words QTL, Marker-assisted selection, Bulked segregants, Trait indexing, BES, Motif, Genome

1 Introduction

Crop geneticists and breeders aim to improve harvestable yield, reduce crop losses, and reduce grower inputs by reassortment among favorable haplotypes (1–3, 12) and the incorporation of new genes for new traits (4, 5). Rapid genetic gains can be made by incorporating genes for resistance to biotic and abiotic stresses. The positive effect of stress resistance is so reliable that journals rarely accept yield data for publication (6). However, each stressor has an economic threshold, a point at which significant yield loss

Stella K. Kantartzi (ed.), *Microsatellites: Methods and Protocols*, Methods in Molecular Biology, vol. 1006,
DOI 10.1007/978-1-62703-389-3_20, © Springer Science+Business Media, LLC 2013

occurs, which is often contentious. For each stress it is important to determine that threshold and use resistance loci only when the threshold is exceeded because many stress resistance genes reduce yield potential in the absence of disease.

Molecular mapping and marker-assisted selection have been cost and labor intensive endeavors until recently (7). Therefore, stress resistance traits with clear easily scored phenotypes are often selected without markers. However, traits with a bigenic or oligogenic inheritance pattern or moderate heritability or with high trait variability are most efficiently selected with DNA markers (8, 9). Traits with a low to moderate heritability due to significant interactions with the environment are most efficiently selected with DNA markers (9) and yield indexing, a weighted mixture of marker and trait data has been particularly useful for selections.

Many marker systems are available in crops (10). Microsatellites have been used extensively for the past two decades as the markers of choice for most breeders. They are codominant and so can be used to identify and protect new loci, study dominance, identify hybrids, track contamination among lines, identify cultivars, and select alleles. They are very abundant in all genomes if both type I and type II microsatellites are considered they approach one potential marker per 5–10 kbp per cultivar pair. Microsatellites tend to show a high frequency of polymorphism, even among pairs of related cultivars. They may have more than two alleles which makes them useful in many populations and for the selection of specific alleles. Since microsatellites use standard PCR conditions and separation techniques that are scalable from dozens to thousands, they are both economical and technically easy to use. The selective genotyping approaches, including genotype ranking (indexing) based on partial phenotype data combined with favorable allele data and bulked segregation event (segregant) analysis (BSA), will be increasingly important uses for microsatellites. Examples of the methods for developing and using microsatellites derived from genomic sequences are presented for monogenic, oligogenic, and polygenic traits. Examples of successful mapping, fine mapping, and gene isolation are given. When combined with high-throughput methods for genotyping and a genome sequence, the use of association mapping with microsatellite markers will provide critical advances in the analysis of crop traits.

2 Intellectual Property

A major use of microsatellite markers is to define regions encompassing resistance genes that can be used to claim and protect intellectual property (11–15). Early patents in this area were quite broad, claiming large tracts of genome because they contained a single resistance gene. More recent actions by the patent office try to

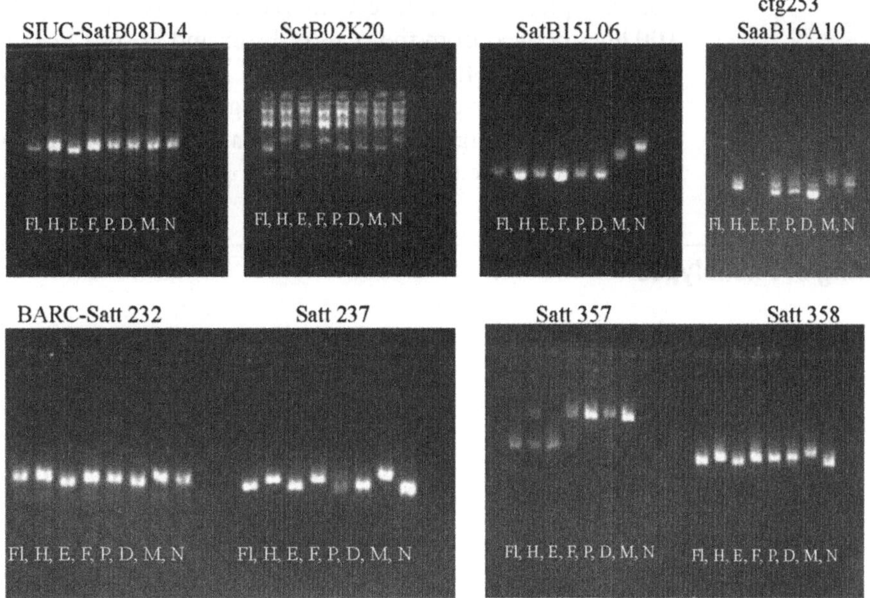

Fig. 1 Screens for polymorphisms among soybean germplasm with BES-SSR markers. Eight Population Parents Flyer (Fl), Hartwig (H), Essex (E), Forrest (F), Pyramid (P), Douglas (D), Minsoy (M), and Noir1 (N) classified by their genotypes comparing SIUC-BES-derived and BARC-SSR microsatellite markers

restrict the claims to single markers for single genes making this type of patent more difficult and expensive to obtain. Some companies holding broad early patents have sought to broaden the claims still further to encompass selection for ANY locus in the region where a stress resistance gene is found (11, 12, 16, 17). Here they seek to limit "unintentional" selection of linked traits. Therefore, each new project should start with due diligence, an examination of intellectual property related to the trait and regions to be selected.

3 Marker Development

For most crops there is a continued distributed community effort to add new markers to the existing linkage maps. The markers of choice include microsatellites and SNPs (Fig. 1; (3, 18–20)). Many efforts have been focused on regions that currently do not have many markers. Current work relies heavily on the genome sequences, resequences, and physical maps (19). In future the development of new markers will rely on the genome sequences of many cultivars and the haplotype maps that will be developed from it (20–22).

For map development renewable genetic resources will be necessary. There are many renewable collections of recombinant inbred lines. However, most lie in private hands and are not widely available. There is a community-wide effort to release certain key populations.

For example, in soybean recently released were two populations of 100 lines derived from the cross of Essex and Forrest (23) and Flyer by Hartwig (24). In addition there are immortal collections of near isogeneic line pairs, each pair capturing the heterogeneous regions derived from a single RIL (25, 26) that are available. Resources of equivalence will be released in many crops over the next decades.

4 Scoring of Phenotypes

In order to avoid garbage in garbage out (GIGO) map, stress resistance traits must be measured under controlled and/or highly replicated assays with clear easily scored phenotypes. Traits with a bigeneic or oligogenic inheritance pattern are most efficiently scored on a percentage scale (8). Traits with a low to moderate heritability due to significant interactions with the environment are most efficiently selected in greenhouse assays IF the assay can be shown to correspond to the field results (9). In pathogen assays the use of low doses of standardized inocula was particularly important to many assays (5, 8). Trait distributions should be analyzed before mapping to determine skewness and kurtosis, indicators of major gene effects. Major morphological effects like maturity dates and determinacy MUST be measured and correlated with the disease resistance traits to avoid selection of loci underlying such confounding traits (7, 27). Trait data with little replication and marker scores can be combined by indexing preliminary QTL and trait ranks to improve rapid selection of the best lines.

5 Choice of Markers

The choice of marker depends upon the objective of the selection program, tools available to the individual/group researcher, markers available for a particular region or trait, and the information content associated with a particular marker system. Linkage disequilibrium must be considered. Linkage blocks can vary by over 100-fold between and even within cultivar pairs and their progeny (21, 28). Hence, comparatively distant markers work well in some regions but not others.

The first marker system to be used was restriction length polymorphism (RFLP). RFLPs can be adapted to high-throughput markers by determination of the SNP detected. However, the polyploid origin of many crop genomes can lead to the detection of multiple DNA fragments with all RFLP probes. Single RFLP probes have been used to detect up to 19 independent loci (29). The multiplicity of RFLP loci can make the locus identity ambiguous. Other factors that prevent the use of RFLP in mapping and marker-assisted breeding are the low frequencies of polymorphisms observed (30)

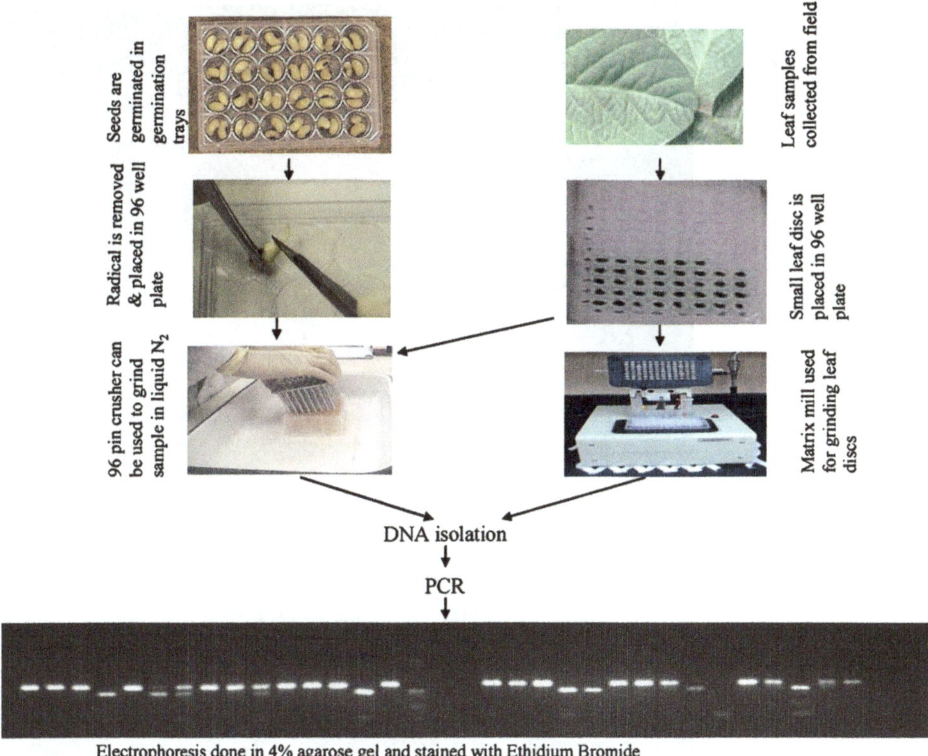

Fig. 2 Steps involved in MAS. Gel electrophoresis can be replaced by other capillary-based electrophoresis techniques, a fluorescent-based assay like TaqMan or invader, or MALDI-TOF can be used for scoring polymorphism at the target allele

and expense of automating the procedure for high-throughput screening.

Microsatellite markers use DNA sequence data and PCR to show polymorphism within 5–9 (class I) or 10–20 (class II) simple tri- or dinucleotide repeat motifs. These class II microsatellite markers have been shown to be highly polymorphic (Fig. 2; (10, 31, 32)). Some highly polymorphic microsatellite loci can have as many as 26 alleles (33–35). Such a high level of allelic diversity increases the possibility of detecting polymorphism between parents of populations derived from the hybridization of adopted soybean genotypes.

Unfortunately, class II SSRs are comparatively rare (1 per 20–50 kbp per cultivar pair) (36). Further, as the maps have become more dense, evidence of clustering has emerged (9, 20, 37, 38). Analysis of genomic DNA has shown the shorter class I microsatellites that contain less than ten simple tri- or dinucleotide repeats (small simple sequence repeats or SSSRs) are more common than SSRs and frequently polymorphic in crops (Fig. 3; (10, 39, 40)). There are about 10–20 such DNA markers (as defined) per 100 kbp

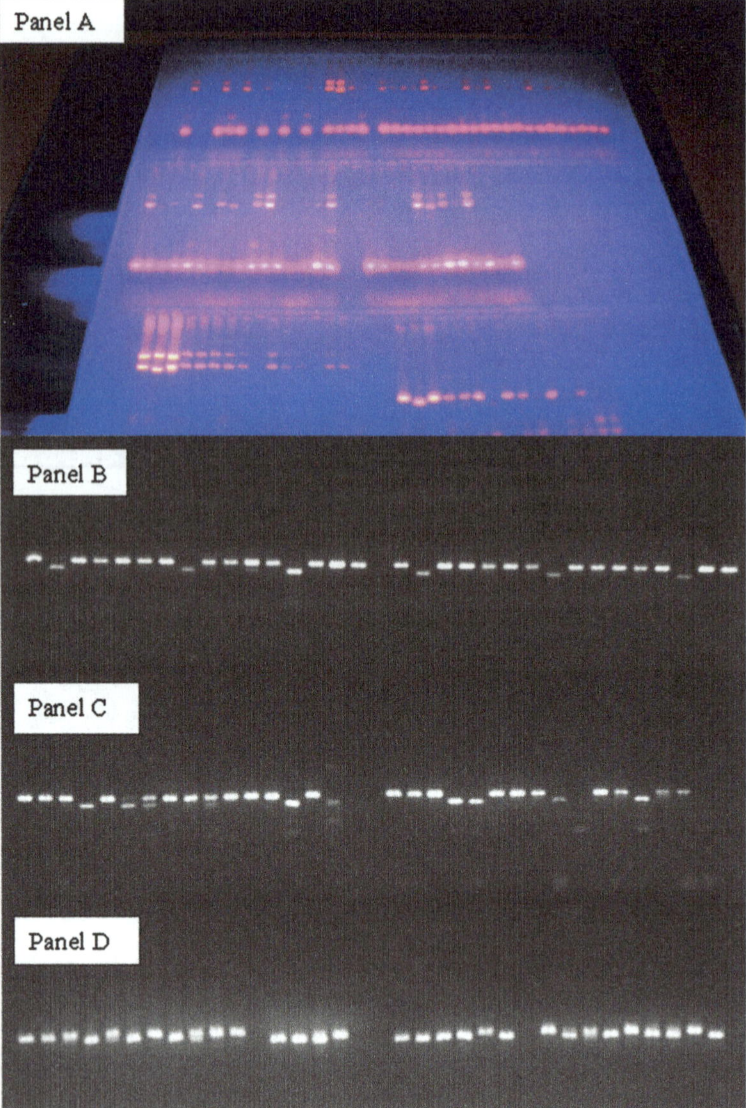

Fig. 3 Segregation of SSR locus, Satt in a population. Individuals can be clearly defined as carrying the alleles of either parent 1 (*P1*), parent 2 (*P2*), or heterozygous. The 10-μl PCR reaction was electrophoresed on a 4 % MetaPhor agarose gel and stained with ethidium bromide. DNA was isolated from radical tissue in a 96-well plate. PCR was carried out in a 96-well plate and samples were loaded on gel with an 8-channel pipette. Panel **a**: A typical gel. Panel **b–d**. SattB54L22 scored among 92 individuals segregating for resistance to biotic stress. First two bands in **a** and last 2 in **c** were the parental DNAs

per cultivar pair. Class I markers can overcome the problem of marker clustering because of their abundance in the genome.

In addition, complex microsatellites such as SSSRs and selectively amplified microsatellite polymorphic loci (SAMPLs) that contain two different tri- or dinucleotide repeats are very common and frequently polymorphic in soybean (10, 41). Selection for class I and II SSRs, SSSRs, and SAMPLs in soybean can be made with standard gel or capillary electrophoresis (7, 42), or electrophoresis-free assays like MALDI-TOF MS (43), and even next-generation sequencing methods. Therefore, microsatellite markers are currently the method of choice in most of the public laboratories based on their distribution in the genome, the polymorphism frequency, and information content attached to each identified locus. The relative ease of automation for high-throughput screening programs is another bonus.

Single nucleotide polymorphisms (SNP) are single base pair changes (or 1–3 bp indels) between individuals or cultivars. They occur about twice as often in noncoding compared to coding DNA. In coding DNA, less than one quarter of SNPs alter amino acid sequence but this depends on the genes examined (21, 44, 45). The observed SNP frequency in crops is about 0.05–0.3 % (21). However, genes under intense selection, like the major resistance genes *Rhg1*, *Rhg4* (resistance to *Heterodera glycines*), *Rfs1*, and *Rfs2* (resistance to *Fusarium solani* f. sp. *glycines*) are encoded by 6–8 alleles defined by SNPs (13, 45–47) but only 3–4 alleles defined by microsatellites (28) reflecting the extra SNP diversity imported by gene introgression. Selection for SNPs in soybean can be made with standard gel or capillary electrophoresis, or electrophoresis-free assays like TaqMan (Fig. 1) (44, 48), MALDI-TOF, and next-generation sequencing. It has been observed that the SNP frequency in the amplicons of microsatellites and AFLP bands is tenfold higher than other regions of the genome (48). Therefore, SNPs and microsatellites can be interconverted if need be.

Deletions and insertions (indels) are multi-base pair changes between soybean cultivars. They occur in noncoding DNA more often than coding DNA. They are the basis of microsatellite and satellite polymorphisms. In coding DNA they alter amino acid sequence (44). The observed frequency in crops is about fivefold lower than for SNPs (44, 48). Some genes under intense selection like the RLK at the major resistance gene *Rhg1*, bear indels in introns, promoter and terminator regions for two of the nine alleles, both encode resistance (13, 45, 46, 49). The indels are found in the 5′ nontranslated region and in both the 5′ and 3′ enhancer region. Selection for indels can be made by non-PCR methods (50) and isothermic assays like Invader (51). The standard PCR-based gel or capillary electrophoresis or electrophoresis-free assays like TaqMan (44, 48) or MALDI-TOF MS (43) or genome sequencing can also be used (Fig. 1). The ease of selection makes indels the most versatile DNA marker available especially when they are also microsatellites.

Rapid advances in the next-generation sequencing (NGS) have provided us the technology with the ability to revolutionize the way the DNA markers are developed and applied in plant breeding. The availability of relatively inexpensive DNA sequencing technology has transformed the way populations are genotyped (52–57), and polymorphisms are discovered (58–61) by genome sequencing (62, 63). However, genomes of major crops such as wheat, soybean, and corn are polyploid and highly repetitive. For the application of NGS in plant breeding and genomics, genotyping-by-sequencing (GBS) approaches that use restriction enzyme-based complexity reduction coupled with DNA-barcoded adapters to produce multiplex libraries of samples ready for NGS have been developed and found to be robust across a range of species and capable of producing tens to hundreds of thousands of molecular markers (53, 56). With the availability of a number of restriction enzymes for complexity reduction, it is possible to increase the coverage of the target genome or increase the multiplexing levels of a target population. The availability of reference genome sequence of soybean (38), the imputation of low coverage marker order data generated through GBS, and the other sequence-based approaches will be straightforward. However, there is still application gap between research laboratories and breeding groups and a knowledge gap among molecular biologists, plant breeders, and bioinformatician necessary for handling large sets of genomic sequence data generated by NGS and its application in plant breeding.

6 Identification of Polymorphism

The first step in applying marker-assisted selection (MAS) to a breeding program is the identification of polymorphic loci between the parents that can be traced in the segregating population. The frequency of polymorphism is related to genetic distance between the parents and the type of marker used (21). When low diversity or a paucity of markers is a problem, genome sequence-based marker identification is an efficient approach (17–19, 39, 40, 42). Screening of markers identified in genome sequences can be made bioinformatically. For instance, if two cultivars have been sequenced from different germplasm bases (e.g., Williams northern USA and Forrest southern USA), it is likely that the useful markers will be polymorphic in other populations. Direct screening of markers without prior knowledge of polymorphism can be efficiently done with 8–10 genotypes selected to encompass 75–90 % of the US soybean crop variability (Fig. 1 and Table 1). Monomorphic loci are discarded and loci polymorphic in many genotypes are selected for (10). BES-derived markers are named for discovering institution (e.g., SIUC-, UMO-, UMN-) and the BAC number from which the BES derive (e.g., B073P06). In a complete sequence, they may be assigned a chromosome number (e.g., Gm18) and a base pair location (1,123,657) as identifiers.

Table 1
BES-SSR derived polymorphic markers in E × F with their Williams 82 genome sequence contig series and numbers of different amplicons found among Essex Forrest Williams 82, Pyramid, Douglas, Minsoy, and Noir1

Marker	Contig series	Amplicon
SIUC_B10A18 SSR	9204	2
SIUC_B02B24 SSR	9102	2
SIUC_B08G14 SSR	9376	2
SIUC_B10P12 SSR	9036	4
SIUC_B08D14 SSR	9376	4
SIUC_B02K20 SSR	9102	2
SIUC_B13L17 SSR	9092	4
SIUC_B20C11 SSR	9320	6
SIUC_B20B01 SSR	9345	3
SIUC_B20K08 SSR	9320	1
SIUC_B27B08 SSR	9319	4
SIUC_B31L10 SSR	9201	5
SIUC_B35F13 SSR	9212	1
SIUC_H32G20 SSR	9284	1
SIUC_B35H07 SSR	8227	1
SIUC_B15I12 (or B15I21)	8131	2
SIUC_H100B10 SSR	8073	2
SIUC_B24H14 SSR	8128	2
SIUC_B32L16 SSR	8047	2
SIUC_B15J11 SSR	8131	6
SIUC_B32K10 SSR	3195	4
SIUC_B13G15 SSR	2562	2
SIUC_B12C13 SSR	2475	1
SIUC_B29K15 SSR	2894	5
SIU_SAT37 SSR	1716 (*G. max*, version 3)	3
E-CCC/M-ATG "Forrest allele" AFLP marker	1716 (*G. max*, version 3)	1
SIUC_B15B14 SSR	1632	4
SIUC_B31I19 SSR	1668	1
SIUC_B30A11 SSR	1536	6

(continued)

Table 1 (continued)

Marker	Contig series	Amplicon
SIUC_B14L17 SSR	1246 (*G. max*, version 3)	1
SIUC_B31B02 SSR	899	1
SIUC_H14J18 SSR	859 (*G. max*, version 3)	1
SIUC_B21C09 SSR	810	4
SIUC_B15L05 SSR	628	4
SIUC_B15L06 SSR	628	2
SIUC_B15P23 SSR	244	3
SIUC_H32P20 SSR	187	1
SIUC_B30I21 SSR	175	1
SIUC_B30J24 SSR	175	6
SIUC_B34L17 SSR	167	3
SIUC_B34N08 SSR	167	3
SIUC_B29H12 SSR	142	1
SIUC_B17E19 SSR	158	6
SIUC_B31A06 SSR	137	4
SIUC-SAT138 SSR	Rhg1 ctg[a]	1
SIUC-100B10-SAG9 SSR	Rhg1 ctg[a]	1
SIUC-SAT96 SSR	Rhg1 ctg[a]	1
SIUC-SAT1 SSR	Rhg1 ctg[a]	1

[a]Ruben et al. 2006

7 Genetic and Association Map Development

There are a wide variety of map build and QTL analysis programs available and many work well (64, 65). At a minimum the map should be built at a high LOD (4 to 5) then expanded at lower LODs to reduce transitive map errors. QTL should be detected by ANOVA, an interval map program and a composite interval map program to identify all possible QTL. Map errors must be removed before IM and CIM. An efficient error removal method is to look closely at the scores inferring double recombination events over short genetic distances. Modern methods for early selections based on an index weighting of marker-identified QTL beneficial alleles at two- to threefold over yields measured with low replications in incomplete sets of environments may be efficiently carried out with 200–300 microsatellite markers.

Across germplasm collections, programs like TASSEL 3.0 work efficiently with more than 1,000 markers, a number easily achieved with genome sequence-based microsatellites or genome resequencing or GBS. False associations may be identified within small groups of related germplasm. The ability to detect more than two alleles can also present problems from over partitioning. Associated loci with several alleles should be subject to a phylogeny analysis to determine their relatedness among alleles that might allow reduction to two clades.

8 Marker-Assisted Recovery of Recurrent Parent Genome

In plant breeding for resistance to biotic stress, the accelerated backcross procedure was often used to transfer favorable alleles from donor genotypes into a recipient elite genotype (4, 66). The donors often have many poor agronomic properties requiring several backcrosses that only partly eliminate them. The goal of this "introgression" is to recover transgressive segregation event (segregants) that yield more than the elite parent. To be worth the effort, yields must be increased by 1–2 % per year for each year necessary to develop the introgression line. Only genes of large effect on stress resistance can be selected in this way. In backcross gene introgression programs, marker-assisted selection has a great impact on increasing the genome recovery of recurrent parent genotype (RPG; (67)). Individuals that are homozygous for the alleles of the recurrent parent at a large number of marker loci covering the entire genome are selected. Population size and marker density required in a background selection program can be determined. The minimum recommended is the use of four markers per chromosome (of ~200 cM length) and a selection strategy for proximal recombinants of the target allele. The number of markers used for RPG recovery varies, about a 20-cM marker density represented an optimal trade-off between percentage recurrent parent recovery and management of data-point throughput requirements (68). Uniformly distributed markers are more effective than equal numbers per chromosome, because smaller population sizes and fewer data points are required with equally spaced markers (68). Therefore, selection of markers from a genome sequence or physical map rather than a less precise genetic map can improve the efficiency of RPG recovery. Genome resequences by NGS provide more markers than needed for the few recombination events present but do allow the identification of rare and valuable events.

9 Marker-Assisted Selection in Recurrent Cross Populations

The majority of crop breeding programs make advances by reassorting the entire genomes of two elite lines with favorable and complementary characteristics (69–72). The goal of shuffling the alleles was to recover transgressive segregants and epistatic interactions that yielded more than either parent. Again, the yield increase target was 1–2 % per year for each year necessary to develop the line. Traditional recurrent selection cycles last from 5 to 7 years from cross to release. The average cultivar is sold for 2–3 years before becoming obsolete. Therefore, marker selection that reduces development time by stacking selections into a shorter time frame will be very valuable. Further, since released cultivars appear to contain more recombination events and be heterozygous at more loci than the population mean, markers can be used to find well shuffled, heterotic, genomes for phenotypic analysis (2, 73).

The method of MAS for stress resistance often competes with field scores, greenhouse assays, and even low-cost laboratory assays of phenotypes for price and effectiveness (8, 74, 75). This competition is not always market driven, as investments in infrastructure made in the past determine economics in the present. Over time though, MAS was widely adopted by all effective soybean breeders. MAS as employed today usually focuses on two to four genes per population with 25–6.25 % of lines "passing the test." Samples are selected after preliminary visual selection for yield potential (at the F3–F5), rarely are F2 selections made. Hence, about 10 % of a program's recombinants will be tested in the MAS lab. Since a single lab will serve dozens of breeders each with a 100,000 lines, the task can quickly reach millions of selections per year. Automation is very important in this situation (3, 7).

10 Examples of Marker-Assisted Selection for Targeted Stress Resistance Traits

A large number of QTL for hundreds of agronomic traits have been mapped in crops like soybean. Databases contain gene class, locus, alleles, phenotypes, and two-point data. The availability of all this information and commercial drive to release new and improved varieties assists the use of MAS for new and improved variety development.

10.1 SCN

One example of MAS in soybean is selection for resistance to soybean cyst nematode (SCN). Soybean cyst nematode, *Heterodera glycines*, is a small plant-parasitic roundworm that attacks the roots of soybeans and causes significant crop losses in the infected fields. SCN resistance was introgressed from non-adapted or weedy types by backcrossing and was associated with both a linkage drag on yield and a genome load affecting yield (Fig. 4; (73)). Two QTL significantly contributing to soybean resistance to *H. glycines*, *rhg*1, and *Rhg*4 were mapped on linkage groups G and A2 (chromosome 18 and 8) and the candidate genes isolated (45, 49, 76). Cultivars selected for *rhg1* provide good resistance to *H. glycines* Hgtype 0 and 7 (race 3). SSR marker Satt309 was mapped at a 0.5–2 cM

Fig. 4 The effects of gene stacking on resistance to SDS. To the *left* of the *white line* is a cultivar with a stack of six resistance genes. To the *right* of the *line* is a cultivar with few or no resistance genes showing the chlorosis and defoliation symptomatic of SDS

distance from *rhg*1 (42). It was shown to be 42 kbp in physical distance from the RLK at *rhg1* the most distal gene (to the telomere) of a three gene cluster (45, 49). Satt309 was widely used for MAS for resistance to SCN race3 (77) but only for research purposes because of patent positions held privately.

With the sequencing of a large region of linkage group G of soybean genome by Monsanto (St. Louis, MO; (13)), and the SIUC patent for perfect markers for SCN (15, 28, 45, 46, 49), it is now possible to select the four resistance alleles among of the nine total *Rhg1/rhg1* alleles with 100 % certainty. There are microsatellite markers for the promoter region, indels in the intron, promoter and terminator, and four SNPs adapted to Taqman in the leucine-rich repeat region of the rhg1 receptor like kinase, plus two in the kinase domain. In total 26 SNP markers are directly located on the RLK gene associated with 50–90 % of resistance to SCN race 3, six alter protein allotype but no SNPs seem to alter cis regulatory elements (CREs). There are indels in the 5′ and 3′ enhancer regions. Selection for recombination events within the gene has been made possible by saturation mapping based on three genome sequences of this region (two susceptible and one resistant cultivar; (49)). Recombinants tend to be susceptible (45) that provide 800 informative markers.

10.2 SDS

A different example of MAS in soybean is selection for resistance to sudden death syndrome (SDS). SDS in the USA is caused by a member of the *Fusarium* sp. disease complex, *F. virguliforme* (e.g., *solani* f. sp. *glycines*) that attacks the roots of soybeans and causes significant crop losses in the infected fields. SDS resistance was found among elite adapted types and was exploited by gene stacking during intercrossing. Resistance to SDS was associated with no linkage drag on yield (Fig. 4); indeed, it is a disease of high-yield genotypes. Among several resistance genes associated with resistance to sudden death syndrome of soybean (SDS; Table 1), three were clustered with SCN resistance *Rhg*1 (46, 69). The first *Rfs2* was formally proven to be pleiotropic with *Rhg1* (28, 49). The second clustered gene, *Rfs1*, was responsible for part of a bigenic root resistance to *F. virguliforme* and can be selected for by a set of BAC-derived SSRs SIUC-Satt122 (10, 44), a RAPD-derived SCAR (SIUC-OI03), or a candidate gene sequence (15). The third locus was near the telomere linked to BARC-Satt163.

The second recently discovered SDS root resistance gene can be selected for by SSR Satt594 (24, 71, 72) and was linked to, but not pleiotropic with *rhg5* needed for race 14 of HgType1.2.5. resistance. Since linkage was in repulsion, the microsatellite markers can be used to identify recombinants of value to breeding programs.

Each of the several resistance genes discovered for SDS resistance has associated a SSR marker (Tables 1 and 2). Each gene has been detected in multiple populations. Therefore, gene pyramids have been developed incorporating up to six genes and gene pyramids of 10–12 genes were produced (Fig. 5). Gene pyramids ave been shown effective in eliminating SDS symptoms from fields. In recent work an expression fingerprint for resistance to SDS has been extended into the proteome (28, 78–80). The expression profile and associated protein changes will be used to develop a very high-throughput assays for resistance to SDS.

However, the SDS success story in Illinois led to complacency among breeders to the north that was misplaced (3). The fungus was still present in the soil and was under strong selective pressure to adapt to the partially resistant germplasm. Consequently the disease has become severe in Iowa and is resurgent in Illinois by 2012. As resistance is increased, the danger of selecting for pathogen races is increased. Further, a second pathogen that causes SDS in South America, *F. tucumanes,* may yet infest US fields (81) and cannot be controlled by the same resistance gene set.

Table 2
SDS QTL found in adapted germplasm and different RIL populations

Linkage group	SoyBase QTL name (linked markers)[a]	Populations	Probable locus names[b,c]	References
A2	SDS 11-1 (Satt 187)	Ripley × Spencer	*qRfs7*	Hashmi (86)
C2	SDS 1-1 (0005-250)	Essex × Forrest	*cqRfs4*[c]	Hnetkovsky et al. (6)
	SDS 4-2 (K455_1)	Essex × Forrest	*cqRfs4*	Njiti et al. (25)
	SDS 7-5 (Satt 371)	Essex × Forrest	*cqRfs4*	Iqbal et al. (69)
	SDS 8-2 (Satt 307)	Pyramid × Douglas	*cqRfs4*	Njiti et al. (75)
	SDS 8-3 (Satt 316)	Pyramid × Douglas	*qRfs9*	Njiti et al. (75)
	SDS 9-2 (Satt277)	Flyer × Hartwig	*cqRfs4*	Kazi (85)
D2	SDS 11-2 (Satt 528)	Ripley × Spencer	*cqRfs11*[d]	Hashmi (86)
	SDS 9-1 (Satt574)	Flyer × Hartwig	*cqRfs11*[d]	Kazi (85)
F	SDS 10-1 (Satt 160)	Essex × Forrest	*qRfs10*	Kassem et al. (9)
G	SDS 6-1 (Bng 122_1)	Essex × Forrest	*cqRfs1*[d]	Meksem et al. (46)
	SDS 3-1 (0G13_490)	Essex × Forrest	*cqRfs3*	Chang et al. (87)
	SDS 6-2 (Satt 309)	Essex × Forrest	*cqRfs2*	Meksem et al. (46)
	SDS 8-4 (0G01)	Pyramid × Douglas	*cqRfs3*	Njiti et al. (75)
	SDS 7-1 (Satt 214)	Essex × Forrest	*cqRfs2*	Iqbal et al. (69)
	SDS 7-2 (Satt 309)	Essex × Forrest	*cqRfs2*	Iqbal et al. (69)
	SDS 7-3 (Satt 570)	Essex × Forrest	*cqRfs1*[d]	Iqbal et al. (69)
	SDS 7-4 (OEO2_1000)	Essex × Forrest	*cqRfs3*	Iqbal et al. (69)
	SDS 4-1 (0I03-512)	Essex × Forrest	*cqRfs1*[d]	Njiti et al. (25)
	SDS 4-3 (Bng122_1)	Essex × Forrest	*cqRfs1*[d]	Njiti et al. (25)
	SDS 5-1 (Satt 038)	Flyer × Hartwig	*cqRfs1*[d]	Prabhu et al. (7)
	SDS 8-1 (Satt 163)	Pyramid × Douglas	*cqRfs2*	Njiti et al. (75)
	SDS 9-3 (Satt427)	Flyer × Hartwig	*cqRfs3*	Kazi (85)
I	SDS 7-6 (Satt 354)	Essex × Forrest	*qRfs5*	Iqbal et al. (69)
J	SDS 10-2 (Satt 285)	Essex × Forrest	*cqRfs8*[c]	Kassem et al. (9)
	SDS (Satt183)	GC87018-12-2B-1 × GC89045-13-1	*cqRfs8*[c]	Sanitchon et al. (88)

[a]QTL numbers are from Soybase where assigned

[b]From 27 QTL detections, there were 15 QTL counting each detection in a separate population once. Assuming QTL detected in common intervals in separate populations represents the alleles there are 11–12 loci

[c]QTL detected in common intervals in separate populations or derived NILs are considered confirmed and suffixed with c under Soybean Genetics Committee recommendations (http://soybase.agron.iastate.edu/nomenclature/QTL.html)

[d]QTL associated with resistance to root infection

a NO YIELD DRAG FROM SDS RESISTANCE GENES

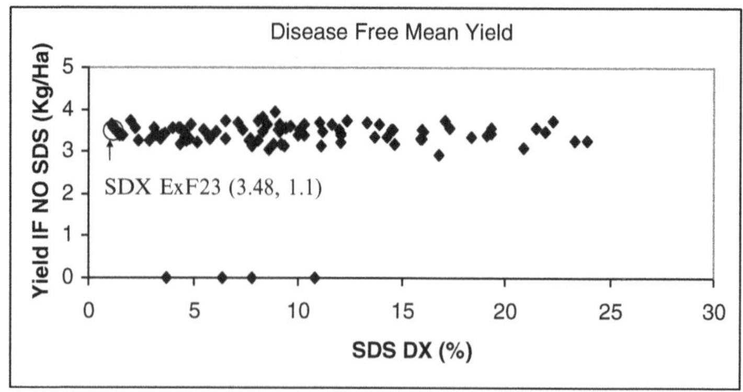

b YIELD DEPRESSION FROM SCN RESISTANCE

Yield Compared to SCN IP

r = 0.38
x = 0.1

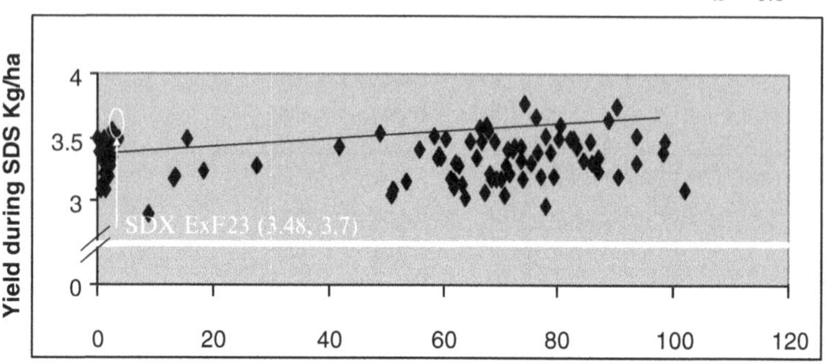

c NO YIELD DRAG FROM SDS RESISTANCE GENES

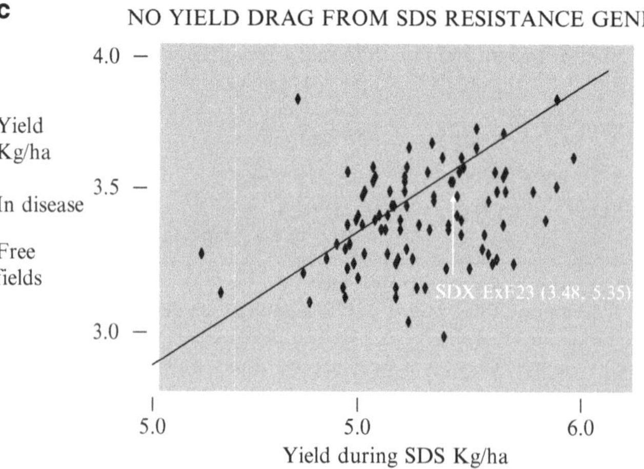

Fig. 5 Yield drag from resistance to SCN but not SDS. (**a**) Correlation between resistance to SDS and yield. (**b**) Correlation between resistance to SCN and yield. (**c**) Correlation between yield during SDS infestation and yield in disease free fields

11 Methods for Marker-Assisted Selection

Screening for markers linked to a trait or recovery of RPG can be carried out with leaf or seed samples. A high-throughput method of MAS is described (Fig. 2; (45, 47)). Briefly, DNA can be isolated from plants or seeds. For sample collection from plants, a 1-cm^2-leaf disc from a young leaf was removed and stored in small note pads and kept on ice or dry ice. Alternately, seeds are placed on small filter paper discs soaked with sterile distilled water for 48 h for germination. About 1–2 mm of the tip of the radicals is carefully removed with sharp scalpel blades and seed stored a 4 °C until selections are ready. For dry seeds a small drill is used to obtain cotyledon tissue from a 1–2 mm hole in the seed. Leaf discs radical tips or cotyledon dusts are placed in 96- or 384-well plates. Samples can be ground either by a matrix mill in the presence of 20–100 μl of 0.5N sodium hydroxide or by 96- or 384-pin crusher (Fig. 2). After crushing, DNA is neutralized by an equal amount of acidic TRIS. Other methods are popular including CTAB and propanol precipitations (82). Alternately, a small sliver of seed leaf or radical tissue (0.1–0.2 mg) can be placed directly into the PCR reaction if appropriate direct PCR polymerases are used.

Once the DNA was isolated, PCR with the selected primers was carried out in 96- or 384-well plates. Products can be electrophoresed in agarose gels when alleles differ by >3 bp (Fig. 3) or acrylamide gel or capillaries if the difference is 1–3 bp. For a higher throughput, SSR or SCAR primers can be labeled with fluorescence tags and multiplexed in the PCR reaction or at the electrophoresis stage. The samples can then be electrophoresed using capillary-based or gel-based DNA sequencers and analyzed by fragment analysis software as provided by the fragment analysis equipment used. The shortcomings of gel electrophoresis can be avoided by developing TaqMan-™ allelic discrimination (Fig. 1) or high-resolution melt methods (83) to polymorphisms within the microsatellite amplicons. Also to consider for use are MALDI-TOF, PCR-OLA, molecular beacons, padlock probes, and well fluorescence assays.

12 Conclusions

The main aim of MAS with microsatellites is the selection of highly desirable lines among thousands of genotypes and dozens of crosses. Selection can only be achieved by high-throughput, rapid, automated procedures for the detection of DNA polymorphisms attached with the desirable traits. With the availability of capillary electrophoresis-based genetic analyzers, thousands of samples can be screened per day per machine. However, the availability of non-electrophoresis-based tools such as TaqMan (84), the invader assay (51), and MALDI-TOF for the detection of polymorphisms further simplifies the screening procedures for large-scale MAS procedures. Moreover, the availability of gene markers or perfect markers such as SNPs directly on the genes of interest makes the MAS procedures more desirable. With the availability of millions of soybean ESTs and more than 50 cultivar genome sequences and genomic sequence and a physical map, crops like soybean have sufficient genomic resources for the rapid development of new markers for resistance to new biotic stresses. The application in marker-assisted breeding of new and improved varieties will become a widely accepted breeding tool.

References

1. Iqbal MJ, Lightfoot DA (2004) Application of DNA markers: soybean improvement. In: L"rz H, Wenzel G (eds) Molecular marker systems in plant breeding and crop improvement. Springer, New York, p 475
2. Stefaniak TR, Hyten DL, Pantalone VR, Klarer A, Pfeiffer TW (2006) Soybean cultivars resulted from more recombination events than unselected lines in the same population. Crop Sci 46:43–51
3. Lightfoot DA (2008) Soybean genomics: developments through the use of cultivar Forrest. Int J Plant Genomics 2008:1–22. doi:10.1155/2008/793158
4. Anand SC (1992) Registration of 'Hartwig' soybean. Crop Sci 32:1060–1070
5. Arelli AP (1994) Inheritance of resistance to *Heterodera glycines* race 3 in soybean accessions. Plant Dis 78:898–900
6. Hnetkovsky N, Chang SJC, Doubler TW, Gibson PT, Lightfoot DA (1996) Genetic mapping of loci underlying field resistance to soybean sudden death syndrome (SDS). Crop Sci 36:393–400
7. Prabhu RR, Njiti VN, Bell-Johnson B, Johnson JE, Schmidt ME, Klein JH, Lightfoot DA (1999) Selecting soybean cultivars for dual resistance to soybean cyst nematode and sudden death syndrome using two DNA markers. Crop Sci 39:982–987
8. Njiti VN, Johnson JE, Torto TA, Gray LE, Lightfoot DA (2001) Inoculum rate influences selection for field resistance to soybean sudden death syndrome in the greenhouse. Crop Sci 41:1726–1731
9. Kassem MA, Shultz J, Meksem K, Cho Y, Wood AJ, Iqbal MJ, Lightfoot DA (2006) An updated 'Essex' by 'Forrest' linkage map and first composite interval map of QTL underlying six soybean traits. Theor Appl Genet 113:1015–1026
10. Shultz JL, Kazi S, Afzal JA, Bashir R, Lightfoot DA (2007) The development of BAC-end sequence-based microsatellite markers and placement in the physical and genetic maps of soybean. Theor Appl Genet 114:1081–1090
11. Webb DM, Baltazar BM, Rao-Arelli AP, Schupp J, Keim P, Clayton K, Ferreira AR, Owens T, Beavis WD (1995) QTLs affecting soybean cyst-nematode resistance. Theor Appl Genet 91:574–581
12. Webb DM (1996) Soybean cyst nematode resistant soybeans and methods of breeding and identifying resistant plants. US Patent 5,491,081
13. Hauge BM, Wang ML, Parsons JD, Parnell LD (2001) Nucleic acid molecules and other molecules associated with soybean cyst nematode resistance. US patent WO 0151627-A 19-JUL-2001
14. Lightfoot DA (2001) Soybean sudden death syndrome resistant soybeans, soybean cyst nematode resistant soybeans and methods of breeding and identifying resistant plants. US Patent 6,300,541
15. Lightfoot DA, Meksem K (2011) Isolated soybean cyst nematode and sudden death syndrome polypeptides. US Patent 7,902,337
16. Webb DM (2000) Positional cloning of soybean cyst nematode resistance genes. US Patent 6,162,967
17. Webb DM (2003) Quantitative trait loci associated with soybean cyst nematode resistance and uses thereof. US Patent 6,538,175
18. Song QJ, Marek LF, Shoemaker RC, Lark KG, Concibido VC, Delannay X, Specht JE, Cregan PB (2004) A new integrated genetic linkage map of the soybean. Theor Appl Genet 109:122–128
19. Shultz JL, Jayaraman D, Shopinski KL, Iqbal MJ, Kazi S, Zobrist K, Bashir R, Yaegashi S, Lavu N, Afzal AJ, Yesudas CR, Kassem MA, Wu C, Zhang HB, Town CD, Meksem K, Lightfoot DA

(2006) The soybean genome database (SoyGD): a browser for display of duplicated, polyploid, regions and sequence tagged sites on the integrated physical and genetic maps of glycine max. Nucleic Acid Res 34:D758–D765

20. Wu X, Ren C, Joshi T, Vuong T, Xu D, Nguyen HT (2010) SNP discovery by high-throughput sequencing in soybean. BMC Genomics 11:469
21. Zhu YL, Song QJ, Hyten DL, van Tassell CP, Matukumalli LK, Grimm DR, Hyatt SM, Fickus EW, Young ND, Cregan PB (2003) Single-nucleotide polymorphisms in soybean. Genetics 163:1123–1134
22. Lam HM, Xu X, Liu X et al (2010) Resequencing of 31 wild and cultivated soybean genomes identifies patterns of genetic diversity and selection. Nat Genet 42:1053–1062
23. Lightfoot DA, Njiti VN, Gibson PT, Kassem MA, Iqbal MJ, Meksem K (2005) Registration of Essex×Forrest recombinant inbred line (RIL) mapping population. Crop Sci 45:1678–1681
24. Kazi S, Njiti VN, Doubler TW, Yuan J, Iqbal MJ, Cianzio S, Lightfoot DA (2007) Registration of the Flyer by Hartwig recombinant inbred line mapping population. J Plant Reg 1:175–178
25. Njiti VN, Doubler TW, Suttner RJ, Gray LE, Gibson PT, Lightfoot DA (1998) Resistance to soybean sudden death syndrome and root colonization by *Fusarium solani* f. sp. *glycines* in near-isogeneic lines. Crop Sci 38:472–477
26. Njiti VN, Myers O, Schroeder D, Lightfoot DA (2003) Glyphosate on roundup ready soybean: effects on root infection by *Fusarium solani* f. sp. *Glycines* and sudden death syndrome. Agron J 95:1140–1145
27. Njiti VN, Lightfoot DA (2006) Genetic analysis infers Dt loci underlie resistance to SDS caused by *Fusarium virguliforme* in indeterminate soybeans. Can J Plant Sci 41:83–89
28. Afzal AJ, Srour A, Hemmati N, Saini N, Shemy E, Lightfoot DA (2012) Recombination suppression at the dominant Rhg1/Rfs2 locus underlying soybean resistance to the cyst nematode. Theor Appl Genet 124:1027–1039
29. Mansur LM, Orf JH, Chase K, Jarvick T, Cregan PB, Lark KG (1996) Genetic mapping of agronomic traits using recombinant inbred lines of soybean. Crop Sci 36:1327–1336
30. Shoemaker RC, Specht JE (1995) Integration of the soybean molecular and classical genetic linkage groups. Crop Sci 35:436–446
31. Akkaya MS, Bhagwat AA, Cregan PB (1992) Length polymorphism of simple sequence repeat DNA in soybean. Genetics 132:1131–1139
32. Morgante M, Olivieri AM (1993) PCR-amplified microsatellites as markers in plant genetics. Plant J 3:175–182
33. Rongwen J, Akkaya MS, Bhagwat AA, Lavi U, Cregan PB (1995) The use of microsatellite DNA markers for soybean genotype identification. Theor Appl Genet 90:43–48
34. Powell W, Morgante M, Andre C, Hanafey M, Vogel J, Tingey S, Rafalski A (1996) The comparison of RFLP, RAPD, AFLP, and SSR (microsatellite) markers for germplasm analysis. Mol Breed 2:225–238
35. Diwan N, Cregan PB (1997) Automated sizing of fluorescent-labeled simple sequence repeat (SSR) markers to assay genetic variation in soybean. Theor Appl Genet 95:723–733
36. Marek LF, Mudge J, Darnielle L, Grant D, Hanson N, Paz M, Huihuang Y, Denny R, Larson K, Foster-Hartnett D, Cooper A, Danesh D, Larsen D, Schmidt T, Staggs R, Crow JA, Retzel E, Young ND, Shoemaker RC (2001) Soybean genomic survey: BAC-end sequences near RFLP and SSR markers. Genome 44:572–581
37. Akkaya MS, Shoemaker RC, Specht JE, Bhagwat AA, Cregan PB (1995) Integration of simple sequence repeat DNA markers into a soybean linkage map. Crop Sci 35:1439–1445
38. Schmutz J, Cannon SB, Schlueter J, Ma J, Mitros T, Nelson W, Hyten DL et al (2010) Genome sequence of the palaeopolyploid soybean. Nature 463:178–183
39. McCouch SR, Teytelman L, Xu Y, Lobos KB, Clare K, Walton M, Fu B, Maghirang R, Li Z, Xing Y, Zhang Q, Kono I, Yano M, Fjellstrom RJ, DeClerck G, Schneider D, Cartinhour S, Ware D, Stein L (2002) Development and mapping of 2240 new SSR markers for rice (*Oryza sativa* L.). DNA Res 9:199–207
40. Rota ML, Kantety RV, Yu JK, Sorrells ME (2005) Nonrandom distribution and frequencies of genomic and EST-derived microsatellite markers in rice, wheat, and barley. BMC Genomics 6:23–32
41. Witsenboer H, Vogel J, Michelmore RW (1998) Identification, genetic localization and allelic diversity of selectively amplified microsatellite polymorphic loci (SAMPL) in lettuce and wild relatives (*Lactuca* spp.). Genome 40:923–936
42. Cregan PB, Mudge J, Fickus EW, Danesh D, Denny R, Young ND (1999) Two simple sequence repeat markers to select for soybean cyst nematode resistance conditioned by the rhg1 locus. Theor Appl Genet 99:811–818
43. Chen CH, Potter NT, Taranenko NT (2003) Detection of trinucleotide repeat containing genes by matrix-assisted laser desorption/ionization (MALDI) mass spectrometry. Methods Mol Biol 217:91–100
44. Meksem K, Ruben E, Hyten D, Triwitayakorn K, Lightfoot DA (2001) Conversion of AFLP

bands into high-throughput DNA markers. Mol Genet Genomics 265:207–214

45. Ruben E, Aziz J, Afzal AJ, Njiti VN, Triwitayakorn K, Iqbal MJ, Yaegashi S, Arelli P, Town C, Meksem K, Lightfoot DA (2006) Genomic analysis of the 'Peking' rhg1 locus: candidate genes that underlie soybean resistance to the cyst nematode. Mol Genet Genomics 276:320–330
46. Meksem K, Doubler TW, Chang SJC, Chancharoenchai K, Suttner R, Cregan P, Rao-Arelli P, Gibson PT, Lightfoot DA (1999) Clustering among genes underlying QTL for field resistance to Sudden Death Syndrome and cyst nematode race 3. Theor Appl Genet 99:1131–1142
47. Triwitayakorn K, Njiti VN, Iqbal MJ, Yaegashi S, Town C, Lightfoot DA (2005) Genomic analysis of a region encompassing QRfs1 and QRfs2: genes that underlie soybean resistance to sudden death syndrome. Genome 48:125–138
48. Meksem K, Hyten D, Ruben E, Lightfoot DA (2001) High-throughput genotyping for a polymorphism linked to soybean cyst nematode resistance gene Rhg4 by using Taqman probes. Mol Breed 77:63–71
49. Srour A, Afzal AJ, Saini N, Blahut-Beatty L, Hemmati N, Simmonds DH, El Shemy H, Town CD, Sharma H, Liu X, Li W, Lightfoot DA (2012) The receptor like kinase transgene from the Rhg1/Rfs2 locus caused pleiotropic resistances to soybean cyst nematode and sudden death syndrome. BMC Genomics 13:368
50. Landegren U, Schallmeiner E, Nilsson M, Fredriksson S, Banr J, Gullberg M, Jarvius J, Gustafsdottir S, Dahl F, Sderberg O, Ericsson O, Stenberg J (2004) Molecular tools for a molecular medicine: analyzing genes, transcripts and proteins using padlock and proximity probes. J Mol Recognit 17:194–197
51. Mein CA, Barratt BJ, Dunn MG, Siegmund T, Smith AN, Esposito L, Nutland S, Stevens HE, Wilson AJ, Philips MS, Jarvis N, Law S, de Arruda M, Todd JA (2000) Evaluation of single nucleotide polymorphism typing with invader on PCR amplification and its automation. Genome Res 3:330–343
52. Baird NA, Etter PD, Atwood TS, Currey MC, Shiver AL, Lewis ZA, Selker EU, Cresko WA, Johnson EA (2008) Rapid SNP discovery and genetic mapping using sequenced RAD markers. PLoS One 3:e3376
53. Elshire RJ, Glaubitz JC, Sun Q, Poland JA, Kawamoto K, Buckler ES, Mitchell SE (2011) A robust, simple genotyping-by-sequencing (GBS) approach for high diversity species. PLoS One 6:e19379
54. Davey JW, Hohenlohe PA, Etter PD, Boone JQ, Catchen JM, Blaxter ML (2011) Genome-wide genetic marker discovery and genotyping using next-generation sequencing. Nat Rev Genet 12:499–510
55. Truong HT, Ramos AM, Yalcin F, de Ruiter M, van der Poel HJA, Huvenaars KHJ, Hogers RCJ, van Enckevort LJG, Janssen A, van Orsouw NJ, van Eijk MJT (2012) Sequence-based genotyping for marker discovery and co615 dominant scoring in germplasm and populations. PLoS One 7:e37565
56. Poland JA, Brown PJ, Sorrells ME, Jannink JL (2012) Development of high-density genetic maps for barley and wheat using a novel two-enzyme genotyping-by-sequencing approach. PLoS One 7:e32253
57. Wang S, Meyer E, McKay JK, Matz MV (2012) 2b-RAD: a simple and flexible method for genome-wide genotyping. Nat Methods 9: 808–810
58. Mardis ER (2008) The impact of next-generation sequencing technology on genetics. Trends Genet 24:133–141
59. Futschik A, Schltterer C (2010) The next generation of molecular markers from massively parallel sequencing of pooled DNA samples. Genetics 186:207–218
60. You FM, Huo N, Deal KR, Gu YQ, Luo M-C, McGuire PE, Dvorak J, Anderson OD (2011) Annotation-based genome-wide SNP discovery in the large and complex Aegilops tauschii genome using next-generation s 639 sequencing without a reference genome sequence. BMC Genomics 12:59
61. Nielsen R, Paul JS, Albrechtsen A, Song YS (2011) Genotype and SNP calling from next-generation sequencing data. Nat Rev Genet 12:443–451
62. Xu X, Pan S, Cheng S, Zhang B, Mu D et al (2011) Genome sequence and analysis of the tuber crop potato. Nature 475:189–195
63. Wang X, Wang H, Wang J, Sun R, Wu J et al (2011) The genome of the mesopolyploid crop species *Brassica rapa*. Nat Genet 43:1035–1039
64. Lander E, Green P, Abrahamson J, Barlow A, Daley M, Lincoln S, Newburg L (1987) MAPMAKER: an interactive computer package for constructing primary genetic linkage maps of experimental and natural populations. Genomics 1:174–181
65. Basten CJ, Weir BS, Zeng Z (2001) QTL cartographer version 2.0. Department of Statistics, North Carolina State University, Raleigh, NC
66. Hartwig EE, Epps JM (1973) Registration of forest soybeans. Crop Sci 13:287
67. Tanksley SD, Young ND, Paterson AH, Bonierbale MW (1989) RFLP mapping in plant breeding: new tools for an old science. Bio-technology 7:257–264

68. Frisch M, Bohn M, Melchinger AE (1999) Comparison of selection strategies for marker-assisted backcrossing of a gene. Crop Sci 39: 1295–1301
69. Iqbal MJ, Meksem K, Njiti VN, Kassem MA, Lightfoot DA (2001) Microsatellite markers identify three additional quantitative trait loci for resistance to soybean sudden death syndrome (SDS) in Essex×Forrest RILs. Theor Appl Genet 102:187–192
70. Yuan Z, Njiti VN, Meksem K, Iqbal MJ, Triwitayakorn K, Kassem MA, Davis GT, Schmidt ME, Lightfoot DA (2002) Identification of yield loci in soybean populations that segregate for disease resistance. Crop Sci 42:271–277
71. Kazi S, Shultz J, Bashir R, Afzal J, Njiti VN, Lightfoot DA (2008) Separate loci underlie resistance to soybean sudden death syndrome in 'Hartwig' by 'Flyer'. Theor Appl Genet 116:967–977
72. Kazi S, Shultz J, Afzal J, Hashmi R, Jasim M, Bond J, Arelli PR, Lightfoot DA (2010) Isolines and inbred-lines confirmed loci that underlie resistance from cultivar 'Hartwig' to three soybean cyst nematode populations. Theor Appl Genet 120:633–640
73. Karangula UB, Kassem MA, Gupta L, El-Shemy HA, Lightfoot DA (2009) Locus interactions underlie seed yield in soybeans resistant to Heterodera glycines. Curr Issues Mol Biol 11(suppl 1):i73–i84
74. Njiti VN, Gray L, Lightfoot DA (1997) Rate-reducing resistance to Fusarium solani f. sp. phaseoli (nee: glycines) underlies field resistance to soybean sudden-death syndrome (SDS). Crop Sci 37:1–12
75. Njiti VN, Meksem K, Iqbal MJ, Johnson JE, Kassem MA, Zobrist KF, Kilo VY, Lightfoot DA (2002) Common loci underlie field resistance to soybean sudden death syndrome in Forrest, Pyramid, Essex, and Douglas. Theor Appl Genet 104:294–300
76. Liu X, Liu S, Jamai A, Bendahmane A, Lightfoot DA, Mitchum MG, Meksem K (2011) Soybean cyst nematode resistance in soybean is independent of the Rhg4 locus LRR RLK gene. Funct Integr Genomics. doi:10.1007/s10142-011-0225-4
77. Mudge J, Cregan PB, Kenworthy JP, Kenworthy WJ, Orf JH, Young ND (1997) Two microsatellite markers that flank the major soybean cyst nematode resistance locus. Crop Sci 37:1611–1615
78. Iqbal MJ, Yaegashi S, Njiti VN, Ahsan R, Cryder KL, Lightfoot DA (2002) Resistance locus pyramids alter transcript abundance in soybean roots inoculated with *Fusarium solani* f. sp. *glycines.* Mol Genet Genomics 268:407–417
79. Iqbal MJ, Afzal AJ, Yaegashi S, Ruben E, Triwitayakorn K, Njiti VN, Ahsan R, Wood AJ, Lightfoot DA (2002) A pyramid of loci for partial resistance to Fusarium solani f. sp. glycines maintains myo-inositol-1-phoshate synthase expression in soybean roots. Theor Appl Genet 105:1115–1123
80. Iqbal MJ, Yaegashi S, Ahsan R, Shopinski KL, Lightfoot DA (2005) Root response to Fusarium solani f. sp. glycines: temporal accumulation of transcripts in partially resistant and susceptible soybean. Theor Appl Genet 110: 1429–1438
81. Aoki T, O'Donnell K, Homma Y, Lattanzi AR (2003) Sudden-death syndrome of soybean is caused by two morphologically and phylogenetically distinct species within the *Fusarium solani* species complex *F. virguliforme* in North America and *F. tucumaniae* in South America. Mycologia 95:660–684
82. Xin Z, Velten JP, Oliver MJ, Burke JJ (2003) High-throughput DNA extraction method suitable for PCR. Biotechniques 34: 820–826
83. Yuan J, Haroon M, Lightfoot DA, Pelletier Y, Liu Q, Bizimungu B, Li XQ (2009) High-resolution DNA melting analysis of allelic expression. Curr Issues Mol Biol 11(S1):1–9
84. Landegren U, Nilsson M, Kwok PW (1998) Reading bits of genetic information: methods for single nucleotide polymorphism analysis. Genome Res 8:769–776
85. Kazi S (2005) Minimum tile derive microsatellite markers improve the physical map of the soybean genome and the Flyer by Hartwig genetic map at *Rhg*, *Rfs* and yield loci. MS Thesis SIUC Carbondale IL, USA, pp 212
86. Hashmi RY (2004) Inheritance of resistance to soybean sudden death syndrome (SDS) in Ripley x Spencer F5 derived lines. PhD dissertation, Plant Biology, SIUC, Carbondale, USA
87. Chang SJC, Doubler TW, Kilo V, Suttner RJ, Klein JH, Schmidt ME, Gibson PT and Lightfoot DA (1996) Two additional loci underlying durable field resistance to soybean sudden-death syndrome (SDS). Crop Sci 36: 1624–1628
88. Sanithchon J, Vanavichit A, Chanprame S, Toojinda T, Triwitayakorn T, Njiti, VM, SrinivesP (2004) Identification of simple sequence repeat markers linked to sudden death syndrome resistance in soybean. Science Asia 30:205–209

Chapter 21

Scoring Microsatellite Loci

Lluvia Flores-Rentería and Andrew Krohn

Abstract

Microsatellites have been utilized for decades for genotyping individuals in various types of research. Automated scoring of microsatellite loci has allowed for rapid interpretation of large datasets. Although the use of software produces an automated process to score or genotype samples, several sources of error have to be taken into account to produce accurate genotypes. A variety of problems (from extracting DNA to entering a genotype into a database) which can arise throughout this process might result in erroneous genotype assignment to one or more samples, potentially confounding the conclusions of your study. Correctly assigning a genotype to a sample requires knowledge of the chemistry you use to generate the data as well as the software you use to analyze these results. In this chapter we describe the critical and more common points that researchers experience when scoring microsatellite loci. More importantly we provide insight from an experienced perspective for these challenges.

Key words Allelic drop-off, Error rate, Fluorescent markers, Genotyping, Null alleles, Polymerase slippage, Scoring microsatellites, Size standard

1 Introduction

Microsatellites, also called simple sequence repeats (SSR) or short tandem repeats (STR), are short repetitive sequences that are prone to rapid mutations that result in sequence length polymorphisms across individuals. The use of microsatellites as polymorphic DNA markers has considerably increased both in the number of studies and in the number of organisms, primarily for population genetics, genetic mapping, studying genomic instability in cancer, forensics, conservation biology, molecular anthropology, and in the studies of human evolutionary history ((5, 55, 59) and references therein). Microsatellite loci can be genotyped by PCR amplification of the microsatellite region, and separation of the products from different samples by electrophoresis. The common detection method is to label one of your PCR primers with a fluorophore that can be detected by laser-induced fluorescence on a capillary electrophoresis system. Amplification produces a pair of fluorescent allelic

Stella K. Kantartzi (ed.), *Microsatellites: Methods and Protocols*, Methods in Molecular Biology, vol. 1006,
DOI 10.1007/978-1-62703-389-3_21, © Springer Science+Business Media, LLC 2013

products (for diploid genes) that will vary in size according to the number of microsatellite repeat units. A suitable choice of fluorescent labeling enables analysis of multiple loci in the same capillary injection. Using different color and size to distinguish between fragments, it is possible to multiplex or pseudoplex >20 markers in a single capillary (14, 15, 33), although it is more common to multiplex only five to eight markers at a time (32).

Fluorescently labeled DNA fragments mixed with an internal size standard migrate through polymer-filled capillaries past a laser beam which excites them. Emission spectra from individual fluorophores are separated by a diffraction grating, and a CCD camera converts the fluorescence signal into digital data that is processed by the instrument data collection software. Allele sizing, scoring, and subsequent data analysis are performed using external software. The automated process of allele scoring allows the analysis of a massive amount of data (number of samples and markers). However, several sources of error have to be taken into account to produce accurate genotypes.

Even if you are adept of the use of your preferred analysis software, correct assignment of genotypes to your sample data is contingent upon first performing PCR using correct chemical conditions. Otherwise, your work may suffer from the computer science adage, "garbage in, garbage out." It is therefore essential that you test each locus to be amplified individually prior to initiating data collection for your project. Many researchers find this step to be difficult, time-consuming, and therefore intimidating. A genotyping error rate of even 1 % (i.e., 1 % of the alleles in an entire dataset are misidentified), which is an uncommonly good value for most studies, can lead to a substantial number of incorrect multilocus genotypes in a large dataset, which in turn will lead you to wrong conclusions (34). In addition to poor amplification, sources of error include incorrect interpretation of stutter patterns or artifact peaks, contamination, mislabeling, and data entry errors (6). In many cases, knowing the sources of error in the genotype data can allow one to correct for it, such as re-genotyping homozygous individuals to catch poorly amplifying alleles. With a few tips, we hope that you can identify and reduce the sources of error, thus improving the allele scoring in your future projects.

2 Materials

For microsatellite scoring by capillary electrophoresis you will need a thermal cycler and reagents for amplifying the desired loci (such as fluorescent label primers), formamide, size standard, access to a capillary electrophoresis-based genetic analyzer, and a computer to run your analysis software and for databasing your results. For fragment analysis on an Applied Biosystems Genetic Analyzer, detectable fluorescent labels are available from Life Technologies,

Table 1
Commonly used fluorophores in microsatellite analysis and their excitation and emission spectra

Dye	Excitation (nm)	Emission (nm)	Analysis color
6-FAM	495	520	Blue
HEX	535	553	Green
TAMRA	555	590	Yellow
ROX	575	607	Red
LIZ (ABI)	638	655	Orange (size std)

though several fluorophore analogues are also available at cheaper prices from companies that offer oligonucleotide synthesis (e.g., Operon, IDT, Table 1).

2.1 Software

There are several programs available to perform fragment analysis of microsatellite electropherograms. Unfortunately, most are not open source and require one to purchase expensive licenses for unrestricted use. This is the case for programs such as GeneMapper from Applied Biosystems, CEQ 8000 software from Beckman Coulter, and GeneMarker from SoftGenetics LLC. Applied Biosystems does provide a simple electropherogram viewer (Peak Scanner) for examining individual samples, but it does not perform comprehensive analyses. SoftGenetics will provide a demo version of GeneMarker to the end user upon request. Freely available software is also able to perform the most important tasks of identifying peak sizes relative to your internal size standard. For instance, the software STRand (60), created at University of California Davis, is available for download free of charge (http://www.vgl.ucdavis.edu/informatics/strand.php). Each software package is different, so we will detail a standard procedure for scoring microsatellites using the popular software packages GeneMapper (Applied Biosystems) and GeneMarker (SoftGenetics, LLC).

3 Methods

(1) *PCR amplification*: Find the optimal conditions to amplify your markers. We find the following concentrations of PCR components to be optimal for most microsatellite analyses: 1× PCR buffer (specific to the manufacturer of your polymerase), 200 nM each dNTP, 200 nM each primer, 0.015 U/μL polymerase, 1.5 mM $MgCl_2$ (or $MgSO_4$), and ~25 ng template DNA. The volume of your reaction matters little, but we find that in the interest of conserving reagents, one can reliably perform 10 μL reactions in a 96-well PCR plate or 4 μL reactions in a 384-well

plate. A 4 μL reaction permits one to check the PCR product on a gel using 2 μL and have 2 μL leftover for capillary analysis (in case of a weak amplification), resulting in no wasted reagent. A thermal cycling protocol we use with standard *Taq* polymerase is 90S °C 1 min; 35 cycles of 90S °C 30 s, 65S °C 2 min, and 72S °C 15 s, though cycling conditions may vary somewhat depending upon the locus being tested and the polymerase being used. Longer annealing time and a higher and therefore more stringent annealing temperature seem to dramatically improve amplification efficiency of microsatellite loci. Our primers are typically designed with a T_m of about 60S °C. For studies involving many loci, it may be more cost-effective to use tailed primers for fluorescent labeling of your amplified PCR products (53). We find the set of universal primers reported by Missiaggia and Grattapaglia (44) to function well in studies of plant population genetics when the fluorescent and reverse primers are included in the reaction at 200 nM each and the forward tailed primer is included at 20 nM.

(2) *Preparing and running the samples in the genetic analyzer (sequencer)*: One microliter PCR product may yield ideal signal strength, though the product may need to be diluted up to tenfold. One microliter of the appropriate dilution is mixed with 10 μL deionized formamide (nucleic acid denaturant) and an internal fluorescent size standard which encompasses the range of product sizes expected for your products. Numerous size standards are available from commercial sources or one can synthesize their own standard (see ref. 8). We routinely use between 0.1 and 0.2 μL GeneScan 500 LIZ Size Standard (Applied Biosystems) per sample with excellent results. To ensure that none of your PCR amplicons possess any secondary structure which may interfere with data interpretation, the samples are then denatured for 5 min at 90–95S °C. Using too little or too much sample can cause problems. Your genetic analyzer instrument can convert a limited range of fluorescence signal into digital values. For optimal results, you should keep the fluorescence signal between approximately 150 and 4,000 relative fluorescent units (RFU). Below this range, the signal-to-noise ratio may be too low to discriminate between sample peaks and background fluctuations. When fluorescence is too intense, the peak may not be sufficiently narrow to accurately assess your allele size.

Size standard: A set of 5′ fluorescently labeled fragments of known sizes. The size standard, possessing a fluorophore distinct from those bound to your fragments of interest, is simultaneously read with the PCR product(s) allowing the software to calibrate fragment sizes within a given sample. In some cases the software may misinterpret one or more of the internal size standard peaks, creating the potential for miscalled alleles due to incorrect calibration. When this occurs, the software will indicate

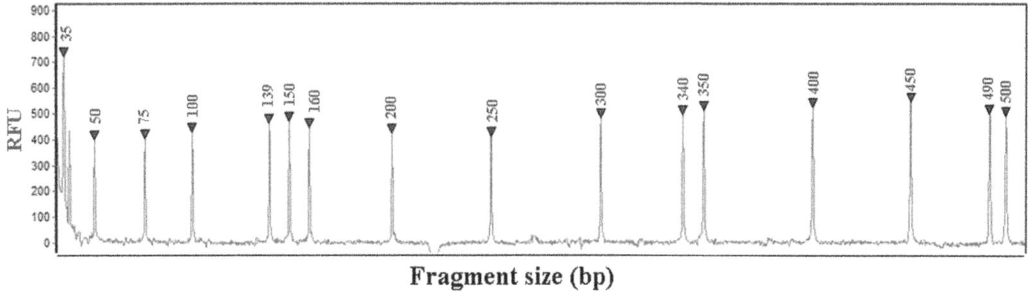

Fig. 1 Pattern of peaks in GeneScan 500 LIZ size standard. The height of each peak corresponds to its relative fluorescence intensity (*y*-axis)

which samples require manual calibration. The manufacturer of the internal size standard will provide information with their product that will enable the researcher to accurately make the necessary corrections. A popular standard is GeneScan 500 LIZ (Applied Biosystems), which ranges from 35 to 500 bp and uses an orange analysis color (Table 1 and Fig. 1).

(3) *Scoring alleles*: Although microsatellite genotyping was first developed in agarose and later in acrylamide gels, capillary electrophoresis (CE) is now the preferred method due to the higher accuracy and increased throughput. Ideally, a peak on the electropherogram (visual representation of a DNA fragment resolved by CE) rises sharply from the baseline, has smooth sides, and is symmetrical in shape.

The steps in data interpretation are the following:

1. Create a panel (see Table 2).
2. Assess peaks of interest versus PCR artifacts.
3. Assess the data from each sample or allele calling.
4. Export a data table.

3.1 Create a Panel

Once your sample files (.fsa, .ab1, .abi, .scf, .rsd, .esd, .smd, or .hid format) have been imported, you need to run your data initially without a panel (use default settings). This permits the software to compare peaks in your sample to those in the internal size standard so that you can begin to build a panel. Once run, enter alleles for each locus using your software's panel editor function. Screening of a locus across several samples should reveal alleles within the expected size range, exhibiting characteristic peak patterns, and any potential scoring problems for that locus. When a large sample size is represented in the analyzed data it is common to observe all expected alleles within the reported size range. For example, if you work with a perfect trinucleotide SSR ranging from 200 to 215 bp, you would expect to observe six alleles at 200, 203, 206, 209, 212, and 215 bp. Such "perfect" results are not always the case however, so one should not get discouraged if an allele remains unrepresented or additional loci are observed within one's dataset (see Subheading 4 for possible explanations).

Table 2
Common terms and definitions

Term	Definition
Bin	For each marker, separate bins (size in bp) are defined by the user for each allele observed. A group of bins is sometimes referred to as a bin set.
Marker	Each marker (or locus) is defined by name, size range (bp), dye color, and repeat length. The size range will include bins for each expected allele.
Panel	A group of markers for simultaneous analysis.
Color channel	Each channel is viewed in the analysis as a separate color defined by the emission spectra of each fluorophore. Different instruments can interpret different dyes, but each will be capable of reading four or five colors simultaneously.
Peak	Visual representation of a DNA fragment resolved by capillary electrophoresis.
Size standard	A set of 5′ fluorescently labeled fragments of known sizes.
RFU	Relative fluorescence units which measure the intensity of a fluorescence signal.

3.2 Get Familiar with Your Loci

Once you have identified all possible alleles in your data you can create bins for each allele (expected size limit for each allele). Bins usually are one bp long to avoid capturing neighboring alleles within the same bin, and to allow for slight variation among called alleles due to sequencer error (approximately ± 0.5 bp). Selection of the fluorophore used is assigned during the panel creation.

When working with only a handful of samples and loci, it may be expedient to simply call each allele one at a time and record the results in a spreadsheet. For larger datasets it will be necessary to instead automate this process by creating a panel of expected allele sizes for each locus against which the software will compare your samples. Though not all software are the same, this process is fairly uniform. Use the panel editor function of your analysis software to identify peaks present in the expected size range across all your samples, and record their positions. Once entered into the panel (specific to an individual project), the software will be able to call peaks observed within each sample into bins which refer to the individual alleles that you designate. Though post-processing editing of automated allele calls may be necessary, this step will greatly facilitate analysis of medium- to large-sized datasets.

3.3 Assess Peaks of Interest Versus PCR Artifacts

The complexity of distinguishing between peaks of interest and PCR artifacts is associated with the complexity of the genome amplified (haploid, diploid, or polyploid) or the number of markers included in a multiplex design. For example, the use of haploid chloroplast or mitochondrial SSRs will yield one allele per sample for each marker, making it relatively simple to perform allele calls. In contrast, working with an organism such as hexaploid wheat will yield between one and six peaks per sample, each of which must be efficiently amplified to be accurately scored.

There are multiple reasons to have confounding peaks in the data. Common causes for such peaks may be PCR artifact(s) created during amplification, incomplete terminal adenylation, mono- or dinucleotide stutter, or pull-up due to spectral overlap of two fluorophores. However, there may be other factors associated with the nature of the marker such as null alleles or the presence of an imperfect repeat. All of these factors can challenge the scoring process. Solutions to eliminate or reduce these confounding peaks are given in Subheading 4.

3.4 Allele Calling

Once your panel has been established you will be able to determine the genotype (sizes of your PCR products) of each sample based on the pattern of peaks or bands on the electropherogram. Rerun your data against your new panel (default settings again), and verify each allele call by hand. You may find that you did not capture all of the alleles present in your sample data; therefore some panel adjustment may be necessary (and subsequent rerunning of your samples) before you finalize your data. Software-automated allele calling will take a few seconds or minutes depending on the size of your dataset.

3.5 Create and Export a Table

The software generates a table with your genotype associated to each sample. For subsequent analysis in a population genetics analysis package such as Arlequin (25) or GenAlEx (48), export your data as a bin table or a genotype table as appropriate.

Though the above protocols are quite similar and may seem straightforward, and the software performs allele calls in an automated fashion, one should never fully trust one's initial data output. Variations among your PCR products may be due to a number of factors including well-to-well variation across your thermal cycler, pipeting errors, and inconsistent quality among your template DNA samples. These factors (and others) can contribute to variation in PCR amplification efficiency that is exhibited as differences in peak height among your electropherograms. Since it is these data that are interpreted by the software and you may restrict allele calling based on peak heights across all samples, the software should be considered fallible. Following the automated allele calling step, scan through the samples and look for obvious errors (e.g., peaks not called, stutter peaks called). You may notice that certain loci yield different characteristic peaks. For instance, a trinucleotide locus may exhibit very clear peaks, each with a small preceding peak while a dinucleotide locus may suffer more from stutter, contributing to a "rooster-comb" appearance (Fig. 2). Mononucleotide repeats will generate peak profiles similar to those of dinucleotide repeats; however there are some strategies that allow reduction of the stutter in these markers difficult to score (see Subheading 4).

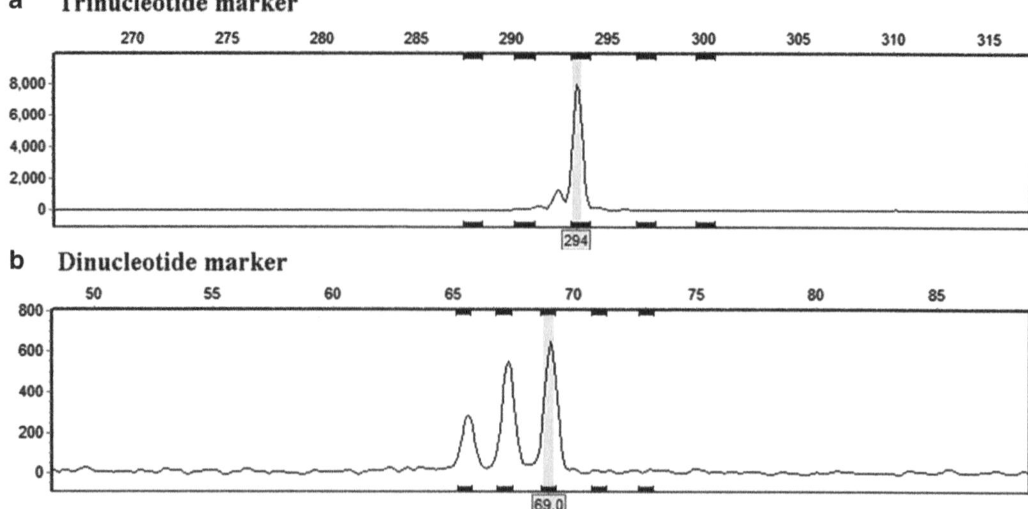

Fig. 2 Peak profiles characteristic of different repeat lengths. Trinucleotide (or longer) repeats (**a**) usually exhibit a very clear major peak with a preceding minor peak. Dinucleotide repeats (**b**) usually exhibit more than one major peak per allele

4 Notes

One should manually check the quality of automated allele calling. Most errors incurred during the allele calling process are derived from poor PCR amplifications which result in low amplification efficiency or production of nonspecific products. You might need to optimize your PCR conditions or even redesign your primers. Re-extracting DNA from difficult samples and re-amplifying questionable genotypes (e.g., heterozygotes with closely sized alleles, faint alleles) is a common practice to increase the accuracy during genotyping. However other factors can cause some troubles during allele calling. In this section we describe common problems that researchers encounter during this procedure and suggest some solutions.

4.1 Previously Unreported Alleles

Occasionally, one may encounter alleles that were not previously reported in the publication from which one derived one's microsatellite markers. This is more common for loci that exhibit many alleles than for loci that exhibit only a few. It will be up to the researcher to determine whether these "new" alleles are valid genotypes or if the result is an error. We suggest that you first look through all your samples to determine if you are observing unreported alleles in more than one sample. If observing a new allele in multiple samples, you can feel more confident that the allele is valid. If instead a new allele is represented in just a single sample, the PCR reaction should be

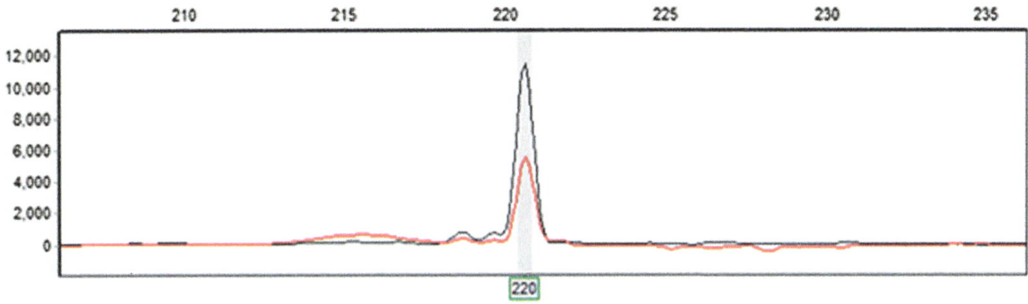

Fig. 3 TAMRA fluorophore (black peak) causing pull-up of the red channel on an ABI 3730xl genetic analyzer

repeated to verify that there was no error in your chemistry. In either case, a subset of samples should be re-amplified to provide a measure of confidence in each allele call.

4.2 "Extra" Alleles

Some loci will exhibit more alleles than expected for a given organism. For example, one would expect a maximum of one allele per locus for a haploid organism or two alleles per locus for a diploid organism. If, when working with a haploid organism and you observe two peaks within your expected allele range, you are likely encountering a locus that has been duplicated within the genome (e.g., ref. 63), yielding amplification of a microsatellite family rather than a single discrete locus. Anderson et al. (2) were the first to report successful utilization of these loci, characterized by their repetitive flanking sequences (ReFS). Though ReFS can be useful for genetic inquiry of populations, such loci are beyond the scope of this chapter and require statistical treatment as dominant markers. Many researchers will choose to simply discard such anonymous loci in favor of properly functioning microsatellites which will better serve to answer their particular research question.

4.3 Pull-Up

If you include too much PCR product on a capillary run, you may experience very strong fluorescence signal (approximately >20,000 RFU) from the labeled fragments therein. High fluorescence signal can prevent the instrument from properly compensating for spectral overlap among the dyes resulting in artifact peaks in one channel derived from the strong signal intensity in another (called "bleedthrough" or "pull-up"). Artifact peaks can corrupt both automated size-calling due to pull-up peaks in the size standard color and the analysis of co-loaded samples when pull-up peaks overlap a bin set for another marker. Certain combinations of fluorophores are more prone to the pull-up effect than others. For example, TAMRA tends to cause pull-up in the ROX channel, but not vice versa (Fig. 3). For this reason, one would choose to use a LIZ-labeled size standard rather than a ROX-labeled standard if you plan to score PCR products labeled with TAMRA.

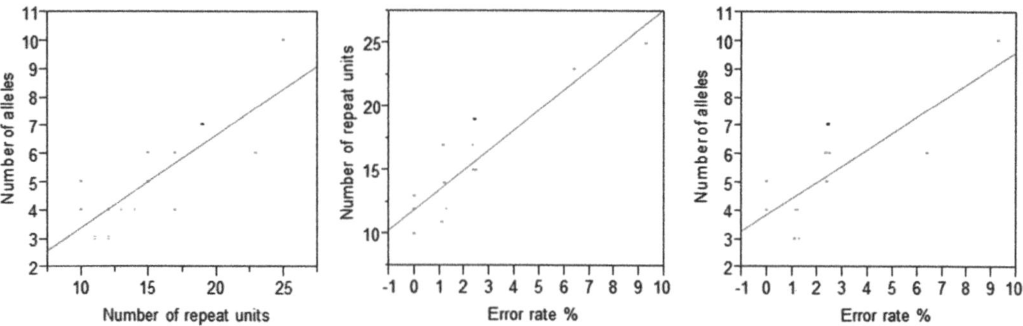

Fig. 4 There is a positive relationship between the repeat size and the number of alleles; however, there is a trade-off associated. Loci exhibiting longer repeat units and high number of alleles will have higher error rates. This is especially true for mononucleotide and dinucleotide repeats. Data generated using 12 mononucleotide SSRs from Flores-Rentería and Whipple (28)

4.4 Stutter in Mononucleotide and Dinucleotide Repeats

Variation in the number of repeats in microsatellite loci is primarily due to polymerase slippage (slipped-strand mispairing) during DNA replication, as well as repair mechanisms during recombination ((62); reviewed in (41)). Slippage can also be generated during PCR reactions making allele designation difficult (40, 43), especially for heterozygotes with adjacent alleles, resulting in high error rates in scoring (18). Mononucleotide repeats are the most common SSRs in the plant chloroplast genome and, due to their high mutation rates, they represent the most variable markers in this organelle (51). According to Guichoux et al. (32) among 100 studies surveyed from 2009 to 2010, none made use of mononucleotide repeat SSRs. This reflects the fact that because mononucleotide repeat SSRs are difficult to accurately assay (58) they are often eliminated at the outset (37). In contrast, dinucleotide SSRs were the most frequently used class of microsatellites (32). Unfortunately, mononucleotide and dinucleotide repeats often show one or more "stutter" peaks arising from multiple PCR products derived from the same reaction template that are typically shorter by one or a few repeats than the full-length product (12). The error rate in allele calling for dinucleotide SSRs is ~5 % with samples amplified by *Taq* polymerase (31), and it could be higher for mononucleotide repeats. Polymerase slippage is positively correlated with the length of the microsatellite ((35, 36); see Fig. 4), making scoring of mononucleotide SSRs >11 bp highly error-prone (18). In contrast, tri-, tetra-, or pentanucleotide repeats appear to be significantly less prone to exhibiting stutter peaks (17, 24, 45). Hence, SSRs with core repeats three to five nucleotides long are sometimes preferred for forensic and parentage applications (17, 38). Note however that stutter bands, when not too strong, can be useful, by helping distinguish true alleles from PCR artifacts (e.g., ref. 54). Note also that a few solutions have been proposed to overcome stuttering problems. The most common solution has been to simply select loci that present the lowest degree of stutter (e.g., refs. 21, 46).

However, mono- and dinucleotide repeats have been used successfully in studies of chloroplast DNA variation in plants (23, 51), SSR-poor fungi (16), or in other circumstances, for assessing microsatellite instability associated with cancer (e.g., ref. 27), where such markers are of special interest.

In addition to the importance of mono- and dinucleotide SSRs mentioned above, there is a methodological relevance in the use of these markers. According to Guichoux et al. (32), focusing on the shortest motifs (such as mono- or dinucleotide repeats) rather than on longer ones (≥trinucleotide repeats) should allow for more dense packing of loci on a given separation system, resulting in larger multiplex designs. This can be important because the capillary electrophoresis-based genetic analyzers used for SSR genotyping make use of no more than four or five fluorophores, thus limiting the number of SSR loci that can be analyzed simultaneously. Given that the allelic range size often reaches up to 50 or 100 bp and that amplicons measuring over 300 bp are rarely used (e.g., refs. 14, 33).

4.5 Reducing Stutter

Stutter bands are typically shorter than the original fragment (56). Thus it has been generally assumed that choosing the largest fragment (bp) will resolve the problem. However, in our experience this is not always right, so improvement during the PCR amplification has to be done in order to reduce stutter. We list the few solutions that have been proposed to overcome stuttering problems:

1. To decrease denaturing temperature to 83S °C (47).
2. Varying the reaction conditions or including additives such as formamide, bovine serum albumin, or dimethyl sulfoxide (9).
3. Adjusting the PCR program by using touchdown or hot start techniques, reducing the number of cycles, or maintaining a stringent annealing temperature (21).
4. To use new-generation polymerases, such as fusion enzymes (26) or PCR kits designed especially for microsatellite analysis (e.g., Multiplex PCR Kit or TypeIT Microsatellite PCR Kit from Qiagen).

Flores-Rentería and Whipple (28) developed a new method to increase the accuracy of scoring mono- and dinucleotide alleles by designing primers that include part of the microsatellite in order to reduce the slippage. This method was tested using primers developed to amplify mononucleotide repeats (≥10 bp) in the chloroplast of *Pinus* spp.

4.6 Null Alleles and Allelic Drop-Out

A microsatellite null allele is any allele at a microsatellite locus that consistently fails to amplify to detectable levels via polymerase chain reaction (PCR) (19). There are at least three potential causes of null alleles or allelic drop-out: (1) poor primer annealing due to mutation on the primer region (e.g., substitutions or indels in one or

Box 1 Improving Scoring by Reducing Stutter in Mononucleotide and Dinucleotide Repeats

For example, if there is a dinucleotide repeat of $(TA)_{7-12}$ that when amplified, the profile generated on capillary sequencer looks like panel A. There is an easy way to reduce the stutter by designing a new primer that contains part of the microsatellite. In this example we redesign the reverse primer.

1. To amplify and sequence the SSRs of interest using the original primers in at least eight samples under normal conditions (samples from the most diverse source the best).
2. To align your sequences in order to find the range in length of the repeat (panel A). In this case the longest repeat is $(TA)_{12}$ and the shortest is $(TA)_7$.
3. Then a new reverse primer should be designed including the flanking region of the microsatellite and part of the microsatellite. The repeat length in the primer should be equal to the smallest microsatellite detected, minus one or two bases, in our example $(TA)_6$ (panel B).
4. The new reverse primer should be compatible with the original forward primer, e.g., no hetero or homo dimer formation. The unequal length of the forward (let us say 24 bp) and the new reverse (let us say 32 bp) primers does not affect the amplification as long as they have similar melting temperature above 50S °C.
5. You can decide to use the forward region to design your primer containing part of the repeat if the flanking region has better conditions than the reverse flanking region (e.g., G+C content).
6. When using the internal primers the PCR should be performed under standard conditions.
7. This method allows a multiplex assay, if similar melting temperatures are used for all primers.
8. If you are using mononucleotides SSR multiplex primer combinations should not mix A and T repeats to avoid primer-dimer formation.
9. In our experience up to six primer pairs can be multiplexed in a single PCR reaction.
10. You can try also to use higher concentration of the reverse primer.

This method requires a little bit of knowledge about designing primers. However, it is more cost-effective in comparison to the use of fusion polymerases, and may require less troubleshooting than the use of additives. However, the benefit achieved through this method of stutter reduction diminishes with longer repeats as the primer may not be able to be designed with an acceptable annealing temperature due to the necessary length. Further, if the repeat-containing 3′ end of the primer is too long, mispriming may occur at other SSR loci containing the same repeat. Though this effect may not directly interfere with the scoring of a locus, it can consume one of your primers, thus reducing the efficiency of your PCR reaction.

If you want to amplify a mono- or dinucleotide SSR longer than 20 bp using a new reverse primer containing only 10 bp of the repeat you might not have a clear peak. In that case you can design an additional reverse primer containing, for example, 16 bp.

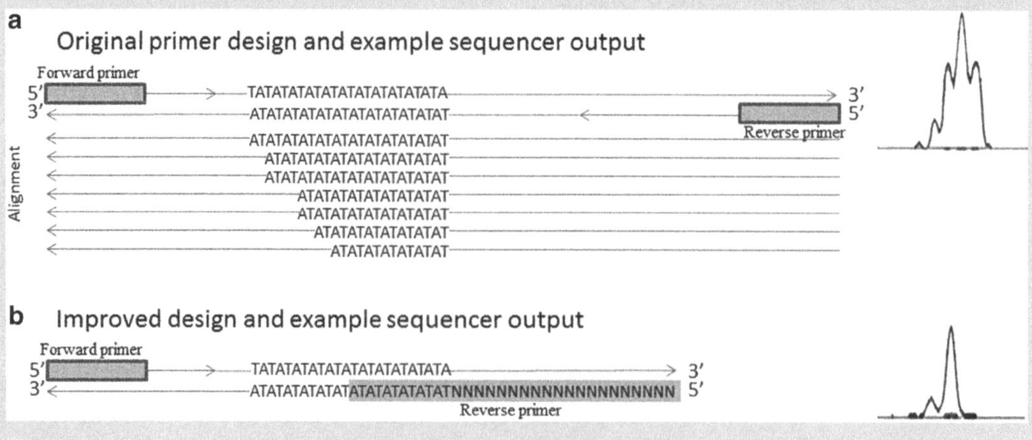

both primer annealing sites). In particular, mutations in the priming site at or near where the 3′ end of the primer anneals are thought to be especially detrimental to PCR amplifications (39) and can contribute to an allele becoming null for a given locus. The allele can be "resurrected" following a redesign of the primers. In most cases internal primers are designed, resulting in a slightly smaller PCR product. Degenerate primers are another alternative, or external primers can be designed when the necessary sequence data is available. (2) Differential amplification of size-variant alleles or "partial nulls" (61). Due to the competitive nature of PCR, alleles of short length often amplify more efficiently than larger ones such that only the smaller of two alleles might be detected from a heterozygous individual. Outcompeted alleles may stochastically amplify more strongly in a second PCR reaction. Alternatively, by loading more sample undetectable peaks become evident. (3) PCR failure due to inconsistent DNA template quality or low template quantity. These problems are insidious because in some cases only one or a few loci (or alleles) fail to amplify, whereas others amplify with relative ease from the same DNA preparation (29, 30). When DNA template at a given locus is poor in some specimens but not others, some samples may appear artificially homozygous rather than heterozygous for the null allele. A potential solution for this is to improve DNA quality by either a further purification step (e.g., ethanol precipitation or column purification) or by re-extracting DNA from the sample in question. Of the above three causes for null alleles, the first one is generally accepted as a legitimate cause of a "true" null allele while causes 2 and 3 are more likely due to technician deficiencies in the amplification process (13).

In addition to these primary causes of null alleles and drop-out, several population genetic phenomena might give the false impression that null alleles are present in a given study. Biological factors such as the Wahlund effect (reduction of heterozygosity in a population caused by subpopulation structure) or inbreeding, for example, can cause significant heterozygote deficits relative to Hardy–Weinberg equilibrium that might be misconstrued as evidence for null alleles (11). Wahlund effect or inbreeding tends to be observed more or less concordantly across loci, whereas the effects of null alleles are locus specific. Therefore the comparison across multiple loci will be useful to discard these possible causes of homozygosity bias.

4.7 Compound Microsatellites

It is often assumed based on a handful of sequence observations that microsatellite loci have a single, discrete repeat sequence (e.g., $(GT)_{14}$). Perhaps in part due to a complex mutational process that leads to variation in microsatellite repeats (22), many microsatellite loci may in fact exhibit variation in the sequence of the repeat structure (e.g., $(GT)_9(GA)_6$; (7, 50)). Such repeats are known as compound microsatellites and are discussed at

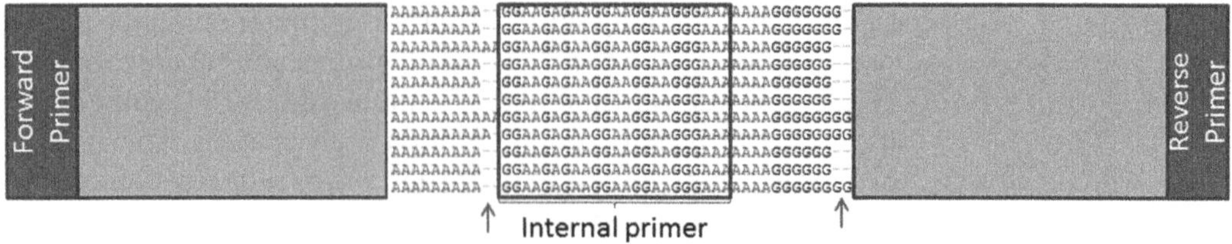

Fig. 5 Compound microsatellite creating homoplasy (*see arrows*). Breaking of the compound microsatellite can be achieved by designing an internal primer; potential region for a redesigned forward or reverse primer is shown in a *black box*

length in Bull et al. (10). Scoring a compound microsatellite may be more challenging than scoring a perfect repeat, as alleles observed within a given population may not be represented in other populations, and since the length of the various repeat motifs may also vary. Unfortunately such complexity can lead to some amount of homoplasy among individuals (e.g., ref. 50), so perfect repeats are desirable. If possible, redesign your primers to only assess one type of repeat (Fig. 5).

4.8 Incomplete Terminal Adenylation

Alternatively, an allele may falsely appear to exhibit such complexity when using *Taq* polymerase and incomplete terminal adenylation results in a peak that is 1 bp shorter than an expected allele (9), regardless of the length of the repeat motif. Most researchers try to ensure complete terminal adenylation when using *Taq* polymerase by using a final extension step (e.g., 60S °C for 15 min) once thermal cycling is complete. Brownstein et al. (9) found that including the "PIGtail" sequence, 5′-GTTTCTT-3′, at the 5′ end of your reverse (non-fluorescently labeled) primer will further facilitate complete terminal adenylation of the fluorescently labeled strand.

4.9 Controls, Confidence, and Error Rate

In the previous paragraphs we described some error sources associated to PCR and the nature of the microsatellites used, and gave potential alternatives or solutions to lower the error rate. Additionally, human error can be introduced directly by DNA contamination, mislabeling samples, or entering wrong data. According to Selkoe and Toonen (55), some amount of error is unavoidable. Regardless of the error source, the error rate within each study should be quantified and reported (reviewed in ref. 49).

Confidence in your scoring procedure can be achieved by including some controls in your data. To ensure that amplification of alleles is consistent throughout the duration of a study, a positive control should be run with every PCR plate, especially any time multiple sequencers are used for genotyping in a single study, or new batches of primers are used (20). According to

Selkoe and Toonen (55) the whole dataset can be genotyped in duplicate or more, as is performed for human parentage or forensics. Conversely, population genetics studies lack the ability to conduct this practical quality check, so accurate reporting of error rates is essential. Fortunately, by keeping track of one's error rate, one can identify and correct the major sources of systematic error in order to bring the overall error rate to an acceptable minimum.

We have established the error rate associated to a marker by repeating marker amplification under same condition in a 96-well plate. Error rate has to be calculated consistently with a simplex or a multiplex design, counting the number of inconsistent genotypes between the first and second attempts. The error rate can then be expressed as either the ratio of incorrect genotypes to the number of repeated reactions (28) or the ratio of incorrect alleles to the total number of alleles (34). Alternatively, the false discovery rate (3) can be employed and has been used to establish and control error rates for a variety of genotyping studies (e.g., refs. 4, 57). By examining the sources of each error, it is possible to determine whether the majority of errors are broadly distributed (such as typographical errors), or biased towards some subset of the data (such as homozygotes in the case of null alleles). For researchers investigating samples of known pedigree as in half-sib/full-sib association mapping studies (e.g., ref. 52), genotyping errors will reveal themselves when one or more alleles segregate inconsistently with Mendelian inheritance patterns. Such an obvious genotyping error will cue the researcher to re-genotype the aberrant samples and can quickly inform whether the trouble is with the PCR chemistry or the initial DNA extraction.

Just a few programs take the error rate into account. One of them is the parentage program CERVUS, which can estimate error rate while also accounting for mutation (42). The effect of error on measures of genetic structure can be estimated using a bootstrapping technique developed by Adams et al. (1). Once the error rate is accounted for, it can be controlled in order to achieve the desired statistical power (3). Due to the potential sources of error incurred during microsatellite allele scoring, we encourage software developers to continue to incorporate error rate into their programs.

Despite the recent advances in DNA sequencing technologies, we expect microsatellites to continue to be utilized as a user-friendly, cost-effective genetic marker system. Such genetic inquiry remains necessary in various research disciplines including ecology and agriculture. With a little experience, it is our sincere hope that a researcher can confidently make use of microsatellites in order to answer their particular research question.

References

1. Adams RI, Brown KM, Hamilton MB (2004) The impact of microsatellite electromorph size homoplasy on multilocus population structure estimates in a tropical tree (Corythophora alta) and an anadromous fish (Morone saxatilis). Mol Ecol 13:2579–2588
2. Anderson SJ, Gould P, Freeland JR (2007) Repetitive flanking sequences (ReFS): novel molecular markers from microsatellite families. Mol Ecol Notes 7:374–376
3. Benjamini Y, Hochberg Y (1995) Controlling the false discovery rate: a practical and powerful approach to multiple testing. J R Stat Soc B 57:289–300
4. Benjamini Y, Yekutieli D (2005) Quantitative trait loci analysis using the false discovery rate. Genetics 171:783–790
5. Bhargava A, Fuentes FF (2010) Mutational dynamics of microsatellites. Mol Biotechnol 44:250–266
6. Bonin A, Bellemain E, Eidesen PB et al (2004) How to track and assess genotyping errors in population genetics studies. Mol Ecol 13:3261–3273
7. Brinkmann B, Klintschar M, Neuhuber F et al (1998) Mutation rate in human microsatellites: influence of the structure and length of the tandem repeat. Am J Hum Genet 62:1408–1415
8. Brondani RPV, Grattapaglia D (2001) Cost-effective method to synthesize a fluorescent internal DNA standard for automated fragment sizing. Biotechniques 31:793–800
9. Brownstein MJ, Carpten D, Smith JR (1996) Modulation of non-templated nucleotide addition by Taq DNA polymerase: primer modifications that facilitate genotyping. Biotechniques 20:1004–1010
10. Bull L, Pabon-Pena C, Freimer N (1999) Compound microsatellite repeats: practical and theoretical features. Genome Res 9:830–838
11. Chakraborty R, DeAndrade M, Daiger SP et al (1992) Apparent heterozygote deficiencies observed in DNA typing data and their implications in forensic applications. Ann Hum Genet 56:45–57
12. Chambers GK, MacAvoy ES (2000) Microsatellites: consensus and controversy. Comp Biochem Physiol B Biochem Mol Biol 126:455–476
13. Chapuis M-P, Estoup A (2007) Microsatellite null alleles and estimation of population differentiation. Mol Biol Evol 24:621–631
14. Chen JW, Uboh CE, Soma LR et al (2010) Identification of racehorse and sample contamination by novel 24-plex STR system. Forensic Sci Int Genet 4:158–167
15. Cherel P, Glénisson J, Pires J (2011) Tetranucleotide microsatellites contribute to a highly discriminating parentage test panel in pig. Anim Genet 42:659–661
16. Christians JK, Watt CA (2009) Mononucleotide repeats represent an important source of polymorphic microsatellite markers in Aspergillus nidulans. Mol Ecol Resour 9:572–578
17. Cipriani G, Marrazzo MT, DiGaspero G et al (2008) A set of microsatellite markers with long core repeat optimized for grape (Vitis spp.) genotyping. BMC Plant Biol 8:127
18. Clarke LA, Rebelo CS, Goncalves J et al (2001) PCR amplification introduces errors into mononucleotide and dinucleotide repeat sequences. Mol Pathol 54:351–353
19. Dakin EE, Avise JC (2004) Microsatellite null alleles in parentage analysis. Heredity 93:504–509
20. Delmotte F, Leterme N, Simon JC (2001) Microsatellite allele sizing: difference between automated capillary electrophoresis and manual technique. Biotechniques 31:810
21. DeWoody JA, Nason JD, Hipkins VD (2006) Mitigating scoring errors in microsatellite data from wild populations. Mol Ecol Notes 6:951–957
22. Dieringer D, Schlötterer C (2003) Two distinct modes of microsatellite mutation processes: evidence from the complete genomic sequences of nine species. Genome Res 13:2242–2251
23. Ebert D, Peakall R (2009) Chloroplast simple sequence repeats (cpSSRs): technical resources and recommendations for expanding cpSSR discovery and applications to a wide array of plant species. Mol Ecol Resour 9:673–690
24. Edwards A, Civitello A, Hammond HA et al (1991) DNA typing and genetic mapping with trimeric and tetrameric tandem repeats. Am J Hum Genet 49:746–756
25. Excoffier L, Laval G, Schneider S (2005) Arlequin ver. 3.0: an integrated software package for population genetics data analysis. Evol Bioinform Online 1:47–50
26. Fazekas AJ, Steeves R, Newmaster SG (2010) Improving sequencing quality from PCR products containing long mononucleotide repeats. Biotechniques 48:277–281
27. Ferreira AM, Westers H, Sousa S et al (2009) Mononucleotide precedes dinucleotide repeat instability during colorectal tumour development in Lynch syndrome patients. J Pathol 219:96–102

28. Flores-Rentería L, Whipple AV (2011) A new approach to improve the scoring of mononucleotide microsatellite loci. Am J Bot 98:e51–e53
29. Gagneux P, Boesch C, Woodruff DS (1997) Microsatellite scoring errors associated with noninvasive genotyping based on nuclear DNA amplified from shed hair. Mol Ecol 6:861–868
30. Garcia de Leon FJ, Canonne M, Quillet E et al (1998) The application of microsatellite markers to breeding programmes in the sea bass, Dicentrarchus labrax. Aquaculture 159:303–316
31. Ginot F, Bordelais I, Nguyen S et al (1996) Correction of some genotyping errors in automated fluorescent microsatellite analysis by enzymatic removal of one base overhangs. Nucleic Acids Res 24:540–541
32. Guichoux E, Lagache L, Wagner S et al (2011) Current trends in microsatellite genotyping. Mol Ecol Resour 11:591–611
33. Hill CR, Butler JM, Vallone PM (2009) A 26-plex autosomal STR assay to aid human identity testing. J Forensic Sci 54:1008–1015
34. Hoffman JI, Amos W (2005) Microsatellite genotyping errors: detection approaches, common sources and consequences for paternal exclusion. Mol Ecol 14:599–612
35. Jakobsson M, Säll T, Lind-Halldén C et al (2007) Evolution of chloroplast mononucleotide microsatellites in *Arabidopsis thaliana*. Theor Appl Genet 114:223–235
36. Kelkar YD, Strubczewski N, Hile SE et al (2010) What is a microsatellite: a computational and experimental definition based upon repeat mutational behavior at A/T and GT/AC repeats. Genome Biol Evol 2:620–635
37. Kim TS, Booth J, Gauch H et al (2008) Simple sequence repeats in Neurospora crassa: distribution, polymorphism and evolutionary inference. BMC Genomics 9:31
38. Kirov G, Williams N, Sham P et al (2000) Pooled genotyping of microsatellite markers in parent-offspring trios. Genome Res 10:105–115
39. Kwok S, Kellog DE, McKinney N et al (1990) Effects of primer-template mismatches on the polymerase chain reaction: human immunodeficiency virus 1 model studies. Nucleic Acids Res 18:999–1005
40. Levinson G, Gutman GA (1987) Slipped-strand mispairing: a major mechanism for DNA sequence evolution. Mol Biol Evol 4:203–221
41. Li Y-C, Korol AB, Fahima T et al (2002) Microsatellites: genomic distribution, putative functions and mutational mechanisms: a review. Mol Ecol 11:2453–2465
42. Marshall TC, Slate J, Kruuk LEB et al (1998) Statistical confidence for likelihood-based paternity inference in natural populations. Mol Ecol 7:639–655
43. Meldgaard M, Morling N (1997) Detection and quantitative characterization of artificial extra peaks following polymerase chain reaction amplification of 14 short tandem repeat systems used in forensic investigations. Electrophoresis 18:1928–1935
44. Missiaggia A, Grattapaglia D (2006) Plant microsatellite genotyping with 4-color fluorescent detection using multiple-tailed primers. Genet Mol Res 5:72–78
45. Nater A, Kopps AM, Krützen M (2009) New polymorphic tetranucleotide microsatellite improve scoring accuracy in the bottlenose dolphin *Tursiops aduncus*. Mol Ecol Resour 9:531–534
46. O'Reilly PT, Canino MF, Bailey KM et al (2000) Isolation of twenty low stutter di- and tetranucleotide microsatellites for population analyses of walleye pollock and other gadoids. J Fish Biol 56:1074–1086
47. Olejniczak M, Krzyzosiak WJ (2006) Genotyping of simple sequence repeats factors implicated in shadow band generation revisited. Electrophoresis 27:3724–3734
48. Peakall R, Smouse PE (2006) GenAlEx 6: genetic analysis in Excel. Population genetic software for teaching and research. Mol Ecol Notes 6:288–295
49. Pompanon F, Bonin A, Bellemain E et al (2005) Genotyping errors: causes, consequences and solutions. Nat Rev Genet 6:847–859
50. Primmer CR, Ellegren H (1998) Patterns of molecular evolution in avian microsatellites. Mol Biol Evol 15:997–1008
51. Provan J, Powell W, Hollingsworth PM (2001) Chloroplast microsatellites: new tools for studies in plant ecology and evolution. Trends Ecol Evol 16:142–147
52. Riday H, Krohn AL (2010) Genetic map-based location of the red clover (*Trifolium pratense* L.) gametophytic self-incompatibility locus. Theor Appl Genet 121:761–767
53. Schuelke M (2000) An economic method for the fluorescent labeling of PCR fragments. Nat Biotechnol 18:233–234
54. Schwengel DA, Jedlicka AE, Nanthakumar EJ et al (1994) Comparison of fluorescence-based semi-automated genotyping of multiple microsatellite loci with autoradiographic techniques. Genomics 22:46–54
55. Selkoe KA, Toonen RJ (2006) Microsatellites for ecologists: a practical guide to using and evaluating microsatellite markers. Ecol Lett 9:615–629

56. Shinde D, Lai Y, Sun F et al (2003) Taq DNA polymerase slippage mutation rates measured by PCR and quasi-likelihood analysis: (CA/GT) n and (A/T)n microsatellites. Nucleic Acids Res 31:974–980
57. Storey JD, Tibshirani R (2003) Statistical significance for genomewide studies. Proc Natl Acad Sci U S A 100:9440–9445
58. Sun X, Liu Y, Lutterbaugh J et al (2006) Detection of mononucleotide repeat sequence alterations in a large background of normal DNA for screening high-frequency microsatellite instability cancers. Clin Cancer Res 12:454–459
59. Thibodeau SN, Bren G, Schaid D (1993) Microsatellite instability in cancer of the proximal colon. Science 260:816–819
60. Toonen RJ, Hughes S (2001) Increased throughput for fragment analysis on ABI Prism Automated Sequencer using a membrane comb and STRand software. Biotechniques 31:1320–1324
61. Wattier R, Engel CR, Saumitou-Laprade P et al (1998) Short allele dominance as a source of heterozygote deficiency at microsatellite loci: experimental evidence at the dinucleotide locus Gv1CT in *Gracilaria gracilis* (Rhodophyta). Mol Ecol 7:1569–1573
62. Weber JL, Wong C (1993) Mutation of human short tandem repeats. Hum Mol Genet 8:1123–1128
63. Zhang D-X (2004) Lepidopteran microsatellite DNA: redundant but promising. Trends Ecol Evol 19:507–509

INDEX

Stella K. Kantartzi (ed.), *Microsatellites: Methods and Protocols*, Methods in Molecular Biology, vol. 1006, DOI 10.1007/978-1-62703-389-3, © Springer Science+Business Media, LLC 2013

Lightning Source UK Ltd.
Milton Keynes UK
UKOW07n2023051115

262161UK00003B/32/P